T0275877

The Concept of the Gene in Development and Evolution

Historical and Epistemological Perspectives

Advances in molecular biological research in the last fifty years have made the story of the gene extremely complex. The gene has become a curiously intangible object, defying any straightforward definition. Relating the gene as a molecular biological unit to the gene as a Mendelian factor produces internal inconsistencies; but genes have been deeply elusive entities even within the traditional Mendelian framework. Philosophers, historians, and working scientists approach the issue from a variety of perspectives. This volume provides evidence of the diversity of scientific disciplines that presently have a stake in the quest for the gene.

The essays collected here offer challenging perspectives on some of the most fundamental concepts of twentieth-century biology. Conceptual perspectives about the gene as it is presently known provide the substance of three contributions. The examination of pre-Darwinian heredity concepts, Goldschmidt's demission of the gene, and Seymour Benzer's work on the fine structure of the gene are also explored. A critique of the "genetic program" is presented as well as modern findings about the functioning of "master genes" during embryogenesis. In the final essay, Raphael Falk reviews the material laid out in this volume and lucidly summarizes the primary themes.

The Concept of the Gene in Development and Evolution is unique in that it is the first interdisciplinary volume solely devoted to the quest for the gene. It will be of interest to professionals and students of philosophy and the history of science, genetics, and molecular biology.

Peter J. Beurton is Senior Scientist at the Max Planck Institute for the History of Science in Berlin.

Raphael Falk is Professor Emeritus at the Hebrew University of Jerusalem.

Hans-Jörg Rheinberger is Director at the Max Planck Institute for the History of Science in Berlin.

Cambridge Studies in Philosophy and Biology

The Concept of the Gene in Development and Evolution

Historical and Epistemological Perspectives

Edited by

PETER J. BEURTON

Max Planck Institute for the History of Science, Berlin

RAPHAEL FALK

Hebrew University, Jerusalem

HANS-JÖRG RHEINBERGER

Max Planck Institute for the History of Science, Berlin

CAMBRIDGE
UNIVERSITY PRESS

PUBLISHED BY THE PRESS SYNDICATE OF THE UNIVERSITY OF CAMBRIDGE
The Pitt Building, Trumpington Street, Cambridge, United Kingdom

CAMBRIDGE UNIVERSITY PRESS
The Edinburgh Building, Cambridge CB2 2RU, UK   http://www.cup.cam.ac.uk
40 West 20th Street, New York, NY 10011-4211, USA   http://www.cup.org
10 Stamford Road, Oakleigh, Melbourne 3166, Australia
Ruiz de Alarcón 13, 28014 Madrid, Spain

© Cambridge University Press 2000

First published 2000

Typeface Palatino 10/13 pt.   *System* MagnaType™ [AG]

A catalog record for this book is available from the British Library.

Library of Congress Cataloging in Publication Data
The concept of the gene in development and evolution: historical and epistemological
perspectives / edited by Peter J. Beurton, Raphael Falk, Hans-Jörg Rheinberger.
p.   cm. – (Cambridge studies in philosophy and biology)
Includes bibliographical references and index.
ISBN 0-521-77187-0
1. Genes. 2. Genetics – Philosophy. 3. Developmental genetics. 4. Evolutionary
genetics. I. Series. II. Beurton, Peter, J. III. Falk, Raphael. IV. Rheinberger, Hans-Jörg.
QH447 .C66   2000
572.8'38 21 – dc21   99-042106

ISBN 0 521 77187 0 hardback

Transferred to digital printing 2003

Contents

Contents

Introduction

Everybody knows about genes. One can read about them in the press. Often we are told that genes are selfish and help themselves rather than the bodies they are housed in. Genes play their role in the nature/nurture debate and in health care. Also, there is an urgent need to conserve the biodiversity around us for future generations that has, of course, to do with genes. There is a big science industry of genome sequencing that is an inventory-taking of all of man's (and other organisms') DNA. Darwin searched a lifetime in vain for the hereditary units, and indeed, genes are indispensable in modern Darwinian evolutionary theory. Early this century, genes were inferred from the Mendelian behavior of traits. The year 1953 marked a breakthrough when Watson and Crick disclosed the double-stranded helical structure of DNA. This suggested an elegant explanation of how genes could replicate themselves from one generation to the next but also serve the purpose of building an individual organism in each generation. Henceforth, the gene came to be viewed as a piece of DNA that coded for a protein or, more generally, a functional or structural product. Genes were seen as inviolable messages passed down the generations (save for occasional mutations) and as the ultimate causal factors lying behind development. Once, these findings were considered evidence for one of the most successful research strategies in the life sciences during the first half of the twentieth century.

Molecular biological discoveries over the last fifty years have made this story vastly more complicated in the details. Somewhat detached from the gene as a public icon, but also unknown to many biologists, these new findings have caused a watershed during the last few decades. The more molecular biologists learn about genes, the less sure they seem to become of what a gene really is. Knowledge about the structure and functioning of genes abounds, but also, the gene has become curiously intangible. Now it seems that a cell's enzymes are capable of actively manipulating DNA to do this or that. A genome consists largely of semistable genetic elements that may be rearranged or even moved around in the genome thus modifying the information content of DNA. Bits of DNA may be induced to share in the coding for different functional units in response to the organism's environment. All this makes a gene's demarcation largely dependent on the cell's regulatory apparatus. Rather than ultimate factors, genes begin to look like hardly definable temporary products of a cell's physiology. Often they have become amorphous entities of unclear existence ready to vanish into the genomic or developmental background at any time.

Indeed, we are thrown back into a position where we must renew the Socratic question, *What are genes?* It has frequently happened in the history of the sciences that basic concepts became so engrained and all-encompassing that their meaningfulness tended to decline. Today, who would seriously ask, What is life?, or, What is protoplasm? But the case of the gene seems to be in a different category, at least when viewed from the present perspective; and this provides the grounds for the present volume. Nearly all modern textbooks on molecular genetics show their authors' implicit awareness that there is a new (if not outrageous) issue of genes. Though it has been dealt with explicitly in more specialized journals by historians, philosophers, and working scientists as well, the issue has not been brought sufficiently into the open and probably has not received the attention it deserves from a broader audience. We want to provide a timely remedy to this situation by devoting a whole volume to the gene as a hot spot of issues. This volume, then, focuses on the gene and its difficulties as a subject of the sciences, not on the gene as a public icon. Though there is no longer any standard view of the gene, our undertaking is not iconoclastic. Far from destroying the image of the gene, this volume attempts to treat the gene as a focal point of interdisciplinary research.

The contributions to this volume have been arranged under four headings: In the first section the notion of genes in relation to traits and the tension this relationship has engendered is discussed. Fogle re-examines the present molecular condition of the gene. He argues that it is impossible to bridge the gap between the gene as a molecular biological and as a Mendelian unit without producing internal inconsistency. Yet, some consensus definition of the gene may be reached within the molecular sphere by comparing patterns of biochemical architecture of actively transcribing DNA regions. Fogle reviews at some length the possibilities and heuristic limits of such a consensus. Schwartz and Gifford show that even within the traditional Mendelian framework genes have been in a sense deeply elusive entities: Because genes interact it is often impossible to specify which trait is caused by a particular gene. However, a phenotypic difference between individuals resulting from gene mutation usually can be traced back to the alteration of the single gene in question. Schwartz shows that already Morgan had a keen awareness of this problem and she develops some of the consequences of this insight. But even the construal of a gene as that which "makes a difference" is bound to reintroduce ambiguities because the expression of such differences is also environmentally – and therefore often population – dependent, as shown by Gifford.

We readily speak of the new issue of the gene, but it is not easy to trace its original sources. New insights also color the past, and in retrospect, a continuity of problems extends backward into the past. The second section contains the bulk of the historical contributions. Gayon traces the concept of heredity in a broad sweep from pre-Darwinian days, when it was seen as a magnitude, through its structural and particulate phase into the molecular biological dissolution of the gene. But he reminds us that even in the most convoluted cases of molecular biological gene editing it always must be true that a material structure is preserved from one generation to the next to provide the material basis for the existence of such a process. This, even Goldschmidt would not have denied. Dietrich gives an historical account of Goldschmidt's demission of the gene more than half a century ago. Yet in view of the new molecular findings this account reveals a remarkable modernity in Goldschmidt's thinking. Holmes reexamines Seymour Benzer's work on the fine structure of the gene in the bacteriophage T4 of *E. coli* at the end of the 1940s and during the 1950s. Shortly after Watson and Crick established their DNA model in 1953, Benzer showed that the units of mutation, recombination, and function, all of which had been used previously to identify genes, were not coextensive in terms of DNA. With hindsight, then, the evidence provided by Benzer may be taken as a proof that there never was an unambiguous definition of the gene in terms of DNA.

The third section deals with questions of genes in development. Keller shows in a focused analysis how in the 1960s – largely as a consequence of the singling out of DNA as the genetic material – the notion of a developmental program was reduced to that of a genetic program. In contrast to this reductionist bias toward genetics, the last two decades have witnessed the unprecedented discovery of highly conserved "master genes" that make sense only in the realm of development. Such developmental genes control similar aspects of body structure across diverse phyla of the animal kingdom. Against this background, Gilbert puts forth an embryologist's perspective on the differences between the classical gene of neo-Darwinian thought and the reasserted developmental gene, while Morange gives a survey of the history of these recent findings. The conclusions are of a notable harmony: Master control genes do not exist independently but only as part of a highly complex regulatory network manifest in development (Gilbert); it is possible that they do not control anything and are no more than a response of molecular components to local configurations when used to build organisms (Morange); but even the whole developmental program lacks a specific location: As an entity it is "everywhere to be found" (Keller in a footnote). Where then are the genes?

In the fourth section, thoughts are presented on the conceptual prospects of the gene in view of the present impasse. Rheinberger, in a countermove against overzealous attempts toward integration, makes a virtue of the gene's fuzziness. In a short historical discourse he argues that it was the gene as a boundary object within a set of biochemical practices that ensured the spectacular success of molecular biology in this field, and so we should rather try to understand why fuzzy concepts work so well. Griesemer makes the novel and unique attempt to reduce genetics to development rather than the other way around. In the process, he introduces a variety of not-yet-familiar concepts that defy any short description (read the paper!) and pursues various novel avenues of thought. One of the important upshots is that it is the stability of reproduction processes that causes the invariance in genetic coding and of genes rather than the other way around. Beurton argues that it is differences among individuals that bring genes into being. According to him, natural selection, in the process of working on adaptive differences among individuals, leads to a precipitation of genic structures in otherwise continuous DNA strings. If genes are secondary products of organisms as can be distilled from Griesemer and Beurton, this may also provide a powerful argument against genic reductionism.

The present state of the gene does not allow for a fully integrated story at the end. The final review by Falk is a one-man attempt at retelling this open-ended theme by picking up once more some of the threads laid out in this volume. It might be helpful for gaining some access to this diversity to read Falk's essay first. This volume is an interdisciplinary probing of perspectives on the gene by philosophers, historians, and working scientists. As editors we have been pluralists, though not necessarily in how each of us handles the gene in his own contribution; here differences will be discerned. Most of the contributions bristle with potential cross references, but we preferred not to press the authors very hard to be explicit. Cross referencing depends crucially on any one individual's own predilections, and more explicitness in this respect would have meant less suggestiveness and may also lower the exploratory initiative of the reader.

Books also have their own causes. This volume is the outcome of a project that was set up when the Max Planck Institute for the History of Science was founded in Berlin in 1994. Due to the initiative of the founding director Jürgen Renn and of Wolfgang Lefèvre, who constituted the bulk of the institute's personnel in the early days, a workshop on "Gene Concepts and Evolution" was organized in early 1995, and a second one on "Gene Concepts in Development and Evolution" for the fall of 1996. (We are grateful to Renn and Lefèvre for this original stage-setting.) All contributors to this volume participated in at least one of these workshops, and together, these workshops provided the indispensable intellectual setting and impetus for what followed. (The workshop material was made available to a broader audience in nos. 18 and 123 of the Institute's Preprint Series.) Yes, we set up these workshops because we wanted to produce a book on genes. It is a major part of this institute's trade to examine fundamental concepts in the sciences.

Peter J. Beurton   Raphael Falk   Hans-Jörg Rheinberger

List of Authors

Peter J. Beurton
Max Plank Institute for the
    History of Science
Wilhelmstrasse 44
10117 Berlin
Germany

Michael R. Dietrich
Department of Biology
Dartmouth College
Hanover, NH 03755
USA

Raphael Falk
Department of Genetics
The Hebrew University of
    Jerusalem
Jerusalem 91904
Israel

Thomas Fogle
Department of Biology
Saint Mary's College
South Bend, IN 46556
USA

Jean Gayon
Centre de Recherches en
    Epistémologie et Histoire
    des Sciences Exactes et des
    Institutions Scientifiques
Université Paris 7-CNRS
37 rue Jacob
75006 Paris

Fred Gifford
Department of Philosophy
Michigan State University
East Lansing, MI 48824
USA

Scott F. Gilbert
Department of Biology
Swarthmore College
500 College Avenue
Swarthmore, PA 19081-1397
USA

James R. Griesemer
Department of Philosophy
University of California at
    Davis
Davis, CA 95616
USA

Frederic L. Holmes
Section of the History of
  Medicine
Yale University School of
  Medicine
P.O. Box 208015
L130 SHM
New Haven, CT 06520-8015
USA

Evelyn Fox Keller
Science and Technology
  Studies
Massachusetts Institute of
  Technology
E51-263
Cambridge, MA 02139
USA

Michel Morange
Ecole Normale Supérieure
Département de Biologie
Unité de Génétique
  Moléculaire
46 rue d'Ulm
F-75230 Paris Cedex 05
France

Hans-Jörg Rheinberger
Max Plank Institute for the
  History of Science
Wilhelmstrasse 44
10117 Berlin
Germany

Sara Schwartz
Department of Philosophy
Program for the History and
  Philosophy of Science
The Hebrew University of
  Jerusalem
Mount Scopus
Jerusalem 91905
Israel

1

Genes and Traits

1

The Dissolution of Protein Coding Genes in Molecular Biology

THOMAS FOGLE

ABSTRACT

The consensus gene, a methodological outcome of the rapid growth in molecular biology, is a collection of flexibly applied parameters derived from features of well-characterized genes. Broad flexibility unites research programs under one umbrella and simultaneously promotes the false impression that the molecular gene concept is an internally coherent universal. This suggests limitations for genomic interpretations of information content in biological systems and for explanatory models that use genes as a manipulative. Genomic referencing, the development of systemic relationships among DNA domains, will more fully interconnect molecular genetics to biology than the molecular gene alone. Advances in understanding regulation and expression of DNA and the current interest in large-scale sequencing will necessarily supervene on much of the attention currently bestowed on molecular genes.

INTRODUCTION

The gene concept, long regarded as a unit of inheritance, undergoes continuous transformation to accommodate novel structures and modes of action. A little more than a decade after the rediscovery of Mendel's work in 1900, new analytical strategies emerged for mapping genes as loci in a linear array on a chromosome. During the 1940s, the one-gene – one-enzyme model revealed that genes act to generate specific cellular products, a precursor to the science of molecular genetics. In the years that followed, the gene underwent further change. First, the double helix model of DNA made famous by Watson and Crick revealed the physical structure for particulate inheritance. Later efforts clarified the biochemistry of gene expression.

3

Today, in the era of genomic sequencing and intense effort to identify sites of expression, the declared goal is to search for genes, entities assumed to have physical integrity. Ironically, the sharper resolving power of modern investigative tools make less clear what, exactly, is meant by a molecular gene, and therefore, how this goal will be realized and what it will mean.

The legacies of particulate inheritance, localization through mapping and the Central Dogma, shape current perceptions of the gene. Although the empirical details are elaborated today, molecular genes retain an imprint from the past. In a previous paper (Fogle 1990) I analyzed the difficulty with continued attempts to bridge the gap between the Mendelian gene as a "unit of inheritance" and molecular genetics. Text-style definitions strain to find coherence when they incorporate language from both eras. Generic definitions, and hence what I termed "generic" genes, lack internal consistency.

Here, I view the problem through a different lens. The identification of a molecular gene does not stem from definitions. It is a methodological process. Genes are recognized by formally or informally comparing elements of structure, expression, and function to those previously documented. Properties and physical elements for the molecular gene concept have broad social acceptance in the community of molecular biologists. For example, detection of an RNA product serves as strong evidence that there exists a site of transcription, a gene, that acts to generate the RNA. RNA is one component from a collection of consensus features found commonly among well-described genes.

The criteria necessary to anoint new genes require research programs to adopt a community structure that places value on particular chemical states, events, and conditions while accepting considerable flexibility on how to apply them. Flexibility is essential because the large (and growing) array of molecular conjunctions prevents a strict application of rules for the molecular characterization of a gene. The need to bring a set of empirical results in line with other claims for genes forces research programs to emphasize different features in different situations or for different purposes. Molecular genes, then, are best understood as a general pattern of biochemical architecture and process at regions that actively transcribe the product of an ongoing development of consensus building in the face of rapidly changing empirical evidence. Hence, I term this shared interpretation to be a "consensus" gene.

At present, there is strong momentum to absorb new molecular revelations into the consensus gene rather than effect a more fine-grained description of molecular parts and processes. The problem is analogous to that of evaluating when a related group of organisms should be clustered into one taxonomic group or splintered into several. The outcome, sometimes contentious, rests on the analysis of shared characters in relation to established taxa. A desirable outcome is to achieve a widely accepted taxonomic solution for the purpose of efficiently characterizing the biology of that group. In taxonomy, lumping different elements into a single taxon may impede deeper biological and/or evolutionary insights. Similarly, forcing diverse molecular phenomena into a single Procrustean bed, i.e., the gene, implies a universal construction. Therefore, the gene as a molecular vehicle for causation is an ambiguous referent. I explore the difficulties arising from the embrace of the consensus gene and discuss heuristic limitations of the gene concept.

THE PROBLEM WITH MOLECULAR GENES

The consensus gene is an abstraction of molecular detail, a socially generated model for what a gene is supposed to be, formed through the expected parts and processes that empiricists associate with it. Genes are identified by seeking a fit, or at least a partial fit, using empirical evidence at hand against the backdrop of an idealized construct, a consensus gene. The process supports genic claims of different entities with shared properties.

The consensus gene, a summary of the cellular route for expression, acts through production of RNA products that may or may not be translated into polypeptides. Function and structure are inseparable. Even when genes are identified strictly from physical readouts of the DNA sequences, functional significance is inferred by analogy to more fully characterized molecular sites that have similar organizational motifs. For example, detection of a common promoter sequence known as a TATA box, a binding site for the enzyme necessary to initiate transcription, signifies a nearby site of expression. By inference, the presence of the TATA box indicates that neighboring

DNA harbors the potential to produce a transcript with functional significance for the cellular system. Hence, the TATA box is a structural component, a consensus feature,[1] contained within a gene. A consensus gene, in its stereotypical format, places importance on the localized segment of DNA that forms the transcribed region. Additional nucleotide strings (elements) may reside externally or internally with respect to the site of transcription. In addition to TATA boxes, a variety of domains are essential for gene activation and regulation. Among other roles, domains bind RNA polymerase, the enzyme that copies one of the two strands of DNA to form a complementary sequence of RNA. Eukaryotic cells can process newly formed RNA by cutting and removing internal sections known as introns. Most eukaryotic genes have introns, sometimes several dozen. Coding regions, termed exons, are spliced into a contiguous piece of mature RNA ready for translation into a polypeptide at a ribosome. Bordering the coding region, or open reading frame (ORF) of the mature RNA, is an untranslated leader sequence at one end and a trailer sequence at the other. Start and stop codons flank the coding message.

The consensus gene implies a high degree of uniformity among genes and seems, at first glance, to be an internally consistent description of parts and action. However, no simple description embodies the breadth of molecular genes claimed by empiricists (see also Carlson 1991; Falk 1986; Fogle 1990; Kitcher 1982; Portin 1993). Therefore, it is impossible to retreat to abstraction about genes without masking the diversity within. The consensus gene is a framework, not a full elaboration of biochemical detail. To what extent does an outline of its principal components and interactions generalize? I will show that consensus mode of molecular biology struggles uncomfortably to unite disparate phenomena under one banner, the gene.

GENES AND THEIR PRODUCTS

The consensus gene embraces multiple products from a single locus. One way this can occur is with sliding edges. Another is through combinatorial splicing of the transcript.

Some genes have two or more staggered promoter sites that form distinct transcripts encoding different polypeptides. The human dystrophin gene (D'Souza et al. 1995) has at least seven promoters; each regulates expression in a tissue-specific manner, leading to production of polypeptides that vary markedly in size. The many products are considered to arise from a single gene, not a set of different genes that share many parts.

In addition to sliding edges on the transcript, multiple polypeptide products can result from alternatively spliced RNA molecules. Many examples of combinatorial splicing among subsets of exons are known (see Hodges and Bernstein 1994).

In deference to the Mendelian tradition, there is resistance among geneticists to subdivide a region into multiple genes when the variant products share functional relatedness and occupy a single locus. By centering the genic claim around localization of a DNA site for expression and functional significance for the cellular system, fuzzy borders or multiple products can be tolerated.

Despite differences in form, loci that produce multiple products share much of their biochemistry for expression. The relationship between DNA coding and polypeptide formation occurs through a recognizable and common set of events. The continuity of pattern and mode binds production of many products under one linguistic construct, the gene. The embedded familiarity reinforces the central framework of the consensus gene.

The consensus gene readily absorbs convoluted twists on the traditional route to production of a functional product, as demonstrated by "inside out genes" (Tycowski, Mei-Di, and Steltz 1996). Usually spliced exons contain coded information and introns are nonfunctional. The transcript of the U22 snoRNA host gene is processed as usual to remove the introns (nine in the human form and ten in the mouse form) and splice the exons into a segment of mature RNA. The spliced RNA, however, lacks coding ability whereas the introns form RNA constituents of the nucleolus, a nuclear structure that participates in the assembly of the ribosome. Unlike all other genes studied to date, processed introns are functional and spliced exons are not.

The consensus gene of molecular biology embraces the "inside out gene" as new in form, not new in kind. It retains nearly all the structural and biochemical activities of protein coding genes except translation, and except the many types of functional RNA that are processed. The "inside out gene" widens the biochemical modes of expression attributed to the molecular gene. As the consensus gene accommodates new molecular events like the "inside out gene," it must incorporate more contingencies into its fold.

SOLUTIONS TO THE ONE-LOCUS – MULTIPLE-PRODUCT DILEMMA

The molecular revelations from multiple products and biochemical novelties suggest two alternative solutions. Either enlarge the constellation of biochemistry for the gene or propose narrower guidelines for genic ascription. Even prior to the discovery of inside out genes, there was no agreement in the literature on whether multiple functional products from a localized segment of DNA should be considered more than one gene. Lewin (1994) argues that we can reverse the usual statement "one-gene – one-polypeptide" to "one-polypeptide – one–gene." He is emphatic in stating that these are "overlapping" or "alternative" genes.

Lewin's claim is a re-evaluation of the meaning of the gene, yet he is uncommitted to pursuing its implications or upsetting the current paradigm. The implications to the molecular genetics community are substantial. Taken at face value, Lewin's proposal would require a revision of the nomenclature system for thousands of loci as a consequence of his call for a more refined relationship between functionality and a gene. It would also profoundly influence estimates of gene number for humans and most other eukaryotes. Lewin does not discuss either the methodological or ontological consequences. He is clearly ill at ease with the consensus gene that readily accepts multiple products. I suspect that he is applying a Band-Aid to a problem, one that he considers worthy of further reflection, but not one that he takes too seriously.

A more widely held perspective is that polypeptide "isoforms," proteins with nearly the same amino acid structure derived from one expression site, originate from a single gene (for example, Strachen and Read 1996). Here, similarity in structure and function of the products suggest a natural grouping into one causal unit. For those cases in which polypeptides are very different, an indicator of functional divergence, some authors recommend subdividing a site of expression into separate genes (Alberts et al. 1994). How different do the polypeptides have to be to split the locus into more than one gene? Molecular biologists do not quantitatively evaluate polypeptide divergence for this purpose. Like Lewin's call for gene splitting of alternatively spliced RNA products, the recommendation to discriminate types using the polypeptide and/or function is an ad hoc solution to situations that do not fit a one-gene – one-product model. The solution is offered more as a helpful suggestion than as a committed proposal to redefine the gene.

Defining "genes" by working backward from the polypeptide is a slippery venture. Many polypeptides undergo post-translational modifications into a functional form. Conventionally, genic identity correlates with the primary product of translation. Post-translational changes in structure are secondary effects of cytoplasmic interactions with the polypeptide. If function becomes a dominant criterion for the task of mapping the locus, as Alberts et al. recommend, then translation no longer serves as a boundary condition. This is not the intended consequence of the proposal. Their hope is to clarify parameters for a gene. Instead, they expand possible interpretations.

Several examples will show how problems arise with their proposal. A variety of post-translational modifications have been documented. After translation, some polypeptides, particularly neuropeptides or hormones, subdivide by proteolytic cleavage. Polyproteins are consistently regarded as products of one gene, whether or not they cleave into identical or divergent forms. For example, the DNA locus for the alpha factor regulating mating behavior in yeast (Fuller, Brake, and Thorner 1986) encodes a translated polypeptide clipped into four identical peptides. In contrast, an ascribed gene in silkworms produces five functionally distinct products (a diapause hormone, pheromone biosynthesis activating neuropeptide, and three other neuropeptides) cleaved from a 192 amino acid precursor (Xu et al. 1995), each an independent functional unit.

Alberts et al. do not distinguish between subdivided polyproteins and polypeptides generated by alternative splicing, yet both can give rise to more than one functional form. The consensus gene is their salvation. By advancing the importance of function, imposing it as a tool for evaluating the expression site when needed, they can side-step the problems that result if one hardens the rules and applies them to every case. They build a molecular case for a gene using a select cluster of consensus components with structural and transcription and/or translation elements. Alternative spliced variation takes place after transcription but prior to translation, two tightly entrenched processes for the protein coding model of the gene. In contrast, post-translationally formed polyproteins lie beyond the physiological boundary of gene-associated biochemistry. For DNA loci encoding polyproteins, translation is a boundary condition that makes functional distinctions unnecessary. Both lumpers and splitters of genes draw the same sharp line in the sand. Polypeptides formed directly from translation are qualitatively different from polypeptides that undergo post-translational modifications. Both views cling tightly to biochemical mechanisms to locate the gene. Function is not a universally important criterion; it gains or loses importance in a particular case against the backdrop of other consensus elements (structural and/or biochemical).

The comparison between genes encoding polyproteins and alternative spliced products suggests a set of parameters for interpreting well-characterized sites of the consensus gene. Three properties with variable weight designate a molecular gene: localization to a transcript-generating segment of DNA, physiological boundaries located at pre-translational (alternative splicing) compared to post-translational (polyprotein cleavage) activity, and an investigator-based assessment of functional divergence among products. Whether a DNA site constitutes a gene depends both on empirical evidence for that case and subjective emphasis of the parameters.

A closer look at expressed sites indicates that this appealing triangulation of conditions provides little help toward rigorously articulating molecular properties for the gene. Translation does not always neatly divide the origin for variation between pre- and post-translation, creating an additional source for a many-to-one relationship between the molecular phenotype and a locus in DNA. The mammalian gene governing S-adenosylmethionine decarboxylase (AdoMetDC) has two ORFs. The short form codes for a hexapeptide within the leader sequence of the larger AdoMetDC coding section (Hill and Morris 1993). The hexapeptide down-regulates AdoMetDC translation in a tissue specific fashion. The investigators avoid confusion about which is the "real" gene by subordinating the smaller ORF as a regulatory element of AdoMetDC. Once again, the chosen rhetoric is consistent with the consensus view that readily accommodates novelties of form and process in the gene concept. In this instance it also forces the investigators to choose referents to identify the gene. That is, the AdoMetDC coding region could be semantically repositioned as the trailer sequence for the production of the hexapeptide. Hill and Morris select the AdoMetDC coding region as the referent for the gene because they place greater functional importance upon it.

Both polypeptides of AdoMetDC are primary products of translation and could be viewed as separate genes. The investigators seem to unpack the consensus gene as follows. The two polypeptide products originate from a common site of transcription and a common transcript that, until translation, unifies their biochemistry of structure and action. The multiple polypeptide products, through their combined presence, effect one functional end. The smaller polypeptide is viewed by the investigators as a product of a regulatory region for the larger polypeptide. The larger polypeptide is, therefore, accorded the role of the principal gene product (a decarboxylase enzyme) through prior recognition of its importance to cellular physiology.

In this instance, the genic claim is, in a classic sense, a unit of function. The mechanics of transcription and translation are sufficiently similar to other claims for genes to warrant support at the molecular level. Consistent with the consensus gene model, Hill and Morris reconfigure the coding site for two polypeptides into one gene. Many other similar cases could be described.

The assignment of one gene for the CCAAT/enhancer binding protein (C/EBP) of vertebrates (Calkhoven et al. 1994) entails an even more careful choreography of semantics. The messenger RNA contains three ORFs, each starting at a different point along the transcript. The product of the shortest ORF regulates the ratio of product from the two longer, overlapping coding regions that have different start sites for translation. When Calkhoven et al. discuss the nucleotide sequence specific for one of the two large ORFs they choose the term *cistron*, a unit of function. This allows them to avoid attaching a gene label to the three coding regions, eliminating the need to declare whether they are working with three genes or one. Once again, unity of function, against the backdrop of a common molecular biology, provides a serviceable means for representing C/EBP as one gene. The interpretative rendering by Calkhoven et al. and others demonstrates how context, a normative mode for the consensus gene, impacts what is reported to be a gene and how difficult it would be to develop an internally consistent systematic taxonomy for genes.

In this section I have attempted to show how multiple products from one locus conspire to force arbitrary decisions about whether one or more genes are represented. The real problem is that there has been a steady creep of new genetic twists that must either be accepted as part of the structure and biology of the gene or abandoned in favor of an alternative description of molecular events. The reluctance to abandon the molecular gene, and instead, work around problems as they arise, erodes coherence. One might ask when told of a newly discovered molecular gene, "what kind? – one that produces a single product? multiple products? multiple products that have very different functions? functional isoforms? multiple products formed during transcription? or processing? or translation?"

GENES AND CODING

The translational assembly interprets the genetic code. After transcriptional processing removes introns and splices exons, RNA is read in tandem triplets of codons. Each codon specifies an amino acid in the growing polypeptide. For some RNAs there are other mechanisms for readout (Gesteland, Weiss, and Atkins 1992). In some cases, the ribosomal assembly skips from one to fifty nucleotides in the RNA, shifting the reading frame before continuing. In other instances, the meaning of the code changes to read, for example, a stop codon, a polypeptide termination signal, in place of an amino acid. A particular physiological state, not just the transcript itself, causes the translational change. Either form of recoding partially shifts informational specificity for the product into the cytoplasmic space, removed from its usual habitat in the sequence of nucleotides of DNA.

The translational machinery is often likened to a computer reading software, an ungainly metaphor with only superficial similarities. The DNA, imagined as a bit stream of computer code read by hardware, sends a template copy of RNA to the site of translation and threads the sequence through the ribosome to read nucleotides in consecutive blocks of three. With cellular recoding, the cytoplasm is rewriting the software program produced by the DNA. The cellular architecture itself contains an information coding ability that becomes apparent during translation.

The coding regions are not the only portion of the transcript that direct the form and function of the product. Leader and trailer sequences that border the ORF of a transcript are crucial in RNA recruitment for translation (Sonenberg 1994) and also regulate activation or rates of translation. For example, the insulin-like growth factor gene (IGF-II) forms two mature transcripts with an identical coding region and trailer sequence but leader sequences of different lengths (Nielsen and Christiansen 1995). The shorter transcript participates in protein synthesis while the longer form complexes with protein to become a ribonucleoprotein particle, a component of a ribosome. One functional product is translated, the other not.

The IGF-II gene produces qualitatively different functional products, an RNA and a polypeptide, that share a common transcript and DNA locus. One might expect that functional divergence of the IGF-II gene products would lead to a claim for two genes since cases of functional relatedness (AdometDC and C/EBP genes) led to a claim for one gene. The IGF-II locus demonstrates a different set of problems and a similar approach to a solution. To unite radically different end products as components of one system requires that other consensus properties compensate. IGF-II becomes a normative gene by ignoring conflicts with the standard outline for translational events and placing emphasis on pre-translational events.

The cellular biochemistry of gene structure and expression consists of a set of contingent statements substantially larger than molecular biologists, such as Lewin or Strachen and Read, or Alberts et al. seem to admit. Gene-splitters run the risk that any post-transcriptional modifications of RNA altering the polypeptide product, any novel variation of translation, any post-translational chemical modifications of a polypeptide, map to a separate gene. Equally difficult to justify are conservative renditions for genic enumeration that de-emphasize function and read different physiological constructs of RNA or polypeptides as members of one gene. The genes of research programs, as opposed to generic descriptions in texts, form a continuum of material forms and processes. There are no discrete functional packets or molecular mechanisms in the protoplasm to serve as guides for delimiting a gene.

From this unsettling outcome, a molecular gene lacks demarcation without at once specifying the temporal and spatial cyto-complex of the system. Accordingly, the dynamics of the system, more than just the sequence and a vague notion about function, characterize a gene. The route geneticists choose to move past this roadblock, as we have seen, is to craft ad hoc solutions to subsets of problems (such as Lewin's one-polypeptide – one-gene proposal). In this way, flexibility is maintained and genic definitions should be read only as statements about common gene patterns, e.g. most genes have introns that are nonfunctional and most have exons that are functional.

One purpose for a flexible gene concept is to link molecular phenomena to Mendelian genes. Of course, Mendelian analysis does not depend on knowledge from molecular biology. During the expansion of genetics as a discipline, when genes lacked physical description of the sort assigned today, one widely held viewpoint was to liken them to beads on a string. The very success of molecular biology in the 1950s and beyond solidified perspectives about the gene to a physical reality of one type – a site on the DNA. Much of the history of molecular biology reified that the Mendelian gene can, in principle and in practice, be described in molecular terms. As details poured forth from an expanding research enterprise in molecular biology, the molecular gene acquired greater contingency without formally abandoning the Mendelian gene.

GENES AND THEIR BORDERS

Although the same outline of molecular biology has been used to argue for a different number of genes, the most common approach is to claim that multiple products due to transcription processing, translational reading, or post-translational interactions are endpoints of a single agent, a gene whose physical origin lies in a section of

DNA. If so, its residence should have property lines. One approach is to fix the position for the genic site through the generation of a transcript. The other is to locate the gene using both the site of transcription and regions that regulate it.

In molecular biology, the term *expression* denotes active production of an RNA transcript and is an indicator for the presence of a gene. But to map the genic site through its DNA complement in RNA is to ignore either post-transcriptional cytochemistry which modifies the transcript prior to translation or pre-transcriptional sites and events that proscribe what becomes the transcript. For example, post-transcriptional changes caused by the cytoenvironment can change a noncoding intron into a coding exon through a change in the splicing pattern, preempting a simple means for crafting a molecular referent for the physical structure of the gene through a primary transcript. Pre-transcriptional influences directed by neighboring DNA elements, such as multiple promoters discussed earlier, extend this concern. The informational locus on the DNA dictating the transcript formed, or its ability to form, resides in both neighboring elements and in physiological conditions at the time of expression.

Many types of elements have been described (for example, silencers and enhancers); each is a short length of DNA that affects the timing or rate of transcription. Their position and number per gene is highly variable. The activation of protein coding genes is a stepwise series of interactions between protein and multiple DNA elements upstream from the start of the transcribed region. An assembly of more than a dozen globular proteins attracts RNA polymerase to the promoter to initiate transcription. The interplay of multiple DNA elements and multiple proteins is a key regulatory mechanism for gene expression. Therefore, many recent descriptions for a molecular gene include DNA elements within their borders, even at the expense of clarity about limits and boundaries. Lodish et al. (1995) state that a gene is the "entire DNA sequence necessary for the synthesis of a functional polypeptide or RNA molecule" (p. G-8). Similarly, Alberts et al. (1994) consider the gene to include the "entire functional unit, encompassing coding DNA sequences, noncoding regulatory DNA sequences, and introns" (p. G-10). Note the juxtaposition of the Mendelian language of "functional unit" and the molecular language of "DNA sequences." The consensus gene results from a struggle to hold on to the past and represent the present. Methodologically, domains are treated in much the same way as other aspects of the consensus model. Despite clear proposals either for or against inclusion of elements as part of the gene, they are included or excluded to fulfill specific needs.

A gene concept that includes all DNA domains connected to the ability to express is not to be taken literally. Surely the inclusiveness does not mean that a substantial fraction of the genome is a domain of each gene because *in vivo* activity of every gene is interdependent on the products of many others. The claim is much more restricted and localized; a molecular gene produces a transcript together with other regional domains. Yet detected interactions at even the local level suggest complex relationships among domains. Individual or joint synergistic effects of elements on expression can act like a rheostat dialed to the lowest active setting to produce transcriptional effects barely above a detectable level or magnify the rate of expression. And empirical limits complicate the proposal; it is unrealistic to experimentally reference every regional genomic domain with respect to all others.

We are left with a sketchy framework for determining when a DNA segment is part of a gene. Elements are often judged by whether they affect expression and act in a local manner. If one looks more broadly at a larger swath of the genome, the problems associated with elements are further evident.

The sharing of regulatory elements contributes to the problem of finding physical borders for genes also. The beta globin gene cluster in humans produces five related polypolypeptides that form part of the hemoglobin protein. The locus control region (LCR), is positioned at one end of the gene cluster and regulates their expression in a developmental-specific manner (Wood 1996). The LCR orchestrates the timing of transcription activation and rate. Embryonic tissue produces high levels of epsilon globin and low levels of beta, A-gamma, and G-gamma chains. Fetal cells have large quantities of A-gamma and G-gamma globin and small quantities of beta globin. By adulthood, a small amount of delta globin can be detected and beta globin production predominates. A genic model that includes regulatory sequences cannot deny the LCR as part of its structure. It is clear from the literature, however, that this is not the case. The LCR is presented as a separate domain, neither a component of any molecular genes nor a gene itself. For multiple local transcripts, like the beta globin cluster, regulatory elements are attributed responsibility for functional coordination of globin production. Isolating the LCR from any one gene more accurately conveys the functional relationships among the domains comprising the cluster. It also contradicts definitions which embed domains within genes and reveals that the physical dimensions of genes once again depend on methodological need.

Neither the edges of the gene, its relationship to function, nor its biochemistry of expression are constants that can aid the formulation of a finely characterized molecular gene. However, that genes do localize is an important part of the genic claim.

GENES, PSEUDOGENES, AND GENIC STATES

Additional modalities of molecular variation further erode prevailing views of the molecular gene as an integrated localized structure. Expression in trypanosomes, protozoan parasites, and nematode worms commonly requires trans-splicing a short and a long RNA transcribed from different regions of the genome. The spliced leader is less than a few dozen nucleotides long and not part of the coding region for a polypeptide. Maroney et al. (1995) find that trans-splicing in nematodes is essential for "translational efficiency," subordinating the smaller entity as a contributor to the effectiveness of the larger, coding RNA. The smaller locus forming the trans-spliced transcript does not have protein coding function, although this is not unusual. Many genes transcribe RNA that does not translate. The DNA site that transcribes the leader has the hallmarks of gene structure and expression without the title. It is treated as a buttress for the integrity of the larger RNA companion.

The bacterial genetic system presents a complementary example. To degrade an abnormal protein formed from a faulty coding sequence, two unconnected transcripts expressed at different sites on DNA will form one translated product. The transcript that encodes the ability to degrade the abnormal transcript joins the ribosomal complex, remains unbound with the incomplete message, and attaches a linker amino acid together with an additional ten amino acids coded by its nucleotide sequence (Keiler, Waller, and Sauer 1996). The eleven amino acid tag signals the cell to dispose of the polypeptide. The authors consider each locus, independently transcribed yet cotranslating a polypeptide, to be a separate gene.

The organizational and functional construction of the bacterial and trypanosome loci do not reveal why the former should be a two gene system and the latter a one gene system. Both transpliced RNA and the bacterial degradation system bring together products from two loci into the translational machinery. The bacterial degrading system and trypanosome loci coding each contain a structural and regulatory domain on the transcript. Each could be interpreted as one gene or two. The genic systems of bacteria and trypanosomes lead the investigators to opposite interpretations to give explanatory flow to the empirical evidence. Trans-spliced sections take on a parochial role as a regulator of translation; the larger transcript of the two becomes the central object of inquiry, promoting its functional importance. On the other hand, the polypeptide degrading system of bacteria serves a global function for the cellular system. Therefore both the RNA that does the degrading and the RNA product that requires degradation, are functionally independent, the products of two genes.

The perceived significance of a DNA locus to the cell can be critical to the case by case interpretation of the presumptive gene. When function is unknown, molecular biologists sometimes postulate a contribution to the cellular system from contextual cues. Functional effect operates through expression, the formation of a transcript. Pseudogenes are categorized separately from true genes because they have low rates of transcription. Alu-sequences, as one example, are short pseudogenes found in large copy number in the human genome. They have the signature of genes, many capable of transcribing small amounts of RNA with ORFs that do not undergo translation. The dividing lines between whether something is or is not a gene can be thin. Pseudogenes point to a minimum level of expression as necessary but not sufficient.

In addition to variable levels of gene expression, the chemical structure of DNA can act like a toggle switch, alternating between two states that influence gene expression. A string of nucleotides that acquires methyl groups on the cytosine bases can activate or, more commonly, repress transcription. In some cases the methylated pattern, known as imprinting, is preferentially inherited through one sex, resulting in the maternal or paternal expression of specific genes. In others, the pattern of methylation changes between tissues or stages of development. Bartolomei et al. (1993) report a domain of methylation surrounding the transcribed region and promoter of the H19 gene, beyond which they did not detect repression. Imprinting of the H19 gene functions similarly to a regulatory element, a domain often positioned within a gene. Methylation patterns differ from regulatory elements in two ways. They are not sequence specific sites and they can spread over the entire face of the locus, as in the case of the H19 region.

Therefore, at least two classes of transcription regulating sites are present – temporal domains of methylation on nucleotides and heritable DNA strings of nucleotides (enhancers, promoters, and the like). Despite their similarities on physiological effect, their differences point to a key property of genes. Alterations in the methylation patterns differ from mutational changes that take place in DNA strings. Mutations have generational stability, reproduced during the process of replication and transmitted to descendants. Methylation patterns change when specific physiological conditions occur that are not well understood. Therefore, methylation can not be absorbed within the consensus gene as another novelty uncovered through molecular biology.

DNA action and function become meaningful in the context of a cellular system. Coding information in the DNA is necessary but insufficient for the operation of living systems. The mutual dependency of DNA and protoplasmic interactions bedevils a simplistic labeling scheme for expressed segments of hereditary information. The more molecular biology that is unpacked, the greater the need to acknowledge the mutuality of the component parts, forcing arbitrary choices about the physical edges or the physiological properties of the gene. The consensus gene devalues mutual dependency in order to locate hereditary units from a loose and changing confederation of molecular constituents. As a result, the research enterprise can successfully search for genes so long as there is no demand for a rigorous underpinning for their specification.

SEARCHING FOR GENES

Much of molecular genetics research focuses on a search for expression units that do not depend on tight matching to a universal construct. Liberal applications of the gene concept weld research programs into a community dedicated to a common mission. For this purpose, a molecular gene is a useful instrumental tool.

There are, however, consequences for vague notions about molecular genes. Talk of genes plays a major role in the intellectual advancement of evolutionary biology and organismal development. If genes are contextually dependent for structural and functional evaluation, then it is unclear how a fully realized, or at least a richly detailed, theoretical presentation would be possible using genes as an explanatory manipulative.

Coding information acts within a codependent cellular setting: localized sites of expression interacting among DNA domains and contingent upon genomic composition. Here, the term genome means more than the collective set of molecular genes of the organism; it refers to the rich tapestry of DNA domains that weave a pattern of expression. Genetic information is layered within ordered, structured chromosomes. Genomic analysis is expanding rapidly, and will unveil integration among domains positioned far apart, an outcome foreshadowed by trans-splicing. From this broader vantage point on the entirety of genetic information in an organism, domains of action and regulation connect locally and distantly positioned DNA loci into a functional network. As sophistication in the understanding of biological relationships of DNA domains increases, explanatory constructs will subordinate genes as instrumental constructs and increasingly emphasize systemic interactions to communicate insight into cellular processes.

There are many examples, too numerous to document here, that demonstrate positional and contextual integration of function at all levels of genomic organization: coordinated regulation of gene families, loops of chromatin that regulate clusters of genes, distinctive sequence patterns within chromosome banding patterns, and regional functions at the tips and centromeres of whole chromosomes. How these levels of organization cooperatively orchestrate information has yet to be explored, largely because of experimental limitations. Cellular activity requires this hierarchy of genomic information in addition to each locally acting hereditary site (a "gene," however defined). This is more than just a problem for the molecular taxonomy of genes; mutual sites of interaction interpenetrate the genome at many levels, much of which is left on the sidelines when gene number is equated with genomes and genomes with information content.

A molecular gene is a coarse parameter for genomic analysis, poorly suited for the future growth of empirical results. It may be possible to count the number of ORFs or the number of alternatively spliced products or the number of DNA sites producing primary transcripts, but it will not be possible to conduct an exact gene count, at least not without resorting to the consensus model.

The goal of the human genome project is to find the 60,000–80,000 genes, a number based on three methods of estimation. Each method plucks some parameter from the consensus gene as a tool for estimation. But, because the consensus gene is a fluid concept, the derived values are themselves a crude statement about the genome. With most of the DNA sequenced at the time of this writing, one can scan this sample of genetic information and search for the structural hallmarks of a gene (ORF, TATA box, etc.) and extrapolate from an average density of one gene per 20,000 bases to arrive at an estimate for the total number of such sites (70,000 genes). A second measure assesses the number of kinds of expressed RNAs (cloned as complementary copies of cDNA) in different tissues to determine gene number (65,000 genes). Since no one tissue expresses all genes and, as we have seen, alternative splicing and other mechanisms can produce more RNAs than local expression sites, this too is a rough estimate. The third method relies on counting the number of CpG islands, regions often surrounding promoters that have a higher density of neighboring CG bases than the rest of the genome. Slightly more than half of all expressed loci have CpG islands. The counting method offers no indication of the number of gene products or their function. Estimates for the number of genes (80,000) are consistent with the other methods. The three methods, collectively, suggest that

21

somewhere between 65,000–80,000 loci in the human genome fit the standards of the consensus gene. Note that each method is successful for the purpose of approximation and cannot be applied to a fully resolved cellular system, because to do so would require many subjective evaluations of the sort discussed earlier. Even genomes with completed sequences (a nematode, yeast, and many species of bacteria) are evaluated for gene number from the DNA readout alone, leaving ambiguities from gene action unaddressed.

Reporting gene counts, particularly for the human genome, is more than an empirical exercise. It is intended as a scale of information content essential for normal function. The thinking goes like this: Genes are functional units; thousands of functional units are present; the expression of the phenotype is significantly impacted by these thousands of units of genetic activity; the set of these genes tightly mirrors what is meant by a genetic contribution to the phenotype. The failure to successfully proscribe universal genic borders or events for expression calls into question the significance of gene counts for higher organisms. What new insight would result from discovering that there are twice as many genes as thought, or half as many? And some genes have no detectable function. On the other hand, knowing how many domains of a particular type are present might be helpful (e.g., how many CpG islands), indicative of the cellularwide importance of a specific mode of molecular interaction.

Mosaic architecture and activity among claimed genes greatly limits meaningful inference about information content, molecular activity, and functional effect. Suppose, for the moment, a complete human DNA sequence were available. It would be possible to scan the genome through a computer search for the number copies of particular domains and collections of domains, some of which might match those DNA strings currently recognized as genes. It would require many hair-splitting choices using the consensus gene as a guide. And to what end? The real advantage to detailed genomic sequencing will be to make sense of the functional contribution from combinations of domains, not to label lots of loci as genes. As valuable as they are for reductionist evaluations of the genetic contribution to a trait, they limit the potential to integrate a conceptual framework for large-scale complexity within living systems. Genomic analysis will lead to further insight about the distribution of expression sites, their relationship to noncoding chromatin, the effects of chromosome architecture on RNA activity, and much more. With a genome-wide perspective, the intact organization takes on a larger and more important biological role than do genes alone. The gene concept receives heavy attention in molecular biology. The analysis of genomes, the next horizon in genetic research, will seek functional interconnections for the entire DNA readout irrespective of size, location, or match to a consensus gene model. The gene could become a quaint term of the past (at least in molecular biology circles) replaced by language that more accurately conveys relationships among domains contributing to phenotypic effects.

There is already an explosion of terminology to bypass shortcomings of the consensus gene. The product of alternative splicing, for example, has been termed a "complex transcription unit." The LCR and globin genes have been described as a "holocomplex." Fogle (1990) proposed that a focus on domain sets for active transcription (DSATs) in their various configurations would be a more refined means to represent regions of expression. DSATs are an assemblage of the structural building blocks that could be used to construct a taxonomy for expression loci. Along a similar vein, Portin (1993) suggested a classification scheme of nine genic subtypes. Both proposals step away from the one-term-fits-all model pervasive today. Each is a partial solution that recommends a more fine-grained nomenclature for the many kinds of transcription sites. Neither provides guidelines for how to map cellular interactions (such as translational recoding or alternative spliced products) to a DNA locale. Nor does either method prescribe a set of specific rules for delimiting the physical borders of genes. However, the inseparable intertwining of causation between the genotype and the biochemical phenotype will always exist as a problem in biology and should not deter attempts to seek a closer match between empirical outcomes and the explanatory units necessary to effectively communicate them.

NOTES

1. A *consensus sequence* in the parlance of molecular biology is one that is highly conserved in evolution. Here, I intend consensus sequence to mean a component of the gene concept retained in methodological renditions about molecular evidence.

REFERENCES

Alberts, B., D. Bran, J. Lewis, R. Martin, K. Roberts, and J. D. Watson. 1994. *Molecular Biology of the Cell.* New York: Garland.

Bartolomei, M. S., A. L. Webber, M. E. Brunkow, and S. M. Tilghman. 1993. Epigenetic mechanisms underlying the imprinting of the mouse H19 gene. *Genes and Development* 9: 1663–1673.

Calkhoven, C. F., P. R. Bouwman, L. Snippe, and G. Ab. 1994. Translation start site multiplicity of the CCAAT/enhancer binding protein alpha mRNA is dictated by a small 5' open reading frame. *Nucleic Acids Research* 25: 5540–5547.

Carlson, E. A. 1991. Defining the gene: An evolving concept. *American Journal of Human Genetics* 49: 475–487.

D'Souza, V. N., T. M. Nguyen, G. E. Morris, W. Karges, D. A. Pillers, and P. N. Ray. 1995. A novel dystrophin isoform is required for normal retinal electrophysiology. *Human Molecular Genetics* 4: 837–842.

Falk, R. 1986. What is a gene? *Studies in the History and Philosophy of Science* 17: 133–173.

Fogle, T. 1990. Are genes units of inheritance? *Biology and Philosophy* 5: 349–371.

Fuller, R., A. Brake, and J. Thorner. 1986. The *Saccharomyces cerevisiae* KEX2 gene, required for processing prepro-α-factor, encodes a calcium-dependent endopeptidase that cleaves after lys-arg and arg-arg sequences. In *Microbiology,* edited by L. Lieve. Washington: American Society for Microbiology.

Gesteland, R. F., R. B. Weiss, and J. F. Atkins. 1992. Recoding: Reprogrammed genetic decoding. *Science* 257: 1640–1641.

Hill, J. R., and D. R. Morris. 1993. Cell-specific translational regulation of S-adenosylmethionine decarboxylase mRNA. Dependence on translation and coding capacity of the cis-acting upstream open reading frame. *Journal of Biological Chemistry* 268: 726–731.

Hodges, D., and S. I. Bernstein. 1994. Genetic and biochemical analysis of alternative splicing. *Advances in Genetics* 31: 207–281.

Keiler, K. C., P. R. H. Waller, and R. T. Sauer. 1996. Role of a peptide tagging system in degradation of proteins synthesized from damaged messenger RNA. *Science* 271: 990–993.

Kitcher, P. 1982. Genes. *The British Journal for the Philosophy of Science* 33: 337–359.

Lewin, B. 1994. *Genes V.* New York: Oxford University Press.

Lodish, H., D. Baltimore, A. Berk, S. L. Zipursky, P. Matsudaira, and J. Darnell. 1995. *Molecular Cell Biology.* New York: W. H. Freeman.

Maroney, P. A., J. A. Danker, E. Darzynkiewicz, R. Laneve, and T. Nilsen. 1995. Most mRNAs in the nematode *Ascaris lumbricoides* are trans-spliced: A role for spliced leader addition in translational efficiency. *RNA* 1: 714–723.

Nielsen, F. C., and J. Christiansen 1995. Posttranscriptional regulation of insulin-like growth factor II mRNA. *Scandinavian Journal of Clinical and Laboratory Investigation* 22 (Supplement): 37–46.

Portin, P. 1993. The concept of the gene: Short history and present status. *The Quarterly Review of Biology* 68: 173–223.

Sonenberg, N. 1994. mRNA translation: Influence of the 5' and 3' untranslated regions. *Current Opinions in Genetics and Development* 4: 310–315.

Strachen, T. S., and A. P. Read. 1996. *Human Molecular Genetics.* New York: Wiley-Liss.

Tycowski, K. T., S. Mei-Di, and J. A. Steltz. 1996. A mammalian gene with introns instead of exons generating stable RNA products. *Nature* 379: 464–466.

Wood, W. G. 1996. The complexities of β-globin gene regulation. *Trends in Genetics* 12: 204–206.

Xu, W. H., Y. Sato, M. Ikeda, and O. Yamashita. 1995. Molecular characterization of the gene encoding the precursor protein of diapause hormone and pheromone biosynthesis activating neuropeptide (DH-PBAN) of the silkworm, Bombyx mori and its distribution in some insects. *Biochemica et Biophysica Acta* 1261: 83–89.

2

The Differential Concept of the Gene

Past and Present[1]

SARA SCHWARTZ

ABSTRACT

The differential concept of the gene was created by geneticists at the beginning of the twentieth century to fill the need for a workable research method given the newly established formal theory that claimed many-to-many relationships between genes and traits. This concept can be reconciled with both one-to-one and many-to-many relationships between genes and traits because it deals with a different ontological level: the relationships between *changes* in genes and *changes* in traits, rather than the nature of the entities themselves. This flexible nature may also allow the term *gene* to survive its replacement by the term *genome*. The differential concept of the gene, except for its predicate, is not specific to genetics. The concept is fundamental, though to some degree intuitive, hence exhibiting power and weakness. Also, it is used continuously in studies of connections between the genotypic and phenotypic levels. In contrast to classical transmission genetics, where the interest in the relations between mutations and alternative appearances of traits is derived mainly from the need for a research method, the present interest is twofold: a research method and a research subject. The present use of the differential concept of the gene is characterized by its refinement: Different alternative states of a gene are to be related to different alternative appearances of a trait. This requires resolution capability at both the genotypic and phenotypic levels and is guided by the reductionist assumption that at least some mutations make a difference in the phenotype.

INTRODUCTION

The differential concept of the gene states that there is a relation between a change in a gene (an allele) and a change in a trait. In order to distinguish between a change in a trait and a trait, I will name the first an alternative appearance of a trait (AAT). The distinction be-

26

tween a trait and an AAT is analogous to W. E. Johnson's now classical distinction between *determinable* and *determinate* properties (Johnson 1921, chapter 11). Determinable properties are, for example, shape and color. Determinate properties are triangularity and redness, respectively.

Johnson drew attention to three features in which determinables relate to determinates that are of interest here. 1) If a particular has a determinate property (AAT), it then follows that the particular has the determinable property (a trait). If a pea is round, it necessarily has a shape. 2) If a particular falls under a determinable (a trait), it then follows that it has one of the corresponding determinate properties (AATs), although the specific determinate property is not entailed. Thus, if a pea has a shape, it necessarily has some particular shape, i.e., round, wrinkled, etc. 3) One particular cannot at the same time represent more than one of the determinates (AATs) which fall under a common determinable (a trait). A flower petal cannot be both red and pink all over. The third feature must be qualified as it does not seem that tastes exclude each other. The same is true for sounds, however, genetics does not often deal with taste (except in breeding for agricultural purposes) or sound.

The differential concept of the gene is distinct from a one-to-one relationship between genes and traits because the emphasis is on the changes in these units. The one-to-one relationship is between genes and determinable traits, while the differential concept of the gene covers the relations between alleles and determinate traits. The concept was formulated early in the history of genetics by E. B. Wilson (no later than 1912 , see Roll-Hansen 1978, 174). The present work will analyze this concept and the reasons for its appearance and will allude to some of its present-day applications. I argue that, with respect to the relations between genes and traits, genetic theory and its associated research method were traditionally guided by different premises. However, these premises coexisted because they did not contradict each other. Furthermore, I suggest that the pragmatic differential concept of the gene allowed the persistence of the problematic concept of the gene.

The differential concept of the gene, like the one-to-one and many-to-many relationships between genes and traits, considers the environment to be a constant. This attitude was reinforced by the well-controlled laboratory conditions that were used at the time the concept was conceived and later. These controlled laboratory conditions created an artificial environment that allowed repeatable phenotypic results with almost no environmentally induced variability. Unless a researcher wanted to pay attention to environmental factors, it could be ignored in the laboratory. Thus a conceptual tool that relates to the environment, such as the *norm of reaction*, was ignored in the Western genetic literature until 1950 (Sarkar 1999). The *norm of reaction* stresses that it is impossible to predict the phenotype from the genotype (Falk 1999). The differential concept of the gene is therefore more appealing to geneticists because it is more fruitful, despite its oversimplification of the etiology of phenotypic traits. Its fruitfulness and the legitimacy that science has given to the isolation of variables makes the differential concept of the gene a useful tool even today.

HISTORICAL CONSIDERATION

The unit character concept was used by geneticists at the beginning of the century. Some of them (W. Bateson, R. C. Punnett) had a preformationist understanding of it, which is not of interest to the present study. Others (W. E. Castle, J. C. Phillips) saw it as meaning a character or trait indicating the presence of an element or factor. For these geneticists the unit character concept entailed a one-to-one relationship between traits and factors – later genes. Shortly thereafter, however, tests of the universality of Mendel's laws and the continual debates such as those over Castle's contamination hypothesis (see Vicedo 1991), brought the community of geneticists to understand that the situation was much more complicated. If the phenotype may be reduced at all to the genotype, it is a many-to-many relationship; many genes may interact in a single trait, and one gene may contribute to many traits. The formal policy of genetics became that the relationship between genes and traits is a multivariate function.[2] A contradiction was created, however, because the research method was (allegedly) based on a one-to-one relationship between traits and genes. Even scientists who dealt with quantitative traits, such as Nilsson-Ehle and East, believed that they could isolate the proportional contribution of each gene to the trait. The tension between the formal policy, regarding the relationship between genes and traits, and the premise that set in motion the research method of genetics had to be resolved; the differential concept of the gene enabled this resolution.

A resolution of this tension was shown by T. H. Morgan (though he was unconscious of the tension itself). His notion was that the alternative appearance of a trait was caused by a group of genes that determined the trait, when one of the genes of the group was changed. Morgan tried to express this understanding through the nomenclature he suggested, but he finally had to give up the attempt (Falk and Schwartz 1993). He designated characters by a formula of several letters, not by one letter as was (and is) the custom. The letters symbolized the known genes which were involved in producing the character. For example, the red eye color of Drosophila was designated by the formula PVO, the mutation vermilion was designated pVO, pink–PvO, and orange–pvO (Morgan 1913). This nomenclature indicates that the eye color pink, for example, is not determined by one factor, but by the *residuum* – the factors left when this factor is missing. Morgan finally gave up the attempt for practical reasons, but his notion did maintain the concept of many-to-many relationship between genes and traits.

The next step toward reducing the tension between the many-to-many notion and empirical practice was suggested by Morgan and his student coworkers. This was to single out the differences at both the genotypic and phenotypic levels and point to them as causes and effects, respectively. This, however, distracted attention exactly from the point that Morgan tried to emphasize. His formulation tried to divert attention from the mutated factor to the remaining ones that were involved in the character, while the new suggestion focused on the single mutated factor. Moreover, this extraction of the cause and effect relationship included a problem that I will address in the next section. The new suggestion was justified in the following quotation, that is brought *in extenso* because of its importance to the present context.

Mendelian heredity has taught us that the germ cells must contain many factors that affect the same character. Red eye color in Drosophila, for example, must be due to a large number of factors, for as many as 25 mutations for eye color at different loci have already come to light. . . . One can therefore easily imagine that when one of these 25 factors changes, a different end result is produced, such as pink eyes, or vermilion eyes, or white eyes or eosin eyes. Each such color may be the product of 25 factors (probably of many more) and each set of 25 or more differs from the normal in a different factor. It is this one different factor that we regard as the "unit factor" for this particular effect, but obviously it is only one of the 25 unit factors that are producing the effect. However since *it is only this one factor and not all 25 which causes the difference between this particular eye color and the normal*, we get simple Mendelian segregation in respect to this difference. In this sense we may say that a particular factor (p) is the *cause* of pink, for *we use cause here in the sense in which science always uses this expression, namely, to mean that a particular system differs from another system only in one special factor.* (Morgan et al. 1915, 208–209, italics mine)

Here, cause is reduced to the factor that makes the difference in the phenomenon, even though many factors contribute to it. This definition allowed coexistence; the formal policy that declared that there is a many-to-many relationship between traits and genes was maintained, but for practical reasons, geneticists could indicate (changes in) particular factors as the causes for specific AATs. This is the differential concept of the gene.

The differential concept of the gene became the hallmark of genetic analysis, and, once the chromosomal theory of inheritance was formulated, it led to the term "marker." A "marker" is a clear-cut AAT, indicating the presence of an allele of a gene that has a known location on a specific chromosome. It is used as a point of reference when mapping a new mutant or when trying to solve a biological puzzle.[3] This concept has an intricate reference; it refers to the changes in both the trait and the gene as well as to the relation between these changes. However, the "marker" does not indicate a one-to-one relationship between genes and traits; it merely promotes one phenotypic change as indicative of the change in a causal factor. The coining of the term *marker* was the last step in reconciling the tension between the formal policy and the research method in classical genetics.

The importance of the differential concept of the gene for the research method in classical genetics is particularly apparent in the 1940s when the experiments of Beadle and Tatum (1941) with *Neurospora* (and to some extent, of Caspari with *Ephestia* [see Harwood 1993, 87–89] and of Beadle and Ephrussi [1936] with *Drosophila*) gave geneticists the feeling that there was after all a one-to-one relationship between genes and their immediate products (enzymes). Even so, the research method based on the relations between *changes* in genes and *changes* in reaction products and not on a one-to-one relationship between genes and enzymes per se was the one that allowed the research to go on. Thus, it seems that neither a notion of many-to-many nor of one-to-one relations between genes and traits were essential for achieving experimental results in classical genetics and that these concepts of relations between genes and traits were only used on the theoretical level. The differential concept of the gene guided the work, even if unconsciously. Classical genetics did not deal with traits with no variation (without AATs).

The formulation of the differential concept of the gene was not precise. It does seem to cause the misinterpretation of classical genetics as Waters (1994, 173) argues. Scientists sometimes phrased it in terms of *changes* in the gene and sometimes in terms of the gene (factor) per se as causing an AAT. The latter is expressed, for example, when Morgan et al. in the quotation cited above stated that "it is only this *one factor* and not all 25 which causes the difference between this particular eye color and the normal." By referring to the gene rather than to its states they obscure, in a way, the fact that the alternative state of a gene (the presence or absence of a gene in Bateson's terminology or the impact of one of the factor's alleles) causes the appearance of the alternative trait. There was no concrete formulation of the differential concept of the gene at the time, but the lack did not seem to bother geneticists in their laboratory work, for which they needed this concept.

The vagueness of the formulation of the differential concept of the gene obscured the fact that from the beginning of genetics, theory and research method were not guided by the same premises. This difference between theory and method did not start with the consolidation of the position that the relationship between genes and traits is many-to-many. A closer look shows that even when formal policy (the theory) still claimed a one-to-one relationship between traits and genes, the research method was guided by the differential concept of the gene. Even Mendel used AATs (e.g., yellow and green peas), and there were also alternative states of elements (alleles or allelomorphs). Thus, the differential concept of the gene can coexist with both one-to-one and many-to-many relationships between genes and traits because the former deals with the relations on a different ontological level than the two latter concepts of relations. The one-variate and multivariate functions deal with the relations between genes and traits, while the differential concept of the gene deals with the relations between their attributes, which are expressed in terms of alternative states of both genes (alleles) and traits (AATs).[4]

CHARACTERIZATION AND GENERALIZATION OF THE DIFFERENTIAL CONCEPT OF THE GENE

Relationships like the differential concept of the gene and the two concepts of relations between genes and traits are not unique to genetics. Change the predicates and they will fit other fields that deal with two-level phenomena and the connections between them (e.g., brain research). The differential concept can be used whenever there is a desire to understand the connections between two levels of phenomena (e.g., developmental genetics, medical genetics) or when one of these levels cannot be reached directly through scientific manipulation (Mendelian genetics). In the latter case, the manipulation of one level produces results that allow inferences to be made on the other level.

If the differential concept of the gene is seen as a special case of a more general notion of a differential concept, Morgan's understanding that a new AAT is caused by a change in one of the group of genes that determine a trait can be seen as an earlier and specific version of Mackie's INUS condition (Mackie 1965).[5] Mackie's INUS condition is the result of his attempts to analyze causation. His line of inquiry extends back to Hume. The conventional interpretation of Hume attributes two claims to him. One is that causal relations are supervenient upon regularities and the other is that causal facts are logically supervenient upon the totality of noncausal facts. Familiar accounts of causation within this tradition use terms of sufficient and/or necessary conditions. For example, one view states: C is the cause of E, if and only if, C and E are actual and C is *ceteris paribus* sufficient for E. Similar views are phrased in terms of necessary conditions and both sufficient and necessary conditions. There are obstacles to these accounts, and Mackie proposed a more sophisticated attempt to analyze causation in these terms. A short overview of these problems in causation can be found in Sosa and Tooley (1993, 1–8).

Mackie thinks that in order to find the law behind the cause, it is necessary to add detail to the description of the event considered. He writes about the ensembles of conditions that constitute a cause for an event. Each ensemble is a sufficient but unnecessary cause for an event. Within an ensemble, he isolates the conditions that are necessary relative to the ensemble, but are themselves insufficient as a cause for the event. In Morgan's phrase, the group of genes that determine a trait is the unnecessary but sufficient ensemble for the AAT, and the changed gene is the necessary but insufficient component within the ensemble. The differential concept of the gene picks out this isolated necessary but insufficient component (the changed gene) as the cause.

According to Mackie, a cause is at least an INUS condition, but from a logical point of view one INUS condition is not preferred over another. The preference is only the result of pragmatic considerations (which may be legitimate but might have a price). A car accident provides a good example. Each of the following factors is an INUS condition and hence a cause: a wet road, a sharp curve, and the speed of the car. However, a court will point to the speed of the car as the cause for the accident, even though the same speed on a dry straight road would not cause an accident. The reason for this decision is that the driver cannot control the weather but he or she can control the speed of a car and take into consideration hazardous driving conditions. Similarly, each of the genes in the group causing the trait is an INUS condition for the AAT. Pragmatically, the changed gene is seen as the cause for the AAT. This approach is fruitful, but only to a certain limit as I show in the following paragraphs.

There is one more aspect that needs to be discussed, the problem of inferring the normal (wild-type) function/phenotype of a gene from the AAT. I have chosen to illustrate this problem with a quotation from two brain researchers, thus making two points – emphasizing the problem of incorrect inference (the analogy to genetics is clear; see also Dietrich, this volume, on the Goldschmidt predicament) and demonstrating that the differential concept is not specific to genetics.

Consider the brain structure named the subthalamic nucleus. Its destruction in the human brain leads to the motor dysfunction known as hemiballism, in which the patient uncontrollably makes motions that resemble the throwing of a ball. Is the normal function of the intact subthalamic nucleus therefore the suppression of motions resembling the throwing of a ball? Of course not; the condition represents only the action of a central nervous system unbalanced by the absence of a subthalamic nucleus. (Nauta and Feirtag 1979, 78)

The problem of incorrectly inferring the normal function of the gene starts by pointing to the changed gene in Morgan's formulation (see above) as the cause for an AAT. This is analogous to pointing to one of the INUS conditions of Mackie as the cause of an event. This way of drawing conclusions, though justified by Morgan's group in the quotation cited earlier, does not take into consideration the influence of the whole ensemble, such as the group of genes that influence a trait or the whole central nervous system, as in the quotation above. Although Mackie allows such a step, apparently one needs to be careful not to push it too far.

CURRENT RESEARCH

In this section I show that the differential concept of the gene is also used in current research. It has the same advantages and disadvantages as in the past, but it has received new consideration that reflects new research methods and techniques as well as different interests. In current research, it is used in attempts to understand the connections between genes and traits – a field the examination of which was postponed by Mendelian geneticists – rather than merely as a research method. For example, genetic counseling may be considered to be based on the pragmatic need to connect specific genotypic and phenotypic changes, which is the essence of the differential concept of the gene. In order to provide reliable counseling, the genetic counselor should be able to, among other things, characterize this connection. One example is the extremely variable phenotype of Gaucher disease. There is a broad range of possible clinical outcomes even within a given genotype, and up to now there has been no way of knowing where a given patient will be on the spectrum. This uncertainty creates a major problem in genetic counseling. Attempts have been made to plot the phenotypes, using criteria of age and severity of symptoms, onto specific alleles of the Gaucher locus, but there is considerable overlap between the severity of disease observed with each allelic genotype (Beutler 1993). This case reflects a refined mode of the differential concept of the gene. Instead of taking into account two states of a gene that are connected to two AATs, there are many changes in one gene, and geneticists hope eventually to specify the phenotypes. This idea is not new to genetics, but molecular methods have given this aspect a significant boost.

The trend of plotting phenotypes onto specific alleles (of the same gene) is contrary to what is required by the *norm of reaction*. These two procedures deal with the connection of the genotype to variability in the phenotype, but the first tries to connect them to different alleles, whereas the *norm of reaction* relates to the difference in environment. Medical genetics usually prefers to exhaust the possibilities given by the differential concept of the gene before considering the role of different environmental factors. In the case of Gaucher, as well as other diseases, it will be an enormous task to characterize the environment of the different patients.

Another area that is dominated by the desire to understand the influences of genes and therefore by the differential concept of the gene is that which uses the technique of targeted gene replacement or gene knockouts. A targeted mutation can be generated in a selected cellular gene by inserting mutated copies of the gene into cells and allowing one copy to take the place of the original gene on a chromosome. This technique has several applications, but generating targeted mutations and checking the resulting phenotype is considered here. Researchers feel that "knockout mice provide an ideal opportunity to analyze the function of individual mammalian genes and to model a range of human inherited disorders" (Melton 1994, 633). Furthermore, sophisticated gene targeting methods are now being used to introduce subtle gene alterations. The trend is similar to that in medical genetics, i.e., to connect different mutations in the same gene to different and specific AATs. But it does not stop there; it is hoped that, "by rewriting parts of the manual and evaluating the consequences of the altered instructions on the development or postdevelopmental functioning of the mouse, we can gain insight into the program that governs these processes" (Capecchi 1994, 34).

By 1994, some ninety targeted mutations had their phenotypes checked in mice (Shastry 1994). Surprisingly, several knockout mice carrying disrupted genes, which were considered to be important in development or in the molecular pathogenesis of certain tissues, showed normal or "minimal phenotypes," or else the phenotypic effects were found in unexpected tissues (Shastry 1994).[6] The hypothesis that accounts for this phenomenon is gene or gene-function redundancy, where it is claimed that another gene replaces the function of a deleted or dysfunctional gene. It cannot conceal the fact that another difficulty is generated in addition to the difficulty of inferring the original function of a gene from an AAT. The latter difficulty is an experimental one, and results from the absence of expression of the targeted mutation. Actually, it precedes the logical problem of inference: If there is no expression, there can be no problem of inferring from it.

The researchers believed that they had some idea of the phenotypic effects of the genes selected for targeted mutations. However, the technique could also be applied to genes whose function is unknown (see Gilbert, this volume). The extensive sequencing work of the last years revealed "anonymous genes," the function of which is not known.[7] Sequence comparisons do not always help to elucidate their function, and one of the problems facing current research is to decipher the function of these genes. The differential concept of the gene is expected to point out a possible approach to solve this problem through a systematic study detecting the qualitative phenotypic effect of deleted individual anonymous genes by targeted gene replacement. This was done, for example, with anonymous genes of yeast chromosome III (Oliver 1996). Tests were carried out on agar plates containing a large number of different growth media, sometimes incorporating specific metabolic inhibitors, checking for distinct phenotypic differences of the mutant and the wild type on this media. Still, as Culp (1997) stresses, the problem with applying the targeting method to anonymous genes is that it is impossible to make phenotypic predictions, except at the level of nucleic acids (DNA and RNA) and polypeptides, that should guide the researchers toward a phenotypic change. The number of possible changes is enormous and some of them are minor; some may occur in unfamiliar functions for which conceptual tools have not yet been created.

Thus, there are several problems in applying the gene targeting method that is guided by the differential concept of the gene to establishing genotype/phenotype relationships. The problem of inferring the normal phenotype from a mutated phenotype is not specific to this method, but nevertheless affects it. Another problem is that a directed mutation does not always result in a mutated phenotype; this is probably connected to the complexity of interactions between genes and compensatory effect. Yet another problem relates to the special case of anonymous genes, where there is only a limited possibility to predict phenotypic change. Without proper prediction the task of finding the phenotypic change caused by a directed mutation becomes a matter of luck.

The differential concept of the gene cannot only coexist with the one-to-one and many-to-many concepts of relations between genes and traits, it could even survive the abandonment of the term *gene,* which might happen if a more holistic approach in genetics is adopted. Although the differential concept of the gene naturally goes with the use of the term *gene,* the concept itself could easily be adjusted if the term gene is replaced by the term genome, as some theoreticians suggest (see, e.g., Kitcher 1982). Genes are encompassed by the genome and mutations are within genes, and thus mutations are included in the genome. From a logical point of view, it is only a matter of a transition to a more inclusive group. This transition would maintain the relationship between genotypic and phenotypic changes, the essence of the differential concept of the gene, but it would also make attempts to infer the functions of what appears to be genes void of content, because the term *gene* will not be used anymore. These attempts may be replaced with attempts to find a more suitable theory for the connections between genotypic and phenotypic changes, one that will take into consideration the whole network of connections, including compensation mechanisms, etc. So far, however, the term *gene* is used deliberately and abundantly in all the works cited in the present chapter and in many others, and I suspect that this will continue to be the case. Nevertheless, there are indications that a genomic conception could be adopted by some, if not all, geneticists, as is implied by the following quotation:

In the future, more refined genetic analysis and *genome, rather than individual gene, alteration* will be achieved by incorporating site-specific recombination into targeting strategies. (Melton 1994, 633, italics mine)

NOTES

1. I wish to thank Raphael Falk, Yemima Ben Menachem, Marcel Weber, and the anonymous reviewers who read earlier versions of the manuscript and helped to improve this chapter.
2. The policy regarding the relation between genes and traits is fundamental to genetics and persists over time and different disciplines of genetics. There are two extreme versions: the one-to-one and many-to-many relations. The many-to-many version is the one that generally prevails.
3. Figuring out the stage of meiosis when crossing over occurs may be given as an example for a biological puzzle solved by the use of markers (and a cytological trick).
4. For example, "eye color in *Drosophila*" is a trait while "red," "white," "eosin," etc., are different attributes of this trait. The distinction at the genotypic level is more difficult and depends on the way in which the alternative states of genes (the alleles) are perceived.
5. The initials INUS stand for Insufficient, Necessary, Unnecessary, and Sufficient.
6. The expression *minimal phenotype* seems to denote disappointed expectation more than reality.
7. *Anonymous genes* are presumptive genes that are defined in these studies from their structural properties alone.

REFERENCES

Beadle, G. W., and B. Ephrussi. 1936. The differentiation of eye pigments in Drosophila as studied by transplantation. *Genetics* 21: 225–247.

Beadle, G. W., and E. L. Tatum. 1941. Genetic control of biochemical reactions in neurospora. *Proceedings of the National Academy of Sciences USA* 27: 499–506.

Beutler, E. 1993. Gaucher disease as a paradigm of current issues regarding single gene mutations of humans. *Proceedings of the National Academy of Sciences USA* 90: 5384–5390.

Capecchi, M. R. 1994. Targeted gene replacement. *Scientific American* 270 (3): 34–41.

Culp, S. 1997. Establishing genotype/phenotype relationships. *Philosophy of Science* 64: 268–278.

Falk, R. 1999. Can the norm of reaction save the gene concept? In *Thinking About Evolution: Historical, Philosophical and Political Perspectives*, edited by R. Singh, C. Krimbas, D. B. Paul, and J. Beatty. New York: Cambridge University Press.

Falk, R., and S. Schwartz. 1993. Morgan's hypothesis of the genetic control of development. *Genetics* 134: 671–674.

Harwood, J. 1993. *Styles of Scientific Thought: The German Genetics Community. 1900–1933.* Chicago: University of Chicago Press.

Johnson, W. E. 1921. *Logic.* Vol. Part I. Cambridge: Cambridge University Press.

Kitcher, P. 1982. Genes. *The British Journal for the Philosophy of Science* 33: 337–359.

Mackie, J. L. 1965. Causes and conditions. *American Philosophical Quarterly* 2: 245–255, 261–264.

Melton, D. W. 1994. Gene targeting in the mouse. *BioEssays* 16: 633–638.

Morgan, T. H. 1913. Factors and unit characters in Mendelian heredity. *American Naturalist* 47: 5–16.

Morgan, T. H., A. H. Sturtevant, H. J. Muller, and C. B. Bridges. 1915. *The Mechanism of Mendelian Heredity.* New York: Henry Holt.

Nauta, W. J. H., and M. Feirtag. 1979. The organization of the brain. *Scientific American* 241 (3): 78–105.

Oliver, S. G. 1996. From DNA sequence to biological function. *Nature* 379: 597–600.

Roll-Hansen, N. 1978. Drosophila genetics: A reductionist research program. *Journal of the History of Biology* 11: 159–210.

Sarkar, S. 1999. From *Reaktionsnorm* to the Adaptive Norm: The Norm of Reaction, 1909–1960. *Biology and Philosophy* 14: 235–252.

Shastry, B. S. 1994. More to learn from gene knockouts. *Molecular and Cellular Biochemistry* 136: 171–182.

Sosa, E., and M. Tooley. 1993. Introduction. In *Causation*, edited by E. Sosa and M. Tooley. Oxford: Oxford University Press, pp. 1–32.

Vicedo, M. 1991. Realism and simplicity in the Castle-East debate on the stability of the hereditary units: Rhetorical devices versus substantive methodology. *Studies in History and Philosophy of Science* 22: 201–221.

Waters, K. 1994. Genes made molecular. *Philosophy of Science* 61: 163–185.

3

Gene Concepts and Genetic Concepts

FRED GIFFORD

ABSTRACT

This chapter explores connections between the concept of a gene and the concept of a genetic trait. Each concept is problematic in various ways, and it might be thought that there are some common roots to these problems, and that we might aid such discussions by clarifying connections between them. There are several reasons to see the *gene* and *genetic trait* concepts as connected, but there are also reasons to be skeptical of this. Some of the ambiguity and complication arises from the fact that there are two different categories of genetic trait concepts, and these are related to the gene concepts to different degrees and in different ways.

INTRODUCTION

In this chapter I explore some relations between two sets of philosophical issues concerning genetics. The first is discussed at length and from several perspectives in the rest of this volume: *the concept of the gene*, how it has changed historically, and whether a clear, coherent, and general account can be given of it in light of our present knowledge and practice. The latter task is especially challenging, particularly because of the increasingly detailed and precise knowledge of the genetic material, including recent findings about the wide variety of ways in which genes are structured and the complexity of genetic regulation and interaction (Beurton, this volume; Burian 1985, 1995; Kitcher 1982, 1992). The second set of issues concerns the question of *what it is for a trait to be genetic*, what underlies our classification of traits as genetic or not. This latter issue occasionally emerges in the gene concept discussion, but is typically not the focus of direct attention there, and there has been no systematic examination of this connection. There appear to be some parallels and connections between these questions, suggesting the possibility of fruitful interaction. Yet it must also be kept in mind that there are important dissimilarities between these topics, so we shall need to take care before assuming that insights from one domain will map onto the other. Still, on the hypothesis that there is something to be learned from this, the purpose of this paper is to clarify the relationship between these questions and begin to explore how they might illuminate each other.

GENETIC TRAITS

Background concerning the gene concept issue is provided elsewhere in this volume. In this section, I will describe what is in this context the less familiar issue by sketching briefly the nature of the genetic trait problem and various proposals concerning criteria for a trait being genetic (see Gifford 1990).[1]

What is meant by calling a trait genetic? What criteria are and ought to be used to judge whether a phenotypic trait is genetic? The question is not straightforward. The central complication is that all traits are the result of an interaction between genes and environment. Thus defining a genetic trait as a trait with a genetic cause would simply yield the conclusion that all traits are genetic traits.

But perhaps more discriminating criteria can be specified that do not label a trait genetic whenever there is any sort of causal involvement of some gene, but that instead require that the genetic cause be somehow salient, thereby capturing our language of causal citation and capturing something explanatorily significant. In fact, I believe that we can say that there are two such criteria, linked to two different ways in which we pick out causal factors as especially explanatory (Gifford 1990). The first and main factor is the *differentiating factor* criterion (DF), that states a trait is genetic insofar as it is genetic differences that make a (phenotypic) difference in the population. In contrast, the proper individuation criterion (PI) requires that, more than just this gene having a causal influence on the trait, it is this trait specifically that is caused by the relevant gene or genes.

DF (Differentiating Factor)

The DF criterion can be understood with the aid of the following example (Burian 1981–82; Gifford 1990). In PKU disease, a defective gene at a certain locus fails to produce phenylalanine hydroxylase, leaving homozygotes unable to convert phenylalanine into tyrosine, resulting in mental retardation and other symptoms. Having a certain level of phenylalanine in one's diet is as much a causal factor for the phenotypic expression of the disease as is being homozygous for the defective gene. But this can still be a paradigm case of a genetic trait, because all such environmental factors are implicitly ruled out as "mere conditions." They fall into a shared causal background or *causal field* (Mackie 1974) on the grounds that they are constant throughout the population and thus do not account for any of the differences between individuals. Note that segregation or heritability analyses operate in this way implicitly, often keeping the environment constant, and labeling as genetic those traits which "Mendelize" under such circumstances.

It is important to see that, for this differentiating factor perspective, claims that a trait is genetic are *population relative.* To dramatize this, suppose there was a population in which all the individuals have what we call the PKU genotype – that is, they lack the gene coding for the necessary enzyme – but where phenylalanine in the diet is a rarity. Since it would then be the presence or absence of the phenylalanine in the diet that made the difference between PKU and non-PKU individuals, the disease would be conceived of as environmentally induced. And yet note that the causal processes involved in the etiology of the disease in each affected individual would be just as they are in PKU individuals in our population. Thus, on the differentiating factor conception, genetic traits are not being picked out as such due to some fact about the causal story in each individual organism; it is not that there is some special sense in which this individual gene causes this individual trait. Rather, a certain genetic difference causes or explains a given phenotypic difference, and that difference is a feature of (variation in) a population. It is in this sense that only some traits are called genetic even though all traits have some genetic causes. Still, this criterion does appear to capture the majority of our usage of the term.

PI (Proper Trait-individuation)

The claim "P is a genetic trait" may be made more discriminating in another way as well. This concerns how we pick out or describe the effect of the gene, that is, how we individuate the phenotype. Even if it satisfies the above criterion in terms of differentiating factors, P may be rejected as not a genetic trait on the grounds that P is not an accurate or appropriate description of what it is that the gene causes (or genes cause) specifically. The trait picked out may be only one part of the effect, or, conversely, what the gene causes may be only part of the trait of interest. For example, suppose a gene could correctly be said to control the ability to perform some specific subtask of cognitive activity. It would not be correct (and it would be quite misleading) to interpret this as controlling intelligence per se. The problem at issue here is not that there might well be variation in other causal factors as well. For even if there was in fact no variation in the population with respect to the other components of intelligence, so that DF picked out this feature uniquely, we would still hold back from saying that intelligence was genetic (even though it would be true that these genes "had an influence on" intelligence). Instead, we would simply say, that particular component of intelligence was genetic.

A consideration of various analogous cases reveals four possible sorts of mistaken attributions grounded in improper trait individuation, resulting from two dimensions. We can pick out a trait which is too narrow or one that is too broad, and we can do each of these things in each of two ways: with respect to range of physical extension (parts and wholes), and with respect to different degrees of generality of types.

Part-individuation

An example of a trait which, with respect to the part–whole relationship, is individuated too *broadly* for the gene to be a specific cause, is the case just described where some gene (or genes) affects some particular cognitive task and this is mistaken for being a gene for intelligence.

An example of a trait individuated too specifically or *narrowly* (or where what the gene causes is too broad) with respect to the part–

whole relationship is the following: Even in the possible world in which all variation in intelligence was due to genetic variation, this would not show that the ability to do neurosurgery was genetic. Presumably the genes don't specifically affect neurosurgical ability; the latter might be seen as a sort of side effect. Indeed, the phrase *side effect* doesn't fully capture how the claim simply focuses at the wrong level of analysis or breadth of description, or how it picks out the wrong aspects of the phenotype.

Type-individuation

An example of a trait individuated too *narrowly* (too specifically) with respect to type is the following: Suppose that the ability to learn and speak a language is genetic. American scientists, looking at the difference made in their population (i.e., using the differentiating factor interpretation), might conclude that the ability to learn and speak English was genetic, but this would be to individuate that trait incorrectly. In that population, ability to speak a language is manifested in ability to speak English, but such individuals *could* have learned German instead, and would have done so with the same genes in a different environment; that difference is explained completely environmentally.

Finally, an example of a misleading categorization as genetic of a trait that in fact is individuated too *broadly* with respect to type would be the claim that hypercholesterolemia is genetic. For it is actually only the subclass of cases termed familial hypercholesterolemia that can be said to be genetic. It is thus misleading to claim this about the whole, heterogeneous class of cases (Brown and Goldstein 1974).

These are all cases where the differentiating factors approach yields an unintuitive answer as to whether something is a genetic trait.[2] Sometimes this misleadingness due to any of these sorts of lack of alignment between the gene as cause and the trait as effect is quite socially important, as in cases where we are insufficiently careful about the complexity and heterogeneity of traits such as schizophrenia, aggression, intelligence, or race.

This second sense of genetic, tying it to proper individuation, or specific causation, might be said to connect up more directly (than does the first criterion) with a certain cognate notion: Not just "P is a genetic trait," but "gene G is the cause of P" or "G is the P gene." (Perhaps even "G codes for P," but this is more controversial.)

Now, an interesting issue is raised by the following. Dawkins (1982) says that a gene whose alternative allele resulted in dyslexia would quite properly be called a gene for reading. This is unintuitive, a fact explained by the PI criterion. Yet there is a potential rationale for Dawkins' position here, based on his interests as an evolutionary biologist. Dawkins might say that he is simply concerned with what traits are open to selection, not what traits are the result of certain developmental causes. And ability to read might well be the basis of fitness in this context. (Indeed, heritability, which parallels a quantitative DF concept, while often appropriately criticized as misleading, does have a theoretically grounded use in that [in its narrow sense] it gives us a measure of the speed with which phenotypic change can be produced by selection [Feldman and Lewontin 1975].)

Perhaps more generally, we can say that these different criteria are a function of different explanatory interests, and these different explanatory interests can be a function of different disciplinary perspectives. Thus, the approach based on differentiating factors might be used by population geneticists and epidemiologists, who might be described as taking part in a "top-down" approach, starting with correlations in populations. Perhaps it would also be used by medical geneticists; here it might be in part a function of the practical tasks of obtaining and disseminating information relevant to genetic counseling or identifying risk factors. On the other hand, perhaps developmental and molecular biologists would be more likely to use the proper individuation interpretation (though of course the way I have introduced it suggests that it should be an important part of the thinking of more macrobiologists as well). The PI criterion would fit with a "bottom-up" approach, involving laying out the causal story in the individual, rather than looking for patterns in populations. Relatedly, it would seem to fit with the tasks of investigating specific and direct products of a given gene. Perhaps it could be argued that these features don't matter to population geneticists or evolutionary biologists, in that their questions can be addressed without considering such, and they might argue that the misleadingness that can result will not matter very much to their scientific purposes. Their direct scientific concerns are not with the causal processes in individuals. Indeed, such top-down analysis may be used as a first cut, or as a heuristic tactic toward eventual causal analysis, by identifying subgroups for more lower-level analysis (Gifford 1989).

We can see a further aspect of disciplinary relativity by noting that the differentiating factor interpretation itself is a sort of schematic definition which can be applied in different ways: namely by picking out different populations, broader or narrower, natural or experimental. This yields a potentially wider set of possible genetic concepts, ones that might tie to different disciplines or interests.

Now, consider another motivation for having the PI criterion at our disposal in addition to the DF criterion. If direct gene products, such as mRNAs and proteins, are universal in a population, then we will of course not be able to say that a gene (or anything else) is the differentiating factor, for this requires there to be variation in the population. Yet it seems correct to say that proteins are genetic traits par excellence, regardless of variation in the population of interest.[3] Perhaps PI can make sense of this intuition, for in some sense this protein is exactly or specifically what is generated by this (structural) gene. If so, this would count in favor of PI.

Of course genes are not unique causes of proteins in the sense of being the only cause. The idea is rather that these other causal factors (the other necessary conditions in the epigenetic milieu, much of the regulatory apparatus, the availability of amino acids, the transcriptases), would be causing much else besides, such that this gene product would not be properly said to be the specific effect of those causal factors; they would not cause this effect "specifically."[4] This may offer a principled way in which to make the cause/condition distinction, showing structural genes to be salient causes of such things as proteins. (Note that PI is here used by itself, whereas it can also be used, as above, as criterion in addition to DF.)

The question of what counts as genetic becomes complicated and contentious when we examine it carefully. There surely is not a sharp line to be drawn between traits that are genetic and those that are not. Questions are forced upon us about different ways of individuating a given trait or of individuating an organism into traits. There are indeed a variety of sorts of criteria that might be given for how the genetic/nongenetic dimension is to be drawn, and there are some reasons for seeing the choice of criteria as tied to certain explanatory interests or disciplinary perspectives. Laid on top of this is the fact that there are socially charged issues involved – different ways of answering the question may have socially important or controversial implications, especially when calling something genetic can lead some to expect that it is not changeable. (This last point raises some complex issues about the proper role of various pragmatic considerations in how we do and ought to mold our concepts.)

No claim is being made that the differentiating factor and proper individuation criteria proposed here are completely exhaustive. But they seem to capture the majority of usage and tie to these very fundamental explanatory interests.

SOME REASONS FOR OPTIMISM ABOUT CONNECTIONS

Having provided this background on the genetic trait concept, let us now characterize more carefully some reasons for taking questions about the concept of a gene and questions about the concept of a genetic trait to be related. At the same time, we can point out some of the implications there might be from this comparison, the issues concerning which we might gain insight from the interchange.

First, both the gene concept and the genetic trait concept concern conceptual questions relating to genes as causal factors, how we should understand our causal claims and causal language. As a result, examination of each concept potentially raises questions about how certain items (causes and effects) are to be conceptualized, individuated, and described. So it might be hoped that insights in one domain might map onto the other.

Also, the genetic trait question, like the gene question, can be examined from the point of view of scientific concept formation, change, and analysis. And, relatedly, in each case, as one learns more, one may come to believe that we were assuming much too simplistic a picture of the causal processes, and that these assumptions might be infused into our language in ways that need to be reflected upon.

There are also similarities concerning what is at stake. One indication of this is that each has certain connections to genetic determinism, a set of theses about how we conceptualize the relationship between genes, environment, and phenotype, the implications of which are of both theoretical and practical (even political) importance. Another is that the complications, difficulties, and ambiguities that come to light in the process of trying to provide a clear and general analysis suggest that there simply may not be a univocal sense of the concepts examined. Thus, it may well be that there will arise analogous issues of essentialism, pluralism, instrumentalism and the role of different disciplinary perspectives, and of how progress and communication (both between scientists and between scientists and the public) occur in such a situation.

Since the issue of pluralism about concepts arises in each case, this too might be a place where insights from one might be relevant to the other. One such opportunity would be this: If we accept that there are these two different *genetic* concepts – one concerning differentiating factors and another concerning proper individuation – not reducible to one another (along with various subcases of the differentiating factor interpretation), perhaps this could give credence to the thesis that there are two or more independent *gene* concepts. Perhaps it is even plausible that the various perspectives of population genetics, evolutionary theory, molecular and developmental genetics, and medical genetics might yield or presuppose different conceptions of what counts as a gene, just as they do concerning what counts as a genetic trait. And perhaps features of the gene concept and genetic trait concept will be able to be explained by reference to the other. In any event, each domain (gene, genetic) would certainly be one in which to explore issues concerning how different explanatory and disciplinary perspectives play a role in the shaping of our concepts.

Further, in each case the possibility is taken seriously that we should "give up the question" – that things are not just complicated, but hopelessly confused, and that rather than try harder and harder to produce a more complicated but adequate answer, we should simply avoid relying on these concepts and translate what we want to say into different language. In the case of "genetic trait," one may conclude that the question not only leads to confusion and misinterpretation, but is ultimately not especially important from a scientific point of view, and that we should go on instead to ask better questions about the relations between genes, environment and phenotype, namely, the shape of the norm or reaction curves (Lewontin

1974), and the nature of the biosynthetic pathways. In the case of "gene," our being confronted with the intractable difficulties in specifying the boundaries of genes, along with the complex interactions of the different parts of the genome, has led to claims such as "when we reach full molecular detail we are far better off to abandon specific gene concepts and to adopt, instead, a molecular biology of the genetic material" (Burian 1995, 50). (See also Kitcher 1982; 1992; Fogle 1990, and this volume.) Perhaps in each case we can say everything we want to say, and explain everything we want to explain, without worrying about these matters of how things are to be labeled or categorized.

The possibility is then raised that perhaps the concept should be abandoned for roughly the same reasons in each case – that basically these are two sides of the same set of issues (the cause side and the effect side). And if so, then surely it would be very fruitful to carry out these discussions together. Looking more specifically, we might ask questions such as: Does anything about the lack of a sharp line of demarcation, the difficulty of coming up with a general definition, and the population relativity of the genetic trait concept (along with the thesis that it might be best to give up the question in favor of better questions about genetic causation) give credence to the thesis that we should give up on the gene concept, in the sense that we would just talk of the genetic material rather than of individual genes?

A final reason for exploring the connection is the fact that certain other authors have written on issues that cross over between gene concepts and the genetic concepts, touching on theses that relate, to varying degrees, to the proposals about "genetic" described here. For example, Schwartz (this volume) analyzes the differential concept of the gene, and discusses the contrasting perspectives of theory and methodology that parallel the perspectives I have highlighted of causation in the individual and differences in a population. Indeed, there is a substantial parallel between the themes presented in this paper and those presented in Schwartz's paper. And Waters (1994) describes two gene concepts, the classical gene concept and the molecular gene concept: ". . . the molecular gene concept is centered on the idea that genes are for linear sequences in products whereas the classical concept is centered on the idea that genes are units whose mutations result in phenotypic differences" (Waters 1994, 182–183). Thus the classical concept ties to my differentiating factor interpretation. Now, his linear sequence account is not the same as my proper individuation account, though they may dovetail to a large extent, for the linear sequence coded would seem to be a paradigm case of something "specifically caused," as suggested above.

Surely more could be said to indicate a more complete taxonomy of potential connections, or of ways in which discussions in one might presuppose views in the other. But this will suffice to establish that there are some reasons to hope that the connections will be fruitful.

CHALLENGES TO CONNECTION

However, if one had the impression from this that the gene and genetic traits contexts would simply and obviously be two sides of the same coin, with each insight having a direct implication for the other domain, one would be mistaken. Indeed, once one sees some of the differences between these two sorts of questions, or the blocks to the inference from one to the other, one may become skeptical of whether there is much connection at all.

A first challenge to the connection is simply that the one case is about genes, the other about phenotypic traits. One way of seeing how these are very different is to note that one is about causes, the other about effects. A second is to note that the gene case involves entities, while the trait case involves features or properties. One will at least need an argument connecting these.

Genes and traits, while distinct, are closely related by causation. Of course, lots of things are causally related without being closely connected conceptually (baseballs and broken windows, socio-economic status and health), so if we think the connection is tighter, we will need to say why.

But there are some more specific reasons for being skeptical of the connection between conceptual questions about the genetic trait and about the gene. Perhaps the central objection to presuming a close connection between the questions is this: Whether a trait is genetic or not seems to be a matter of whether the cause is (or the causes are) genetic *as opposed to environmental*, not as opposed to caused by some other gene, or by some part of the genome that we don't say counts as a gene. And when we ask this question about a trait being genetic as opposed to environmental, it would not seem to matter whether the genetic causes are one or two or three-and-a-half, or where the borders of those genes lie. From this point of view, the genetic trait concept does not involve matters of individuation, or how to individuate the genes. On the other hand, many of the gene concept questions do center on such matters of individuation. For instance, does a certain segment of DNA count as some one gene, or is this true only of a more inclusive segment of DNA? Or should it be viewed as a gene at all? Should regulatory genes that don't get transcribed count as genes? Can we give an account of what a gene is, in general, that will answer these questions? From this perspective, the gene and genetic questions appear to be of completely different sorts.

To elaborate this apparent indifference of "genetic" to individuation, note that no matter how we individuate the genetic material into pieces, whether or not we call regulatory elements far upstream (Fogle 1990, 360) part of a given gene, surely they are still part of the genetic material. Hence, presumably their causal influence on a trait is relevant to determining whether something is a genetic trait. Even if the variation in the DNA were outside what we were willing to call genes, it would still be part of the genome. And even if we used the massive complexity of gene individuation as part of an argument for saying we should give up talk of genes and just talk about the genetic material, presumably we would still say these were genetic differences, that we had given a genetic explanation of any phenotypic differences that arose.

This suggests one way to state how the gene and genetic questions are *not* tightly linked. Even if we decide to take the ultimate deflationary step and give up on the term *gene* and just talk about the genetic material (even say that there are no genes, that there exists only the genetic material), we are not required to do the same with the term *genetic trait*. It is not clear we are really even encouraged, or given any reason, to do so. We were able to pick out traits as genetic, on the basis of Mendelian ratios, before having any idea of the nature of the gene, let alone having a view about exactly what part of the DNA should be labeled the gene. Even where the gene is radically messy and indeterminate, this would not make *genetic trait* so. Of

51

course, it might still be true for genetic trait, but if so, it would be for wholly different reasons.

It goes the other way around as well. Even if we decide that there is no deciding on one concept of genetic trait for all purposes or disciplines, and even if we judge that, on these and other grounds, there are good reasons to reject the question and go on to better questions instead, it does not follow that there couldn't be a very clean gene concept. It is not clear why the term *genetic* couldn't be radically indeterminate even if genes were simple beads on a string. We might, for instance, say "genetic" is radically indeterminate on grounds of population relativity, but population relativity does not appear to have implications for the nature of the gene. So, again, if "gene" is radically indeterminate, it is for its own reasons.

At this point we should make a distinction concerning kinds of relevance. One could insist that the disanalogies suggested here put a block in the way of drawing a direct inference from one to the other domain, yet one could at the same time assert that we have here two domains with some very important analogies, such that several of the same questions arise and it is useful to compare answers. Even though certain facts which, for example, make the term *genetic* ambiguous and difficult to apply may have no *direct* relevance to the way in which the term *gene* is ambiguous and difficult to apply, nevertheless there may be analogies between these features. Indeed, it would be surprising if there were not at least this ground for a useful comparison between the gene and genetic cases. After all, surely there will be something to be learned by comparing each of the gene and genetic cases to the ambiguities and difficulties of specifying a unified analysis to other quite different biological terms, such as *species*.

Connection by way of mere analogy would ground a less deep and compelling argument than connection by way of being two different sides of the same question, or two different ways of looking at the same set of processes. So while even the connection by way of analogy would be of interest, and would justify those examining one topic making use of ideas from the other, we are most interested here in whether there are tighter, deeper connections, ones that might make proper understanding of the one set of issues directly relevant to or perhaps even necessary for proper understanding of the other.

Would the Proper Individuation Account Fare Better?

Note that the grounds for this skepticism about contact between "gene" and "genetic" may be especially strong when we assume the differentiating factor sense of genetic, where a genetic trait is one where genetic factors explain the phenotypic differences in the population. Perhaps when we consider instead the context of application of PI, where genetic traits are ones where the trait is properly individuated such that it is exactly or specifically what the gene causes, the situation may be different. For instance, when we said above that the genetic trait concept surely did not involve questions of individuation, this was with the differentiating factor conception in mind. Surely PI is concerned with individuation, with the boundaries of the causes and effects. The PI criterion might require some answer to the question, "What is the gene?" in a way that DF does not. So this will make the situation more complicated. We will explore PI further below.

In fact, I believe that PI does suggest deeper connections to the gene concept than does DF. In consequence, while it is true that we must take care not to assume an immediate and unproblematic connection between the gene and genetic concepts, what we really see here is the need not to assume an immediate and unproblematic connection between the DF and PI accounts of a genetic trait.

GENES DEFINED BY PHENOTYPIC TRAITS

Probably the central argument for there being not just an analogy, but a deep and significant connection between these concepts, is that genes are picked out by some phenotype. In the context of transmission genetics, genes are defined by the traits they cause. This is of course how we first notice them: by their effects, for example, in segregation analysis. At the same time that we are identifying these traits as genetic traits, we are identifying the genes (or gene differences) as whatever it is that causes these differences in phenotype. This still describes gene differences being picked out by phenotypic differences, but even when taking the "causation in the individual" point of view, or the direct gene product point of view, there is clearly a close connection between picking out a gene and some phenotypic trait that it causes.

It is worth tying this notion to a related philosophical issue. There has been a long discussion in the philosophy of biology literature, the question of whether and how Mendelian or transmission genetics can be reduced to molecular genetics. Even though that question is about two theories or fields about genetics, or about genes as described by two perspectives, the attempted reductions (in order to establish "connectability" of the terms, to allow derivation of one set of statements from the other) have to make reference to, and deal with complexities of, phenotypic traits (Rosenberg 1985, 100). A conceptual issue about how to describe and classify traits can have relevance to, and even be a prerequisite for, the conceptual question of what counts as a gene.

A central corollary of this is the fact that what counts as a gene is dependent upon, and gets altered with, what phenotype is chosen. Choice of phenotype has a number of aspects. First, as a general point, we may be interested in certain broad categories of traits, such as: all the proteins or direct gene products, all the diseases we notice, all the features which are adaptations. Second, what counts as the trait we identify can change as our interests change and as we learn more. For example, we may redefine a disease or divide it into subtypes, as in the above-mentioned case of hypercholesterolemia, which is subdivided into familial and nonfamilial forms, or in the case of diabetes or cancer. We might make these divisions in part by the ability to intervene, and the knowledge that intervention will be differentially successful in different subtypes. And when there is genetic heterogeneity, the narrower disease category may be correlated to a narrower set of mutations. Third, it is also worth noting that we can gradually alter our focus on the phenotypic trait of interest, and thus the genes that cause it, as we move our attention away from direct gene products along the biosynthetic and developmental pathways to more remote effects.

Let us explore a bit more carefully this last case – the fact that traits that are at various degrees of directness to or remoteness from the genes will individuate the genes differently. We might in a particular case choose as the phenotype of interest a certain amino acid sequence. Or we might instead choose the ability to carry out some function – even the very function carried out by the protein of that amino acid sequence. Now, certain amino acid substitutions will not alter the function of the protein. As a result, whether an alteration at the nucleotide level will count as the same gene will depend upon whether we define the trait of interest as the amino acid sequence, or, instead, as a protein with a certain function. For if we define the gene as that which causes that amino acid sequence, then a given nucleotide substitution will count as creating a different gene, whereas, if we define it as the gene that creates a protein with that function, then it will not. We may generate a still broader characterization of the gene (that is, a wider range of nucleotide sequences would count as being the same gene) by moving further downstream in the biosynthetic pathway, and by broadening further the description of the phenotypic trait. Such phenotypes will also tend to be more macro and more complicated, and there will be more opportunities for other causal factors to have an effect on the trait. As a result, different disciplines and explanatory questions that focus at various points along the biosynthetic or developmental pathways will view genes differently.

One thing to keep in mind as we ponder this picture of phenotype description determining gene individuation is that defining genes in terms of traits is not simple or unidirectional. While it is true that we often pick out a certain segment of the DNA as a (single) gene on the basis of the phenotype we have in mind, it is also true that we sometimes modify what we count as a given trait as we learn more about its causes (including genetic causes) (Burian 1995, 50). Thus, we engage in a sort of back and forth triangulation. We may sometimes just notice the trait because it segregates, but often we may initially be interested in some trait that we pick out independently of this. For instance, we may be interested in some disease state due to its practical consequences. We might then examine whether or to what extent there is genetic causation of this disease. (This is not to say that we would be expecting it to be "completely" genetic.) But upon discovering some genetic cause, we might reclassify or reindividuate the trait of interest, as in the hypercholesterolemia case.

In fact, it is very common for there to be a variety of forms of a given genetic disease resulting from different mutations. The case of PKU illustrates this in that there are several different genetically discernible hyperphenylalanemias or disorders that involve the inability to convert phenylalanine to tyrosine (Rimoin et al. 1996, 1867).

More generally, genetic heterogeneity of disease can involve either different mutations of the same gene (different alleles) or mutations at different loci which generate the same phenotype. In the case of Alzheimer's disease, there are cases of both sorts (Rimoin et al. 1996, 1812).

So in various contexts, genes and traits codefine one another. (Of course, other traits can be similarly defined in terms of some environmental factor.) Note that none of this denies that individuation is important; the point is simply that it is not a one-way constraint. Note further that there will be different interests affecting how a trait is picked out, some of them practical and some theoretical; as a result, there may be no reason to believe that we will eventually land on the one unique or best classification. This may be a source of messiness or indeterminacy.

Type and Part Again

In relation to these various claims about individuating genes by individuating phenotypes, one should keep in mind the distinction – generated from our discussion of different aspects of our proper individuation criterion – between part- and type-individuation. The focus of the examples just given of different conceptions of the gene has been on how broad a *type* counts as the same gene (whether the change in nucleotide sequence makes this a different gene), not how extensive a *part* counts as the gene. But part individuation of genes can also be altered by how we choose the phenotype. For instance, by changing the phenotype from "carrying out a certain function" (e.g., creating the products that digest lactose) to "carrying out that certain function whenever there is a certain environmental stimulus" (e.g., when there is lactose in the environment), one would enlarge the gene to include regulatory genes, sequences which might otherwise be viewed as separate genes or not genes at all.

So the distinction between type- and part-individuation can be useful in our discussions of what counts as a gene. It may help us to make important distinctions between different sorts of issues about what counts as a gene. The case above where amino acid substitutions don't affect protein function concern type-individuation – the question is whether a gene so altered would count as changed to a different gene type, that is, to a different allele. In contrast, questions about recons, mutons, and cistrons are centrally about *how wide an entity* the gene is, what constitutes the extension of a single object (part-individuation). This is also true of questions about whether to include introns and certain regulatory entities in a given gene, and of some questions about the unit of selection in evolutionary theory.

It would be worth exploring whether type- and part-individuation should play rather different roles from one another in our gene concept discussions, and if so, it is important to tease these apart as we think about this issue. Further, should part-individuation be seen as in some way more fundamental, more an issue of "the nature of the gene," or are part- and type-individuation simply two different issues, neither more fundamental? The case of a mutation that doesn't alter protein function will not change our view of whether some particular object in the world counts as a gene, but only whether different genes count as of, or are to be classified as, the same type. It thus might be argued not to raise the issue of ontology in as deep a way as cases of part-individuation do.

Does explanatory interest play itself out in the same way in the gene concept and genetic trait concept cases, and will these features of the two cases be connected? It may well be that there is relativity to explanatory interest in both cases. But I presume, for instance, we wouldn't want to say that the purpose relativity of part-individuation of genes followed from the purpose relativity of their type-individuation, or vice versa.

I conclude that the connection between the identification of genes and that of traits central to the Mendelian tradition is a compelling reason to take seriously the gene–genetic parallel. The fact that there is a mutual back and forth triangulation is one thing that will make the matter complicated, but it does not obliterate this parallel. Seeing the relevance of part- and type-individuation to both cases sets up another parallel between *gene* and *genetic* concepts. But again, the connections are not complete, and there remain questions about whether we should see the reasons for comparing gene and genetic concepts as simply resting on analogies. Finally, it should be noted that the PI perspective is central, and the DF perspective less so, in making these connections.

PI AND ITS CHALLENGE

As described in the previous section, the PI criterion, identifying a trait as genetic only if it is individuated in such a way that it is caused specifically by the gene, appears to be an important source of insights in relation to the gene concept questions. So some of the utility of the connection depends on this PI analysis of "genetic trait."

Still, it might also be questioned whether this PI criterion is a sufficiently coherent and precise analysis of "genetic trait." It might be questioned whether it will help us in the gene concept discussion to clarify various issues and forge a connection between the gene and genetic concepts, or whether it will instead lead us astray.

Recall that there are at least two important claims I make concerning PI. First is the claim that it captures something explanatorily relevant in the underlying idea of specific causation and this does indeed need to be added as a criterion for "genetic." Second is the thesis concerning "direct gene products" – namely, that many direct gene products (mRNAs and proteins) could turn out to be importantly genetic in this PI sense, on grounds of causing the protein specifically.

I would like to comment on a criticism that has been made of PI and its implications for calling mRNAs and proteins genetic traits (Falk 1995; Smith 1992). I hope in the process to clarify some points and raise some questions. Falk here follows Smith (1992) in making two claims (Falk 1995, 26). The first is that the PI criterion "does not properly differentiate between causes and conditions." Presumably it is an implication of this that it is not a clear and coherent analysis, and that any insight it appeared to yield for our questions would in fact be misleading. And the second claim is that PI "embodies an implicit appeal to the genome as a privileged causal factor in morphogenesis." There is a suggestion here both of an inaccuracy or bias that will lead us astray, and also of a genetic determinism whose acceptance would have socially bad effects.

While Smith makes a number of interesting points in his article, he does not show that PI fails to make the cause/condition distinction properly. The charge that the PI criterion leads us to privilege the genome is an interesting and important one. There are important issues concerning the idea that the gene is to be privileged over other parts of the cellular milieu, and (relatedly but not identically) over environmental factors – and thus whether we are led to a sort of genetic determinism. However, I believe this criticism of PI to be mistaken as well, or at least very inconclusive.

Can we say that the protein coding gene can properly be picked out as a cause, whereas such features of the epigenetic milieu as the availability of transcriptases and tRNAs cannot?

First, Smith simply does not apply the PI criterion correctly. Rather than ruling out, as DF does, those factors that are shared throughout the population, PI rules out factors such as those that not only cause the trait in question, but are causally relevant to *much more besides*. For instance, the various parts of the regulatory system and general epigenetic milieu – whether or not they take on varying forms in the population – would have a causal influence on essentially all traits. But when Smith takes himself to apply PI, what he considers is whether the factors can be ruled out because they are universal (shared throughout the population of cases), not whether their effects extend to lots of other parts or aspects of the phenotype. "It does not seem then, that the universality of factors provides a workable means of making the cause/condition distinction – at least not one meaningfully distinct from DF" (Smith 1992, 341). But this is the DF criterion, not the PI criterion. So we have no test here of whether PI works.

There are various ways in which the PI criterion and its application to the direct gene products issue here might be wrong, and it might help clarify PI to describe them. First, it might turn out that PI cannot be given a coherent and sufficiently precise explication. Surely more work would need to be done to show that it can, elaborating the notion of "specifically causes." Second, it might be that, even if clear, it does not in fact apply to genes in the way I'm suggesting. Applying PI may not yield the conclusion that genes (at least in many cases) cause proteins specifically. This is just to say that it might turn out that a given gene will typically be properly said to cause much else besides the "direct gene product" that we were intuitively labeling a genetic trait. Finally, it could turn out that applying PI *does* yield the conclusion that other causal factors do do this – that is, that other factors besides those genes (many other parts of the genome, and also much else besides genetic material) might be properly said to specifically cause mRNAs and proteins. That is, perhaps they in fact cause those things without causing much else besides. But it does appear to be characteristic of most or all of the factors in the epigenetic milieu that they do in fact have much more general effects. For instance, the presence of the various tRNAs is necessary for all traits. It would take us quite far afield to evaluate these matters further. In any case, Smith fails to carry out this test.

The Privileged Genome

Let me now take up the privileged place argument – the claim that PI is to be rejected on the grounds that identifying traits as genetic that are specifically caused by genes leads to an inappropriate privileging of the genome as a causal factor. This concern is a serious one. A biased picture of the role of genes and environment in the generation of organisms can have bad effects, as those who have discussed biological determinism have shown. Falk illustrates well both the fact that such a biased and simplistic picture is common, and the difficulty of correcting it. He quotes a set of statements in the draft of a National Academy of Science report recommending that scientific concepts be taught in primary and secondary school (Falk 1995, 29). They are quite common sorts of statements ("Each organism requires a set of instructions for specifying its traits." "An inherited trait . . . can be determined by either one or many genes."). But an uncritical reading of them yields too simplistic a picture, encouraging a genetic determinist picture. If acceptance of PI were to increase the number of traits properly called genetic, and if this were to have as an effect the further entrenchment of simple and genetic determinist conceptions of the relations of genes, environment and phenotype, this would be worrisome.

If the PI criterion works in the way I have suggested, I suppose it does give a privileged place to genetic material, or certain parts of it. That is, it gives structural genes a privileged place as causes of mRNAs or of proteins. It specifies a sense, suggested to be meaningful and explanatorily important (indeed, one of the two central criteria), in which genes are picked out as the causes rather than mere conditions of these traits. If we agree that proteins are significant traits, then these structural genes are picked out as the cause, in this sense, of significant traits. This may seem to be a mistaken approach in light of, first, all the discoveries about the complexity of the factors involved in transcription and translation and, second, our attempt not to fall into genetic determinism. But is this so and what exactly is wrong with giving such privilege in this context?

Let me first make clear a sense in which the genome is not privileged by the proper individuation criterion. The idea was to pick out a criterion that would make sense of instances of calling a trait genetic. Certain traits are picked out that people have been intuitively calling genetic, an account is given, and this account (in the sense discussed) gives priority to the genome (or, actually, to particular [structural] genes) for those traits. But this is not for traits in general; it is (perhaps) for mRNAs and proteins. Those traits are not the only traits (or trait descriptions) there are. Other traits may be specifically caused by some environmental factor (or by both, or, more likely, neither). Therefore, this is not a way of privileging genes or the genome for traits in general, but for making sense of the intuition that proteins should be said to be caused by genes regardless of whether there is variation in the population.

Let me elaborate on the claim that it won't make everything genetic. First, as stated above, it is reasonable to expect that few of the more macro or complex traits further downstream (where, after all, most traits of interest are) could count as genetic in this sense, for at that point the paths from genetic and environmental sources merge more and more. But second, even if we focus our attention on traits that concern proteins, still rather few of the whole variety of ways we could pick out traits will count as genetic in this PI sense.

Consider the wide variety of possible such traits at the protein level: the exact makeup of this protein (say, the amino acid sequence), the amount of this protein, the timing of when there is a substantial amount of the protein, the ratio between the amounts of proteins A and B, etc. It seems plausible that some of these could well turn out to have a unique, specific genetic cause, while none of the others do. This illustrates the importance of how we pick out the traits in the first place. As a result of such considerations, I believe that there is a large caveat to the claim that PI gives priority to the genome; it should not lead us to the claim that everything is genetic. The fact that many traits downstream may be dependent specifically on such timing and ratios shows one particular way in which it will be incorrect to say that that trait is correctly identified as what a given gene specifically causes.

Suppose there is in fact some significant sense in which PI gives priority to the genetic material. Let me suggest the following way in which the worry about this may be an overreaction.

Consider first the DF interpretation of "genetic trait." Suppose we first naively call a trait genetic (whether because of heritability or segregation analysis or the overconfident extrapolations of workers in the Human Genome Project). Then we come to understand and take seriously that all traits are the result of an interaction between genes and environment. We might then say that nothing is genetic, that the claim doesn't make sense, that it is both conceptually misleading and socially harmful, and conclude that all traits should be viewed as on a par in this way.

On further reflection, we may consider various analyses of our causal citations and judge that an analysis like the differentiating factor view has certain virtues. There are certain virtues in noting that certain traits are genetic in an interesting, but incomplete, sense – namely, all the phenotypic variation in the population is due to genetic variation. Part of giving this analysis would involve pointing out – and making explicit – the limited nature of the claim (it being population relative, not a deep description of the causal processes in individuals, etc.). Rather than just saying this causes or explains that, we specify senses in which it does and does not. Particular claims can then be evaluated more precisely and objectively. It might reasonably be hoped that this would help to curb a naive genetic determinism, not enhance it.

Now consider the proper individuation interpretation of "genetic trait" – especially its application to fairly direct gene products. Can analogous things be said here? Suppose people are initially disposed to say of the presence of particular mRNAs and proteins that they are genetic traits. We then come to understand that the process is not that simple, for there are several stages to be gone through from gene to protein, with large numbers of macromolecules playing various roles. We might then infer that it is a mistake to call them genetic traits (or to say that gene G codes for P, or that G is the P gene) and see this as only due to reductionistic bias. This, it is pointed out, is not only conceptually misleading, but socially harmful. Perhaps we are moved by this and other arguments to hold that there are no genetic traits, or perhaps even to say that there are no genes.

But it is then suggested that we may be able to make sense of this intuitive idea that these are genetically caused traits in some non-trivial sense by means of the PI criterion, concerning what is caused specifically. This yields some intuitive implications about proteins (even universal proteins, about which the differentiating factor inter-pretation would say nothing), and also about the improperly indi-viduated cases discussed earlier. Is this now abetting some form of privileging of the genome in a sense that we should find worrisome? Or that generates tendencies of thought that are not only concep-tually misleading, but socially harmful? Why not view it as before? Such an analysis points out and makes explicit the limited nature of the claim. For example, it does not pretend to pick out all the causal factors, i.e., all factors that are necessary in the circumstances for the trait to emerge, and it says nothing about which causal factors are "stronger" or more "potent" or even which ones explain more of the population variance. Particular claims can then be evaluated more precisely. Why couldn't this again help to curb a naive genetic deter-minism, rather than enhancing it?

It might be naive to expect all science writers, members of the public, and even scientists themselves to always keep carefully in mind the limited nature of these claims. But what exactly should we infer from this fact? After all, getting everyone to stop using the term *gene,* or to stop thinking as if there were genes, or stop using the term *genetic trait,* is a rather tall order as well. Educating all and getting everyone to understand these matters is very difficult. Hence the argument that we should not countenance the proper individuation analysis, on grounds that it gives too much prominence to the ge-nome, is going to have to be a longer one, involving weighing a variety of educational strategies.

This does not by any means resolve the question of whether PI yields some objective and helpful means of identifying genetic traits. If it turns out not to, perhaps saying why this is so would be il-luminating. Perhaps that analysis could help illuminate the gene concept. It might be worth examining whether the gene concept case will involve us in pragmatic and social issues to the same extent as, and in analogous ways to, the genetic trait concept.

I don't think that PI has been shown not to have a way to distinguish causes and conditions. While more thought will need to be given to this important issue of how our gene and genetic concepts connect with ideas of genetic determinism and the priority of the genome, I don't yet see a clear problem for PI in this regard, either. I conclude that the PI criterion is legitimate and thus able to be used as it was earlier in the paper to exhibit and clarify a variety of significant connections between the gene concept and genetic trait concepts.

CONCLUSION

This is just the beginning of an exploration of the gene–genetic trait connection, but there are many parallels between questions concerning gene concepts and those concerning genetic concepts. The two sets of questions have some differences in structure, due to different kinds of concerns that motivate them, such that they will be separable. As a result, it would surely be possible to abandon one concept but not the other. Even so, I believe that the combination of connections and analogies should make interaction between the two sets of questions very fruitful.

NOTES

1. *Genetic* as discussed here, is always a modification of *phenotypic trait* (genetic as opposed to environmental). It should not be confused with *genic*, which is used in other ways, such as to denote the level or unit being referred to concerning selection (genic as opposed to organismic, or as opposed to the whole genome). We often say, "the genetic material," referring generically to the genome or parts thereof. My analysis of the term *genetic* is not about this usage.
2. Schwartz (this volume) discusses the case of the motor dysfunction hemiballism to illustrate an analogous methodological problem, that of "incorrectly inferring the normal function of the gene" from what phenotype (behavior) results when the gene is not there. This brings in another aspect – function. But the problems of improper individuation, as well as how they result from unthoughtfully applying a criterion about what difference one sees in the actual population, are quite parallel.

64

3. Note that Gilbert (this volume) cites the ability to detect and isolate genes without reference to their variability as one of the criteria of the Developmental Synthesis.
4. Note that when applying this criterion, it does not count against the claim that the gene causes the protein specifically (that is, without causing much else in addition), that it of course causes things on the causal chain between the gene and this protein, and causes by transitivity those things that the protein causes.

REFERENCES

Brown, M., and J. Goldstein. 1974. Familial hypercholesterolemia. *Proceedings of the National Academy of Sciences USA* 71: 73–77.

Burian, R. M. 1981–82. Human sociobiology and genetic determinism. *Philosophical Forum* 13: 43–66.

Burian, R. M. 1985. On conceptual change in biology: The case of the gene. In *Evolution at a Crossroads: The New Biology and the New Philosophy of Science,* edited by D. J. Depew and B. H. Weber. Cambridge, Mass.: MIT Press, pp. 21–24.

Burian, R. M. 1995. To many kinds of genes? Some problems posed by discontinuities in gene concepts and the continuity of the genetic material. In *Gene Concepts and Evolution (Workshop).* Preprint no. 18. Berlin: Max Planck Institute for the History of Science, pp. 43–51.

Dawkins, R. 1982. *The Extended Phenotype.* Oxford: W. H. Freeman.

Falk, R. 1995. The gene: From an abstract to a material entity and back. In *Gene Concepts and Evolution (Workshop).* Preprint no. 18. Berlin: Max Planck Institute for the History of Science, pp. 21–30.

Feldman, M., and R. C. Lewontin. 1975. The heritability hang-up. *Science* 190: 1163–1168.

Fogle, T. 1990. Are genes units of inheritance? *Biology and Philosophy* 5: 349–371.

Gifford, F. 1989. Complex genetic causation of human disease: Critiques of and rationales for heritability and path analysis. *Theoretical Medicine* 10: 107–122.

Gifford, F. 1990. Genetic traits. *Biology and Philosophy* 5: 327–347.

Kitcher, P. 1982. Genes. *The British Journal for the Philosophy of Science* 33: 337–359.

Kitcher, P. 1992. Gene: Current usages. In *Keywords in Evolutionary Biology,* edited by E. F. Keller and E. A. Lloyd. Cambridge, Mass.: Harvard University Press, pp. 128–131.

Lewontin, R. C. 1974. *The Genetic Basis of Evolutionary Change.* New York: Columbia University Press.

Mackie, J. L. 1974. *The Cement of the Universe.* Oxford: Oxford University Press.

Rimoin, D. L., J. M. Connor, and R. E. Pyeritz, eds. 1996. *Emery and Rimoin's Principles and Practice of Medical Genetics.* 3rd ed. New York: Churchill Livingston.

Rosenberg, A. 1985. *The Structure of Biological Science.* Cambridge: Cambridge University Press.

Smith, K. 1992. The new problem of genetics: A response to Gifford. *Biology and Philosophy* 7: 331–348.

Waters, K. 1994. Genes made molecular. *Philosophy of Science* 61: 163–185.

2

Extracting the Units of Heredity

4

From Measurement to Organization: A Philosophical Scheme for the History of the Concept of Heredity[1]

JEAN GAYON

ABSTRACT

This paper proposes a general scheme for the historical analysis of the concept of heredity covering the last one and a half centuries. During this period heredity became a major biological concept that was purposefully subjected to experimental and quantitative methods of investigation. The conclusions drawn here are twofold. First, there were basically two alternative quantitative approaches to heredity: Either treat heredity in terms of a magnitude similar to the concepts of the physical sciences, or treat it as a material structure. After some initial pending and leaning toward the first alternative, biologists became increasingly committed to the second alternative. Second, I show that the three major phases in the history of the science of heredity (biometry, Mendelian genetics, and molecular genetics) corresponded to three quasi-explicit conceptions of the cognitive status of scientific theories, namely: phenomenalism, instrumentalism and realism. This periodization does not pretend to account for the entire complexity of the history of biological conceptions of heredity. These should be taken, rather, as ideal types that shed some light on important philosophical aspects of the broad history of the science of heredity.

INTRODUCTION

According to a common understanding among physicists, the concepts of a scientific theory consist fundamentally of magnitudes. Mass or force, for instance, have the status of quantitative concepts or magnitudes. Duhem (1981[1914], chapter 2) proposed that the definition of magnitudes and their measurement was the first of four successive steps in the formation of a physical theory. The three other steps were: the connecting of the magnitudes through a limited number of propositions that serve as principles, or "hypotheses," of deductions; the combination of the principles according to mathematical rules; and the comparison of the theory arrived at with a set of empirical laws (which would also be a set of quantitative relations between measurable properties). In such a context, the classical formula, "all science is of the measurable," has a strong meaning: It is not only required that the phenomena should be accessible to measurements, but also that the theoretical entities should be magnitudes, and therefore measurable. However, this is probably a condition peculiar to the physical sciences. In other empirical sciences, fundamental concepts are not necessarily magnitudes; in fact, they most often denote classes of genuine *things*. To be sure, these things (like the chemist's atoms and molecules, or the organs, tissues, and genes handled by the biologist) display a variety of measurable *properties*, but they are not magnitudes.

These philosophical observations suggest a general scheme for a broad historical analysis of the concept of heredity. Like nutrition or growth, heredity refers to a fundamental *property* of organisms. Indeed, in the nineteenth and twentieth centuries, heredity came to be treated as the most fundamental property of living beings. After giving a short survey of early qualitative conceptions of heredity, I will consider the evolution of quantitative treatments of heredity that began during the 1850s and continues to the present.

With respect to the quantitative phase, there are two intertwined dimensions involved in my argument: First I will show that, having been originally treated as a measurable magnitude, heredity later became a structural property, or the property of an organized object. In fact, while quantitative methods were ever more developed in the experimental science of heredity, the concept of heredity itself became increasingly less like a magnitude. Rather, it was defined in relation to newly discerned levels of organization present in living beings. Secondly, building on this view, I will argue that three phases can be distinguished in the modern history of heredity: the biometric phase, the Mendelian phase, and the molecular phase. Moreover, I will show how these three phases correspond to an implicitly phenomenalist, operationalist (or more widely instrumentalist), and realist interpretation of the status of the science of heredity, respectively.

FORCE VERSUS STRUCTURE IN EARLY QUALITATIVE CONCEPTIONS OF HEREDITY

In this section, I present an overview of conceptual approaches to natural inheritance entertained prior to the introduction of quantitative methods in the field of animal and plant breeding. This period was largely characterized by a contradistinction of notions of heredity as a force and those of heredity as resulting from the distribution of particles.

In the nineteenth century, the dominant belief was unequivocally of heredity as a force. By employing the expression *hereditary force,* naturalists described a power active in transmission and though unknown in its intimate mechanism, it was cognizable through its effects (just like the classical notion of force in mechanics). Louis Vilmorin, a French seedgrower and biologist, played a pioneering role in this story. In 1851 he declared that hereditary forces were not amenable to measurement (Vilmorin 1851). Nevertheless, in 1856, fifteen years before Galton, Vilmorin proposed the first quantitative treatment of a hereditary force, i.e., of the capacity of beet-root to produce sugar (Vilmorin 1856, 873).

However, in the mid-nineteenth century, most theoreticians of animal or plant breeding thought primarily of the hereditary force in terms of qualitative analysis. A good example is furnished by a doctrine known as "constancy of race" which emerged in various parts of northern Europe (especially Germany) among the theoreticians of animal breeding. In a fascinating story, Berge (1961) related that around 1840–1850 this doctrine had been built on the assumption that the hereditary force was of variable intensity that changed over time. The longer the character had been transmitted without crossbreeding, the more powerful its hereditary force became. Later, the same doctrine was developed mainly by Galton and the English biometricians, figuring under the name *ancestral heredity.* Ancestral heredity implied that the hereditary endowments of the ancestors tended to accumulate and act in an additive manner in the progeny (see next section 3).

In his *Variation of Animals and Plants under Domestication,* Darwin (1875) rejected the interpretation of heredity as a force the intensity of which is proportional to the duration of transmission. He presented his argument in a section with the revealing title, "Fixedness of Character." There, Darwin examined "a general belief amongst breeders that the longer any character has been transmitted by a breed, the more fully it will continue to be transmitted" (Darwin 1972 [1875], volume 2, 37). Darwin attacked this belief, using the breeders' own examples. The everyday practices of breeding and horticulture, Darwin wrote, did not at all confirm that "inheritance gains strength simply through long continuance" (Darwin 1972 [1875]). Darwin accepted that a domestic race would tend to become fixed when maintained by eliminating individuals that deviated from the desired form. But artificial selection, though presupposing the heritability of the characters that were to be fixed, did not imply that heredity becomes stronger over time. For measuring the hereditary force of a given character, what counted was not its age, but rather what happened when an alternative character emerged, either spontaneously or due to a cross. The breeders and horticulturists had shown that a new character can become rapidly fixed, or show major variations, or may not be transmitted at all. Thus, there does not appear "to be any relation between the force with which a character is transmitted and the length of time during which it has been transmitted" (Darwin 1972 [1875], volume 2, 38). In fact, even though he used archaic terms like "tendency to inheritance" (Darwin 1859, 12) or "force of inheritance" (Darwin, volume 1, chapter 12, 446, 448), Darwin's fundamental conception of heredity was not energetic but "material" and "particulate" as shown by his "provisional hypothesis of pangenesis" (Darwin 1972 [1875], chapter 27). This hypothesis implied that at any time in the life of an individual, all cells generated "gemmules." The gemmules were not cells, but small buds that retained the characteristic features of the cells they came from. They possessed the ability to circulate through the body and accumulated in the germinal cells where they finally acquired the status of hereditary elements that combined and reshuffled in every generation. Darwin assumed that the gemmules were able to develop cells corresponding to their own type in the course of embryogenesis. Although this hypothesis may have entailed an extreme form of Lamarckism (since for Darwin all variation tends to be inherited),[2] its representation of heredity was in the language of particles rather than in terms of a force. Darwinian gemmules, whatever their origin may have been, were units of transmission. Therefore, what counted for heredity in the Darwinian perspective of pangenesis was less the *past* (the ancestry) than the present *structure* of these collections of units.[3] Thus, even before the introduction of quantitative methods in the study of heredity, we observe an alternative between a concept of a force of varying intensity and an organizational or structural conception of heredity. I will now consider the effects of the introduction of measurement.

THE BIOMETRIC PHASE (1870–1900)

Putting aside the social ideology of Galton and his followers, we can see that the major commitment of their approach to heredity was to reconstruct this theoretical notion in terms of a magnitude. Galton's methodology for treating heredity is well known. Employing statistical tools, he progressively arrived at the idea that the correlation of characters between offspring and parents, as between offspring and ancestors, provided the appropriate means for measuring heredity. This methodology fit well with the older representation of heredity as a force of varying intensity. For Galton and for the early biometricians who adopted his strategy (e.g., Pearson), the main problem revolved around how to evaluate the "strength" of heredity for a particular character.

Galton's statistical studies of heredity were based on a series of popular observations: (1) Children resemble their parents more than they resemble others; (2) children of the same parents vary; (3) exceptional parents generally have more mediocre offspring than themselves; (4) some characters are able to "jump" over several generations (atavism). Such phenomena could be easily submitted to statistical analysis as it existed in the mid-nineteenth century. The typical procedure was to examine the frequency distribution of a particular character (e.g., size) among children of parents with a known character state (e.g., 70±1 inches high), and compare the outcome with the frequency distribution of the whole population. This procedure allowed a formulation in quantitative terms of the first three kinds of common-sense observations mentioned above. In particular, for children of given parents, the mean value of the character was closer to the mean of the population than that of the parents.

Children tended to return to the statistical mean of the population, or they reversed to the population mean in Galton's terminology. Galton called this phenomenon reversion and considered it as the "statistical law of heredity" (Galton 1877). Unfortunately for his theory, he discovered some years later that the same kind of analysis could also be applied to the frequency distribution of parents of given children: Parents also tended to be more mediocre or meaner, and to regress toward the mean. Consequently, Galton renounced the idea that reversion was the fundamental law of heredity. He recognized that it was a general characteristic of the mathematical tool he had invented for the purpose of analyzing hereditary phenomena and consequently gave this statistical tool a new name: *regression*. For any two parameters x and y, it was equally legitimate to speak of "regression of x on y" or of "regression of y on x" (Galton 1885). This resolution led to the invention of the coefficient of correlation (Galton 1888). The correlation coefficient between parents and offspring, or other coefficients derived from it, became for Galton and his followers the typical measure of the hereditary force. Consequently, Galton came to see ancestral heredity as the key concept. Ancestral heredity implied that a given individual does not inherit the relevant character only from his parents, but also from his various earlier ancestors (for more details, see Cowan 1972, 1977; Gayon 1998; Stigler 1986).

A major epistemological characteristic of this approach was its purely descriptive character, in other words, its independence of any hypothesis about the mechanism of hereditary transmission. Galton was probably not fully aware of this, but his followers were. Karl Pearson's treatment of ancestral heredity was exemplary in this respect. In 1898, two years before the rediscovery of Mendel's laws, Pearson proposed a mathematically respectable formulation of the law of ancestral heredity. For Pearson, there was only one possible mathematical interpretation of this law: It had to be an equation of multiple regression (or multiple correlation), linking the character of an individual to the average character of each ancestral generation. Such an equation took the form: $y = b_1 x_1 + b_2 x_2 + \ldots + b_n x_n$, in which y is the character of the progeny, x_n the average character in generation n, and b_n the coefficient of partial regression of y on x for generation n. In Pearson's mind, this treatment of heredity was purely descriptive. In particular, the regression coefficients could not be determined a priori, and it was absolutely not necessary for them to take on the values $1/2, 1/4, 1/8, \ldots 1/2^n$, as Galton had thought. Only actual measurement could determine them in particular situations. Thus the equation of regression did no more than summarize the observable differences between ancestors and progeny (Pearson 1898). Pearson insisted that this approach was independent of any hypothesis on the physiological nature of heredity. "The law of ancestral heredity in its most general form is not a biological hypothesis at all, it is simply the statement of a fundamental theorem in the statistical theory of correlation applied to a particular type of statistics. If statistics of heredity are themselves sound the results deduced from this theorem will remain true whatever biological theory of heredity is propounded" (Pearson 1903, 226). In fact, if one looks at the details, Pearson considered the intensity of heredity as the result of a balance of forces involving all factors of evolution that affected lineage in the past: selection, mating system, etc. (see Pearson 1898).

Pearson was one of the major advocates of the phenomenalist conception of scientific theories around 1900. For this English disciple of Ernst Mach, scientific theories could do no more than abbreviate empirical data in a mathematical language. This was the central thesis of *The Grammar of Science*, first published in 1892. With Pearson, we have the most highly developed version of the first phase of the philosophical history of the concept of heredity: the phenomenalist view of heredity. It is worth observing the agreement between Pearson's general conception of science, and his attempt to construe the scientific concept of heredity as a magnitude.

THE MENDELIAN-CHROMOSOMAL PHASE (1900–1950)

Let us first characterize the methodological shift between biometry and Mendelism. In fact, there were two major shifts. One is related to the role of pedigrees, the other concerns the role of mathematical analysis of data.

For Mendelian genetics, a pedigree was a tool, and no longer a fundamental concept. In previous theories built upon the concept of ancestral heredity, heredity was the sum total of influences received from the ancestors. It meant, rigorously speaking, that a child did not inherit from his parents alone, but from the whole series of ancestors included in his lineage. In this context, heredity was nearly synonymous with descent, or lineage, or else "pedigree." For Mendelism, the origin of characters was an irrelevant issue. The pedigrees had nothing to tell us about the nature of heredity; they were only tools for inferring the genetic structure of individuals. This is why they could be employed in both directions: from parental to filial generations, and conversely, from filial to parental generations. This difference can be illustrated by a comparison of the standard representations of generations in biometry and in Mendelian genetics. In the context of the biometricians' conception of ancestral heredity (Figure 4-1), a succession of ancestral or parental generations ($P_1, \ldots , P_3, P_2, P_1$) converged toward the unique F-generation (filial generation). This kind of representation connoted a vision of heredity as an accumulation of influences. Since all individuals will have roughly the same ancestors when going back far enough, heredity was in the long run confounded with the racial legacy. Ancestral heredity was a modern form of the old notion of racial inheritance. The Mendelian representation of generations was quite different (Figure 4-2). It recognized only one parental generation, and instead of going backward, it went forward toward a series of filial generations ($F_1, F_2, F_3 \ldots$; note that in the case of sexual reproduction, the diagram of Figure 4-2 should be slightly modified: in every generation, new crosses are realized). The important point here is to examine the disjunction of the hybrid generation (F_1) in the course of the follow-

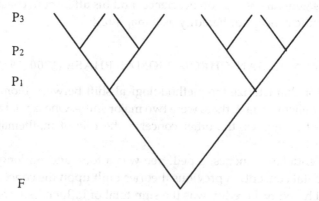

P_3

P_2

P_1

F

Figure 4-1   See text for explanation.

P   X   .....

F_1

F_2

F_3

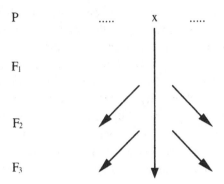

Figure 4-2   See text for explanation.

ing filial generations. This disjunction allowed one to retrospectively infer the structure of the parental generation. The parental generation was, by definition, the generation that by its genetic structure was to be expressed in the context of specific crosses. Thus, for Mendelian genetics, the concepts of descent and heredity did not stand on the same level, as they did for the supporters of ancestral heredity of the English biometrical school. Heredity was not the sum total of ancestral influences; it was a question of structure in a given generation. What happened to the progeny did not depend on what happened to the ancestors of its parents, but only on the genetic makeup of its parents. For instance, if both parents were recessive homozygotes for a certain character (or *aa*), the progeny will also be *aa*, whatever may have been the story of the parents' parents, grandparents, and so on. Accordingly the mathematical analysis of data took on a completely different meaning. For the biometricians, heredity was a *magnitude* – a statistical magnitude. For the Mendelians, the task of statistical analysis was to reveal individual genotypic *structures*, and the dynamics of recombination between units. Algebra played a role very different from that which was conceived in the biometrical treatment of heredity. Mendelians used symbols, but these symbols did not denote magnitudes. They denoted hypothetical units that allowed for a calculus that was not reducible to an abbreviated treatment of data. Mendel's reasoning relied on a calculus of differences, and required finite algebra. This has been repeated again and again for nearly a century: The kind of mathematics involved in Mendelian genetics is basically combinatory. In fact, behind the Mendelian symbols for alleles and genotypes stood hypotheses formed in the context of anatomy and cytology (biological hypotheses about the role and behavior of germinal cells and chromosomes in the reproductive process). This was already conceived by Mendel as shown by his 1866 memoir (see the paragraph on the sex cells of the hybrids). It became much clearer in the 1900s, when the Mendelians reinterpreted Mendel's laws in the language of meiosis and chromosomes. Mendelian genetics did not interpret heredity as a magnitude, but as an expression of a certain level of organization.

Having characterized the Mendelian break, I will now characterize the Mendelian-chromosomal paradigm as a whole, with a particular concern for the relation between measurement and organization. Officially consecrated by the publication of *The Mechanism of Mendelian Heredity* (Morgan et al. 1915), the chromosomal theory of inheritance introduced new kinds of measurement, such as genetic distance. However this quantitative tool was developed in order to provide a cartography of genes. In other words, it aimed at giving an anatomical picture of the genome its proper spatial organization. The chromosomal theory provided a metric for understanding the overall topography or spatial organization of the presumptive hereditary elements (or genes). Thus the new kinds of measurement that were introduced led to a weakening of the old view of heredity as a magnitude, and reinforced its interpretation in terms of structure and organization.

In this new theoretical context, genes had no clear status. In the 1930s nobody knew about the physical characteristics, nor about the chemical constitution and properties, or about the physiological action of genes. There were even disputes about the continuity of their physical existence and whether the gene was a genuine molecule (or part of a molecule) or rather a physiological state (a very common idea before the mid-1940s). Against this background, geneticists developed a philosophical interpretation of the status of genes: Within the theoretical and experimental contexts that defined them, genes could be just the hypothetical units needed for the success of the geneticists' predictions. In his 1933 Nobel lecture, Thomas Hunt Morgan expressed this common epistemological conviction in the crudest way: "... at the level at which the genetic experiments lie it does not make the slightest difference whether the gene is a hypothetical unit or whether the gene is a material particle" (Morgan 1935, 347–348). Most geneticists before World War II shared this operationalist interpretation of genetics.

THE MOLECULAR PHASE

I will not expand on the origins and development of molecular genetics (for a general account, see Judson 1979; Cairns, Stent, and Watson 1966). It is needless to recall the main feature of this new approach to heredity: the elucidation of the physicochemical nature of the gene, and of the machinery that allows it to replicate and code for polypeptides. However, I will discuss two stages of development, beginning with the dogmatic doctrines that dominated around the 1960s and 1970s, and ending with some remarks on the present state of the discipline.

Just as in the previous phases of the history of the science of heredity, measurement and quantitative methods have been crucial for the emergence of molecular genetics. Refined physical methods were necessary for the elucidation of the molecular constitution, conformation, and physiology of the genes, identified, in the dogmatic period, to stretches of chromosomal DNA. But more than ever, measurement remained outside the sphere of the theory of heredity. In molecular genetics, measurement is everywhere in the peripheral instrumentation (ultra-centrifugation, X-ray spectrography, electrophoresis, radioactive markers, etc.), but it is excluded from the concept of heredity. Molecular biology has indeed completed the dissolution of the concept of heredity as a physical magnitude. It has done so in two ways. First, it has provided a clear material picture of the anatomical structures involved in the storage and expression of genetic information. Nobody would have ever imagined before the 1950s that the macromolecules identified as genes (as well as proteins) had so precise and definite an atomic structure and a spatial configuration. Second, the extensive use of informational and linguistic concepts had a deep philosophical implication: It involved a return to "form" at the most elementary level of description of organic beings. Thus, genes happened to be describable both in the mechanistic language of shape (geometrical form) and of information (imposition of a form, in Aristotle's sense). This renewed link between form as both shape and information in biology is epitomized well in the doctrine that the three-dimensional structure (or shape) of the double helix of DNA can be interpreted as a means to the function of this molecule. The three-dimensional structure makes it possible to replicate the linear or one-dimensional sequence that bears the information ("It has not escaped our attention that the specific pairing we have postulated immediately suggests a possible copying mechanism for the genetic material." [Watson and Crick 1953, 738]).

At the beginning of this paper, I stated that the molecular phase of the science of heredity developed in the philosophical mood of realism. This is perfectly clear for the early period of molecular genetics. Indeed, if genes can be identified as fragments of chromosomal DNA, these fragments being both replicating units and functional units coding for amino-acid sequences, it seems reasonable to interpret genes philosophically as a class of real physical things. Genes, on this account, turn out to be robust "natural kinds," just like organisms, tissues, cells, or chromosomes. They are clearly located somewhere, they have a well-identified molecular nature (as stretches of DNA), and they perform well-defined functions (replicating and coding for polypeptides).

But an issue remains open: Given our present knowledge, is there still room for some kind of realism in our philosophical representation of genes? It has been said often that the present advances in molecular biology make such a representation impossible, because genes can no longer be equated, in many cases, with clear-cut fragments of chromosomal DNA. If one thinks of nontranslated regulating genes (operators, promoters), overlapping genes, genes-in-pieces, RNA splicing, and other phenomena, the only fair philosophical conclusion is that there is absolutely no hope to have a type-to-type correspondence between Mendelian genes and some unambiguous molecular entity (or disjunctive list of entities). The relation between a Mendelian gene and corresponding molecular entities is a many-to-many relation. The best one can have is a case-by-case or token-token relation. Thus a realistic physical interpretation of genes as theoretical entities is not possible. In fact it is quite tempting to say that the qualification of something as a gene depends on how the experimenter chooses to manipulate the genome (for a discussion of this point of view, see Beurton this volume; for a very complete review of modern challenges to the classical molecular view of the gene, see Portin 1993). Beurton's observation is illuminating. It amounts to saying that the recent development of genetic engineering has led to a restoration of the instrumentalist view of genetics. I would like, however, to advocate some sort of realism, although I am aware that I will use this philosophical category in a sense that does not fit rigorously with the commonest use of the word in contemporary philosophy.

Let us portray the situation. True, we can no longer pretend that genes constitute a robust class of physical material entities, these entities being fragments of DNA endowed with the properties of coding or polypeptides playing some role in regulation (e.g., operators). However, if we want to preserve the concept of a gene as a natural category, we must accept that the modes of storage of information cover a wide variety of anatomical and physiological situations. Sometimes a gene will be a mere continuous fragment of chromosomal DNA; sometimes it will overlap with other genes (e.g., operator and promoter); sometimes it will have to be defined through developmental process (leading for instance to a mature mRNA). As Hans-Jörg Rheinberger reminds us, this developmental process can sometimes go very far into the metabolism of the cell (Rheinberger, this volume). Furthermore, since there is always a functional aspect in the concept of a gene, what counts as a gene will change according to what functions we think of (Burian 1985).

I am not thoroughly convinced by the argument that this situation forces us to come back to an instrumentalist view of the gene. Rather, it seems to me that we are back to the theoretical situation of comparative anatomy in older times, but at a totally new scale of observation. In comparative anatomy, the same organ can be developed and organized in many ways; similarly, the developmental and organizational schemes leading from DNA to RNAs and to polypeptides may be highly variable, within and between organisms. Sometimes we will be satisfied to call a mere strand of chromosomal DNA a gene, because it is just there that the information necessary for the making of a protein (or regulating its fabrication) is stored. Sometimes a gene will not be identifiable before the occurrence of various processes of editing. Then the information will not be simply stored in a permanent piece of nucleic acid, but rather involve complex metabolic and regulatory processes. Consequently, we must admit that a gene needs not to be thought of as having temporal continuity. This entity can, and will indeed most often, be endowed with temporary and discontinuous existence, and it will often require a developmental process at its own level of organization for functional expression. This does not mean however that there is no temporally continuous material basis for what will function as a gene. Something structural, of course, must be retained: Even in the most complex cases of gene editing, contemporary molecular biology has not discredited the idea of a material structure that is continuously preserved and transmitted in the course of generations (a certain strand of DNA or RNA). Whether or not we decide to call this continuously existing piece a gene, we still believe, however, that something like this must exist as a necessary basis for the existence of a gene and for its identification (Burian 1995, and personal communication). Or, to put it in an older vocabulary that reminds us of Erwin Schrödinger (1945), we need some autocatalytic structure, although this structure need not be the gene.

In such a theoretical context we are back to the ordinary situation that has dominated in biology for many centuries. Biologists are spontaneously prone to realism, because they are committed to studying existent beings, and not only processes (as physicists do) (Piaget 1950, volume III, 5). The knowledge of the fundamental diversity of these beings, and of their components, can reinforce this philosophical conviction. Today's knowledge of the organizational basis of heredity confirms this old tendency of biological sciences. In other words, there seems to be no privilege for heredity: Just as for all other biological functions, it is organization and diversity all through.

A NONRELEVANT OBJECTION: HERITABILITY

The general trend of the argument that has been developed in previous sections can be summarized as follows: Heredity as a scientific concept was first theorized as a magnitude, but this conception was increasingly rejected in the course of the development of genetics in the twentieth century. It is true that the development of Mendelian genetics involved the emergence of new kinds of measurements (e.g., genetic distance, mutation rate), and that molecular biology implied the development of sophisticated material techniques and quantitative methods to establish the structure and function of the molecules involved in the storage and the expression of hereditary traits. But the more genetics developed, the more the idea of heredity as a quantitative concept became obsolete, and the more a structural, organizational, and finally developmental view of the hereditary units (genes) emerged from biological discourse. One important objection could be raised against the present analysis, namely, that under the term of *heritability* the genetics of quantitative traits has developed a modern concept of heredity that is explicitly one of a measurable magnitude. I will now show why this objection is not admissible.

Heritability is a statistical index that, for a given quantitative character and against a fixed environmental background, measures which portion of the observable differences can be attributed to "heredity." This concept corresponds quite well to what the nineteenth century scientists had in mind when they characterized heredity as a force of varying intensity. However, it is only in a Mendelian framework that a satisfactory mathematical solution to this problem was finally found. In the context of early biometry (Pearson), the strategy for determining the respective shares of heredity and environment in a given character consisted in contrasting a parent-offspring correlation coefficients on one side, and an open list of measured correlations between the character and many environmental factors on the other side. This is exactly the kind of work to which Pearson devoted a huge amount of time and money in his Galton Laboratory in London (see Pearson 1910; Norton and Pearson 1976). Generally, the hereditary correlation was quite strong when compared with various correlations with individual environmental factors. However, as noted by Fisher (1918) in his famous paper "The Correlation between Relatives on the Supposition of Mendelian Inheritance" – the paper that really laid the foundations of quantitative genetics – this method did not allow any rigorous statements about the respective causal responsibility of heredity and environment. The concept of heritability originated in Fisher's paper, although the word was introduced only later. In his rich paper, Fisher introduced the words *variance* and *covariance* for the first time. He also invented the classical analysis of variance, which renders it possible to make causal inferences from statistical data. Fisher proposed to analyze the total phenotypic variance for a given character into four parts: additive genetic variance, variance due to interallelic interaction (dominance), variance due to genetic interaction (epistasis), and variance due to environmental causes. Schematically, this can be written:

$$V_p = V_a + V_d + V_{ep} + V_{env}$$

Heritability in the narrow sense is the portion of variance that can be attributed to transmissible traits. It is commonly defined as the ratio V_a/V_p. This ratio measures the part of variance attributable in a population to the additive effects of genes, independently of interaction effects between alleles, between loci, and between the genetic makeup and the environment. In a different, broad sense heritability is used as the portion of the total variance that is due to genetic differences, whatever their origin (additive effects of genes, or genetic interactions). Heritability in the broad sense is then measured by the ratio V_g/V_p ($V_g = V_a + V_d + V_{ep}$). For a population geneticist, however, the only important parameter is heritability in the narrow sense. Although heritability had been widely discussed in the 1920s and 1930s, it was only in 1940 that Jay L. Lush brought a clear distinction between heritability in the narrow sense and heritability in the broad sense (Lush 1940).

Now it is essential to understand that this statistical construction does not permit the conclusions that are commonly drawn on the responsibility of heredity in this or that character. Heritability (whether in its narrow or broad sense) is not a measure of heredity. It is a meaningless concept as long as one does not specify the particular population and the particular environment in which such a measure is carried out. For instance, in a population of individuals genetically identical for a particular character the heritability (narrow sense) of that character would be zero, since the genetic variance equals zero. Many similar paradoxes could be mentioned. They are well-known among population geneticists (see Hartl 1980, chapter 4). In reality, the heritability of a certain character has not much to do with the issue of whether a certain character depends on genetic causes from a physiological point of view. As animal and plant breeders knew from the beginning, heritability is not so much a measure of heredity as an efficient index of the efficiency of selection (artificial or natural), and, more generally, of the evolutionary potential of a population. If there is no heritability for a certain character, there is no room for a biological evolution of the population for this character. But this does not mean that character has no genetic causes.

Therefore, the only magnitude that modern genetics may appear to some to offer as a measurement of heredity is not a good candidate for this status. True, heritability is measurable magnitude, but it is incorrect to say that it would measure heredity.

CONCLUSION

In this paper, I have proposed a reappraisal of the history of the concept of heredity at a relatively wide historical scale. This scale is wider than that usually covered by the history of the disciplinary field called genetics though there is no intrinsic reason not to broaden the history of this field too. After all, the idea of hereditary traits is almost as old as European medicine. Actually, before the nineteenth century, heredity as a naturalistic concept was mainly a medical issue, not a concept of natural history by and large (David 1971; López 1992). The historical scale I have chosen corresponds to a period in which heredity had obviously become a major biological concept, and purposeful attempts were undertaken to study it with the aid of experimental and quantitative methods. For this modern period, which began approximately in the 1850s, I have proposed a philosophical interpretation of the development of the debates. The conclusions of my inquiry are twofold.

First, I have shown that biologists, after having remained undecided for some time which of two general approaches to adopt in the analysis of heredity (measurable force versus material structure), became increasingly committed to the second alternative. Biologists no longer thought of heredity as a magnitude that would play in biology a role similar to that of fundamental quantitative concepts in physics. Such ideas were superseded by structural and organizational representations of heredity: Mendelian genetics, the chromosomal theory of heredity, and molecular genetics (in its successive versions), have increasingly developed and deepened a structural representation of heredity. A constant feature of the twentieth century, heredity (genetics) is the role this science has conferred to quantitative methods in the treatment of its object: in this science, quantification is everywhere (methods, instruments, etc.) except in the very concept of heredity. The more genetics developed in the course of this century, the more it has privileged a structural, then an organizational, then a developmental concept of this fundamental property of living beings. Therefore, heredity is the central concept of a quantitative science without being itself a magnitude. And correlatively, genetics, although it was the first biological discipline that was genuinely mathematized, stands today as a remarkable example of the autonomy of biology: chromosomes, genes, introns and exons, and the related functions and processes, are genuine biological concepts.

The second conclusion of this paper is closely related to the previous one. Those who in the nineteenth century emphasized the treatment of heredity as a measurable magnitude, were the biometricians. Although they took over a very old and popular idea of heredity as a hidden force, they probably never seriously believed that they would succeed in building an abstract mathematical and axiomatized theory of heredity that would be comparable to major physical theories. This is why some of them (especially Pearson) insisted on the purely phenomenalist signification of their treatment of heredity in terms of statistical regression and correlation coefficients. Heredity coefficients were no more than condensed or abbreviated data. Later on, when the materialist conceptions of heredity (e.g., Darwin's pangenesis, Weismann's and de Vries's speculations on the hereditary material) were reformulated in the symbolic language of Mendelian genetics, biologists also felt the necessity to make explicit the philosophical significance of their theories. In the first half of the twentieth century, a majority of geneticists agreed upon the idea that the kind of theory they were developing had essentially an operational value. This meant that they admitted unobservable theoretical entities (especially genes) that allowed them to make good testable predictions, but that were unresolved as to their physical signification and existence. It is only with molecular biology that a realistic interpretation of the hereditary material emerged. Thus I have shown in this paper that the periodization of the history of the science of heredity could be enlightened by considering the spontaneous philosophical conceptions developed by scientists about the cognitive status of their theories. This periodization is certainly not the only one possible, and it should indeed complement others, founded on other criteria (methodological, conceptual, or social). Nevertheless, I find it remarkable that the history of this typically modern biological discipline called *genetics* can be reconstructed along such philosophical lines. I know of no other history of a modern biological discipline that could be reconstructed in terms of such successive epistemological shifts. Of course, the alternative between phenomenalism, operationalism, realism, and other philosophical categories concerning the cognitive status of scientific theories are a constant preoccupation for all sciences. But it is not so frequent to observe these philosophical positions as ideal types shaping massively successive states of a certain area of scientific knowledge. One important reservation must be added at this point. The narrative I have proposed here does not take into account the cytological studies that, in the nineteenth and early twentieth centuries, contributed to shape the modern science of heredity. Cytology, of course, induced a number of biologists, both before and after the rediscovery of Mendel, to adopt, or at least express, the wish of a material and therefore more realistically oriented view of heredity. Therefore, the periodization I have proposed should not be taken as the only one possible. Nevertheless, the material conceptions of heredity that were formulated in the second half of the nineteenth century (e.g., theories of germ plasm in the nineteenth century) or Muller's "material conception of the gene" had a strong speculative, not to say, a metaphysical taste (see Gayon 1997; Muller 1926; Weismann 1889). The present paper has been directed primarily at the cognitive status of theories of heredity that had a strong empirical basis in their own time.

Finally, I do not want to suggest that genetics should be now irreversibly installed in "scientific realism." Many developments in molecular biology since the 1960s suggest that genes can no longer be viewed as unambiguous classes of natural entities in the sense that there would exist a nonambiguous molecular definition of the category of gene. Nevertheless, within the general scheme I have

proposed for the history of the concept of heredity, these developments go in the same overall direction as previous phases of genetics.
Our understanding of genes and allied concepts, although stated in a
molecular language, is more and more akin to what biologists used
to call anatomy and physiology.

NOTES

1. Peter J. Beurton, Richard M. Burian, Marjorie Grene, and Evelyn Fox
   Keller are thanked for their stimulating observations on this paper and for
   their linguistic corrections.
2. "When a new character arises, whatever its nature may be, it generally
   tends to be inherited, at least in a temporary and sometimes in a most
   persistent manner" (Darwin 1972 [1875], volume 1, chapter 12, 416). A
   similar declaration can already be found in the first pages of the *Origin:*
   "Any variation which is not inherited is unimportant for us . . . Perhaps
   the correct way of viewing the whole subject, would be, to look at the
   inheritance of every character whatever as the rule, and the non-
   inheritance as the anomaly" (Darwin 1859, 12–13). Contrary to what
   many historians say, I do not think that Darwin treated heredity as a force
   opposed to the force of variation, whether in the *Origin* or in *Variation.* For
   him, all variation was in principle heritable. This is of course related to the
   extreme form of Lamarckian inheritance he put forward under the name
   of *pangenesis.*
3. Many historians have insisted on the role of blending inheritance in Darwin's thinking. I do not deny this aspect, which played an important role
   in the history of conceptions relative to the efficacy of selection (Fisher
   1930; Provine 1971). However, it is the material and structural aspect
   conveyed by the hypothesis of pangenesis that played a major role in the
   history that finally led to the notion of gene. Weismann's and de Vries's
   speculation on the hereditary particles were directly inspired by Darwin's
   pangenesis. These two biologists agreed both on refusing the Lamarckian
   aspect of the hypothesis (the conjecture that cells produce gemmules that
   accumulate in the germ cells) and on stressing the idea of hereditary
   material particles (Weismann 1889; de Vries 1889).

REFERENCES

Berge, S. 1961. Die geschichtliche Entwicklung der Tierzüchtung.
Schriftenreihe des Max-Planck-Instituts für Tierzucht und Tierernährung.
Special Issue: 127–148.
Burian, R. M. 1985. On conceptual change in biology: The case of the gene. In
Evolution at a Crossroads: The New Biology and the New Philosophy of Sci-

ence, edited by D. J. Depew and B. H. Weber. Cambridge, Mass.: MIT Press, pp. 21–42.

Burian, R. M. 1995. Too many kinds of genes? Some problems posed by discontinuities in gene concepts and the continuity of the genetic material. In *Gene Concepts and Evolution (Workshop)*. Preprint no. 18. Berlin: Max Planck Institute for the History of Science, pp. 43–51.

Cairns, J., G. S. Stent, and J. Watson. 1966. *Phage and the Origins of Molecular Biology*. Long Island, NY: Cold Spring Harbor Laboratory of Quantitative Biology.

Cowan, R. S. 1972. Francis Galton's statistical ideas: The influence of eugenics. *Isis* 63: 509–528.

Cowan, R. S. 1977. Nature and nurture: The interplay of biology and politics in the work of Francis Galton. *Studies in the History of Biology* 1: 133–208.

Darwin, C. 1859. *On the Origin of Species by Means of Natural Selection, or the Preservation of Favoured Races in the Struggle for Life*. London: John Murray.

Darwin, C. 1972 (1875). *The Variation of Animals and Plants under Domestication*. 2nd ed. Facsimile: New York: AMS Press.

David, B. 1971. La préhistoire de la Génétique – Conceptions sur l'hérédité et les maladies héréditaires des origines au XVIIIe siècle. Ph.D. dissertation, Paris, Hôpital Broussais.

de Vries, H. 1889. *Intracellular Pangenesis*. Jena: Gustav Fischer.

Duhem, P. 1981(1914). *The Aim and Structure of Physical Theory*. Princeton: Princeton University Press.

Fisher, R. A. 1918. The correlation between relatives on the supposition of Mendelian inheritance. *Transactions of the Royal Society, Edinburgh* 52: 399–433.

Fisher, R. A. 1930. *The Genetical Theory of Natural Selection*. Oxford: Clarendon Press.

Galton, F. 1877. Typical laws of heredity. *Nature* 15: 492–495, 512–533.

Galton, F. 1885. Regression towards mediocrity in hereditary stature. *Journal of the Anthropological Institute of Great Britain and Ireland* 15: 246–263.

Galton, F. 1888. Co-relations and their measurement, chiefly from anthropological data. *Proceedings of the Royal Society of London* 45: 135–145.

Gayon, J. 1997. Le temps des gènes. In *Les figures du temps*, edited by L. Couloubaritsis and J. J. Wunenburger. Strasbourg: Presses Universitaires de Strasbourg, pp. 335–349.

Gayon, J. 1998. *Darwinism's Struggle for Survival. Heredity and the Hypothesis of Natural Selection*. Cambridge: Cambridge University Press.

Hartl, D. L. 1980. *Principles of Population Genetics*. Sunderland: Sinauer.

Judson, H. F. 1979. *The Eighth Day of Creation. The Makers of the Revolution in Biology*. New York: Simon and Schuster.

López Beltrán, C. 1992. The Construction of a Domain. Ph.D. Dissertation, University of London.

Lush, J. L. 1940. Intra-sire correlations or regressions of offspring on a dam as a method of estimating heritability of characteristics. *American Society of Animal Production. Record of Proceedings.* 33: 293–301.

Mendel, G. 1866. Versuche über Pflanzen-Hybriden. *Verhandlungen des Naturforschenden Vereins, Brünn; Abhandlungen* 4: 3–47.

Morgan, T. H. 1935. The relation of genetics to physiology and medicine. *Smithsonian Institution. Annual Report* (Publ. No. 3365): 345–359.

Morgan, T. H., A. H. Sturtevant, H. J. Muller, and C. B. Bridges. 1915. *The Mechanism of Mendelian Heredity.* New York: Henry Holt.

Muller, H. J. 1926. The gene as the basis of life. *International Congress of Plant Sciences* 1: 897–921.

Norton, B. J., and E. S. Pearson. 1976. A note on the background to, and refereeing of, R. A. Fisher's 1918 paper "On the correlation between relatives on the supposition of Mendelian inheritance." *Notes and Records of the Royal Society of London* 31: 151–162.

Pearson, K. 1892 (1900). *The Grammar of Science.* 2nd ed. London: Scott.

Pearson, K. 1898. Mathematical contributions to the theory of evolution. On the law of ancestral heredity. *Proceedings of the Royal Society of London* 62: 386–412.

Pearson, K. 1903. The law of ancestral heredity. *Biometrika* 2: 211–229.

Pearson, K. 1910. Nature and nurture, the problem of the future. *Eugenics Laboratory Lecture Series,* VI. London: Dulau.

Piaget, J. 1950. *Introduction à l' épistémologie génétique.* 3 vols. Paris: Presses Universitaires de France.

Portin, P. 1993. The concept of the gene: Short history and present status. *The Quarterly Review of Biology* 68: 173–223.

Provine, W. B. 1971. *The Origins of Theoretical Population Genetics.* Chicago: University of Chicago Press.

Schrödinger, E. 1945. *What Is Life?* New York: Macmillan.

Stigler, S. M. 1986. *The History of Statistics: The Measurement of Uncertainty before 1900.* Cambridge, Mass.: Harvard University Press.

Vilmorin, L. 1851. Sur un projet d'expérience ayant pour but de créer une variété d'ajonc sans épines se reproduisant par graines. Paper read at Communication lue à la Société Industrielle d'Angers le 7 juillet 1851, pp. 253–261.

Vilmorin, L. 1856. Note sur la création d'une nouvelle race de betteraves à sucre. – Considérations sur l'hérédité dans les végétaux. *Comptes Rendus hebdomadaires de l'Académie des sciences* 43: 871–874.

Watson, J., and F. H. C. Crick. 1953. Molecular structure of nucleic acids. *Nature* 171: 737–738.

Weismann, A. 1889. *Essays Upon Heredity and Kindred Biological Problems.* Edited by E. Poulton, S. Schönlans and A. E. Shipley. Oxford: Clarendon.

5

From Gene to Genetic Hierarchy: Richard Goldschmidt and the Problem of the Gene[1]

MICHAEL R. DIETRICH

ABSTRACT

This paper examines Richard Goldschmidt's opposition to the classical concept of the gene as a combined unit of structure, function, mutation, and recombination. To replace the classical gene Goldschmidt articulated a genetic hierarchy drawing on the diverse strands of genetic research that had been bound together previously in the classical gene concept. As such, Goldschmidt's genetic hierarchies represent the possibility of retheorizing genetics without a unifying gene concept.

INTRODUCTION

Throughout the 1930s and 1940s Richard Goldschmidt took great pleasure in announcing the demise of the corpuscular gene. The resulting controversy surrounding Goldschmidt's opposition to the corpuscular gene is well known (Allen 1974; Burian 1985; Carlson 1966; Dunn 1965; Gilbert 1991; Maienschein 1992; Richmond 1986). While Goldschmidt relished claiming that the corpuscular gene was dead, his opponents did not refrain from stating, often in his presence, that he had gone crazy (Goldschmidt 1960, 323; Stern 1967, [1980, 83]). The maelstrom of rhetoric surrounding Goldschmidt and the problem of the gene has, however, obscured the history of the development of alternatives to the corpuscular gene. Biologists, historians, and philosophers have characterized Goldschmidt's position as the chromosome-as-a-whole hypothesis. Such characterizations of Goldschmidt's views do not reflect the development of Goldschmidt's thought as much as the fact that most commentators have failed to come to terms with Goldschmidt's views after 1940.

Elof Carlson, for instance, conflates Goldschmidt's early views (late 1930s) of the chromosome-as-a-whole with his later views that Carlson characterized as view of "the genetic continuum of the chromosome" (Carlson 1966, 125, 126, 128). What Goldschmidt actually articulated in the passages that Carlson quoted was a notion of genetic hierarchy of which the chromosome-as-a-whole was merely a constituent. From the mid-1940s until his death in 1958, Richard Goldschmidt articulated and refined multi-leveled genetic hierarchies to replace the classical, corpuscular gene concept. This paper will examine, first, the development of Goldschmidt's opposition to the classical, corpuscular gene as a particulate entity, and second, his articulation of genetic hierarchies as an alternative.

Despite the diversity of genetic units that a genetic hierarchy introduces, Goldschmidt maintained a unified understanding of genetics by advocating a theoretical genetics in which results from cytogenetics, transmission genetics, and physiological genetics were integrated. Through at least 1940, the gene acted as a boundary object, drawing together diverse strands of genetic investigation (Star and Griesemer 1988; Rheinberger 1995). The loss of the gene as a powerful unifying entity created a need, in Goldschmidt's eyes, for a reconceptualization of genetics that would integrate and unify those diverse aspects previously associated with the corpuscular gene. As Pierre Duhem noted, when faced with negative results, a scientist can choose to tinker with the parts of his theoretical system or he or she can choose to question the very foundations of his or her thought (Duhem 1914, [1981, 217]). Goldschmidt sought to maintain a unified theory of genetics by rethinking its foundational entity, the gene. As such Goldschmidt's genetic hierarchies represent the possibility of retheorizing genetics without a unifying gene concept.

THE DEMISE OF THE PARTICULATE GENE

According to his own account of the development of his thought, Richard Goldschmidt began to question the existence of the particulate gene in 1932 when Theodosius Dobzhansky convinced him that position effects should be taken seriously (Goldschmidt 1944a, 185). A position effect occurs when the location of a gene alters the phenotypic effects of that gene. Position effects had been discovered in

1927 by A. H. Sturtevant in his now famous set of experiments on the Bar eye effect in *Drosophila*. Changing phenotypic effects with position raised questions of whether genes were functional units in the sense of whether or not they carried their function with them. The particulate theory of the gene represented genes as indivisible beads on a string, each a unitary structure and sufficient to fulfill its function. The particulate gene or the classical gene was a unifying entity in that it was simultaneously a unit of structure, a unit of function, a unit of mutation, and a unit of recombination. In the mid-1930s, however, the particulate gene was, in L. C. Dunn's words, showing "some signs of disappearing in a cloud of position effects" (Dunn 1937). Indeed, by 1934 with the first report of the results from H. J. Muller's group in the Soviet Union concerning mutation and rearrangements at the *scute* locus in *Drosophila*, Goldschmidt was ready to declare that the demise of the classical gene was at hand. Goldschmidt's conviction was bolstered by his own work on spontaneous mutation and chromosomal rearrangement, which was becoming his main line of experimental research in 1934 (Goldschmidt 1944a, 185). Goldschmidt's campaign against the gene was, thus, sustained by earlier work on position effects by Sturtevant and others, by Muller's work on X-ray induced mutations/rearrangements, and by his own work on spontaneous mutability.[2]

In his biographical memoir of Goldschmidt, Curt Stern remarks that Goldschmidt waited until he arrived in America, the birthplace of Thomas Hunt Morgan's theory of the gene, to announce in a funereal voice: "The theory of the gene is – dead!" (Stern 1967, 83). As Goldschmidt toured the United States in 1935, desperately trying to find a job and raise money with lecturing fees, he began to articulate his doubts about the particulate gene. At the time, his views drew heavily on the experiments of H. J. Muller's group on position effects and on the ability of X-rays to produce minute rearrangements in chromosomes.

Muller had pioneered the use of X-rays to induce mutations in *Drosophila* in the 1920s. The ability of X-rays to create mutations naturally raised questions about the action of radiation and the mechanisms of induced genetic changes. Muller and most notably Timofeéff-Ressovsky's group in Berlin were actively pursuing these concerns in the 1930s. By 1934, the results of X-ray radiation were recognized visually as changes in the banding patterns of chromosomes in the salivary glands of *Drosophila*. The visual evidence of large chromosomal rearrangements and research on the ability of X-ray radiation to create breaks in the chromosomes led Muller to consider the possibility that some mutations were, in fact, the result of breaks and rearrangements.

Together with Alexandra Prokofyeva and later Daniel Raffel, Muller began investigating the ability of X-rays to create single and double breaks in chromosomes that could result in cytologically visible differences (Carlson 1981, 193). Focusing on breaks in the *scute* region, Muller and Prokofyeva had found "definitive evidence . . . of the correctness of the 'position effect' interpretation of the action of chromosomal breaks" (Muller, Prokofyeva, and Raffel 1935, 253). Instead of claiming that the various *scute* effects were the results of gene mutations located near breakages, Muller and Prokofyeva claimed that the breaks themselves produced rearrangements that had position effects (Muller and Prokofyeva 1934; Muller, Prokofyeva, and Raffel 1935). This raised the question, in their words, "as to what proportion of 'natural mutations' in *Drosophila* may really be minute rearrangements." The only reason that Muller et al. gave for not making the inferential leap that most mutations were position effects was that "rearrangements alone must be far from adequate for any indefinitely continued evolution" (Muller, Prokofyeva, and Raffel 1935, 255).

Even armed with Muller's experimental results, the reaction to Goldschmidt's rejection of the gene was overwhelmingly negative (Goldschmidt 1960, 323). In the summer of 1936, however, new experimental evidence on the mutability of the *Drosophila* genome was introduced and brought to bear on the question of the gene. As the Goldschmidts sailed for Berkeley and a new life at the University of California, Milislav Demerec, Harold Plough, and C. Holthausen announced at the meetings of the Genetics Society of America their independent observations of a high frequency of spontaneous mutation in an inbred Florida stock of *Drosophila*. Just before his departure from Germany, Goldschmidt had also observed a higher than normal number of spontaneous mutations in the same Florida stock. Demerec and Plough and Holthausen and Goldschmidt all published accounts of their experiments in 1937 (Demerec 1937; Goldschmidt 1937a; Plough and Holthausen 1937). Goldschmidt's article was significantly different from the others, however (see Dietrich 1996).

In his short article published in the *Proceedings of the National Academy of Sciences*, Goldschmidt used his observations of increased spontaneous mutability to argue against the existence of the gene. Goldschmidt argued that the sudden appearance of many different kinds of mutations was the result of rearrangements affecting a number of different loci. Specifically, experiments involving crosses of a stock containing the *blistered* mutants would result in the next generation in the disappearance of blistered and the appearance in every individual of the brood of the mutants *plexus, dumpy vortex, thoraxate,* and *purple*. Goldschmidt interpreted these events as chromosomal rearrangements with an insertion or break in the *blistered* locus. (Goldschmidt 1937a, 622) From his brief and admittedly preliminary analysis, Goldschmidt concluded that all gene mutations were in fact rearrangements and that all mutations were in fact position effects. One could still localize the production of a phenotypic effect, but, according to Goldschmidt, it did not follow that there was a wild-type allele corresponding to the site of the mutation that produced the phenotypic effect. Instead, the "whole, wild type chromosome" was the "allele for all 'mutant genes' within this chromosome" (Goldschmidt 1937a, 622). Goldschmidt did promise further clarification of his ideas in a future publication, but it is important to note that while he did claim in 1937 that the chromosome was the "unit," he did recognize that it had to have some internal structure or "texture" in order to ensure normal development. As Goldschmidt's thinking advanced this "texture required for normal development," took form as a multi-level hierarchy of genetic units within the chromosome.

In a letter to L. C. Dunn, Goldschmidt admitted that the 1937 article would make him "an outcast in genetics," since he "now stated in writing, as before only orally, that there is no such thing as a gene. Horror!" (Goldschmidt 1937b). In 1938, Goldschmidt submitted the full report of his experiments to *Genetics* for consideration. In his accompanying letter, Goldschmidt admitted that the experiments were not yet completed, but that he wanted to publish in parts, starting with general comments followed by detailed analysis of particular mutants. L. C. Dunn, acting as editor of *Genetics*, was not willing to go along with this plan. Dunn refused to publish Gold-schmidt's results until he had provided the analysis of results from the study of the specific mutants and their representation on the salivary gland chromosomes.

Goldschmidt took Dunn's advice and stepped up research, but he also made his views known (Goldschmidt 1938a; Goldschmidt 1938b). The next year, 1939, Goldschmidt's article "Mass Mutation in the Florida Stock of *Drosophila melanogaster*: Details of an Old Experi-ment Reinterpreted" appeared in *The American Naturalist* (Gold-schmidt 1939a). Dunn was not at all pleased (Dunn 1940). Despite his serious warnings, Goldschmidt had published something that was evidently very similar to the manuscript rejected by *Genetics*. Gold-schmidt pleaded to Dunn that *The American Naturalist* article was not a report of his new experiments, but merely the interpretation of his old temperature shock experiments from 1929. Dunn felt that Gold-schmidt was promoting useless controversy (Dietrich 1996).

From 1939 to 1945, Richard Goldschmidt's experimental program was devoted to analyzing the spontaneous mutations in *Drosophila* completely (Goldschmidt 1944b; Goldschmidt et al. 1945). Where Muller had been given to caution when interpreting the effects of rearrangements because of the implications for evolutionary biology, Goldschmidt was not. The differences between Muller and Gold-schmidt are complex, but Goldschmidt himself summed up many of them in a letter to Muller in 1939. Goldschmidt wrote:

I just read with great pleasure your admirable paper in *The Collecting Net*. But I was a little disappointed that you were so cautious about the gene. I had actually expected (without knowing your Paris paper, but on the basis of the one with Prokofiewa [sic]) that you would be the first to jump over the fence and to discard unequivocally the classic gene. I know of course that temperaments are different and that I may be less cautious than required. But in times of transition from one basic viewpoint to another – these times I think are at hand with the classic theory in its last convulsions – I prefer the analytic mind to be a small step ahead of the experimentation (Physics proceeds that way, but biologists consider it a crime). I know of course what our differences are. First the small regions of similar action and second evolution. Regarding the first point I shall explain my position soon on the basis of some new facts; and as for the evolution I shall show in a forthcoming book (Silliman Lectures) that it is better understood without genes. (Goldschmidt 1939b)

Goldschmidt's exhortations did not convince Muller to throw caution to the wind and leap ahead of his meticulously gathered experimental results. Goldschmidt's forthcoming views on evolution, published as *The Material Basis of Evolution*, did little to help his cause against the classical gene.

The Material Basis of Evolution (Goldschmidt 1940a) was very hotly contested in its own right. Coming as it did in the middle of his campaign against the gene, Goldschmidt's evolutionary theory was an exploration of the phylogenetic consequences of his view of the demise of the gene. In his words, it would be "typical Goldschmidt with everything I like about him, and some others dislike" (Goldschmidt 1940b). I doubt Goldschmidt was prepared for how much his views would be disliked (Dietrich 1995).

The principal thesis of *The Material Basis of Evolution* is that microevolution and macroevolution are distinct phenomena and that the slow and gradual accumulation of micromutations was not a sufficient mechanism to bridge the gaps between species in an evolving lineage. In order for these bridgeless gaps between species to be crossed, Goldschmidt argued that large-scale systemic mutations were needed. Systemic mutations are complete changes of the primary pattern of the chromosome (the reaction system of the chromosome) into a new, well-integrated pattern (Goldschmidt 1940a).

A crucial part of Goldschmidt's argument was that changes in intrachromosomal structure or pattern are the basis for species level differences. In making this argument, Goldschmidt did not simply marshal his own evidence; he instead focused on the work of the influential neo-Darwinian, Theodosius Dobzhansky. The picture Goldschmidt painted of Dobzhansky was that of a man at a scientific impasse: He must decide whether species formation is a matter of genic differentiation or differentiation in chromosomal pattern. According to Goldschmidt, Dobzhansky knew a decision had to be made when he wrote in *Genetics and the Origin of Species:*

To what extent the differences between such species as *Drosophila pseudoobscura* and *D. miranda* are due to position effects is also a matter of specula-

tion; the greatly different gene arrangements in these species may be responsible for many alterations in the morphological and physiological properties
of their carriers. In any event, position effects show that gene mutations and
chromosomal changes are not necessarily as fundamentally distinct phenomena as they at first appear. (Dobzhansky 1937, 117)

Dobzhansky's failure to take the next step and admit that chromosomal repatternings could be the decisive changes needed for specia-
tion was, according to Goldschmidt, the result of "a dogmatic belief
in the inflexibility of the classical theory of the gene" (Goldschmidt
1940a, 242). Unfettered by the dogma of the gene, Goldschmidt could
promote a theory of evolution that was compatible with the "facts,"
meaning Dobzhansky's and Muller's own results on position effects
and chromosomal differences.

It is not surprising that Dobzhansky became one of Goldschmidt's
harshest critics and took him to task specifically on his attack on the
gene in *The Material Basis of Evolution*. In the second edition of *Genetics and the Origin of Species* published in 1941, Dobzhansky made it
clear that he was not about to give up genes in the face of Goldschmidt's attack, but neither was he willing to advocate the idea of
genes as beads-on-a-string. In his words, "regardless of whether
position effects are interpreted as due to interactions of chromosome
products in development or to changes in these products themselves,
the genes can no longer be thought of as absolutely discrete entities."
Dobzhansky thought Goldschmidt had created a false dichotomy:
"Genes must either be separated by impregnable walls, or else they
do not exist at all." Dobzhansky preferred an intermediate position,
where "the germ plasm may consist of genes and yet have a continuity of a higher order." Dobzhansky felt that his view was similar
to that advocated by Muller and allowed him to consider genes as
parts of functionally integrated systems, although the extent of the
functional interaction of gene products was recognized to be an open
question (Dobzhanksy 1941, 110; see also Dietrich 1995, 441–442,
445).

While Goldschmidt continued to articulate his views during the
early 1940s, the Drosophilists lined up their evidence against him:
Most notably at the 1941 Cold Spring Harbor Symposium where
Plough and Demerec sought to set the record straight. As they saw it,

the spontaneous mutations they observed were not associated with any gross rearrangements.

Plough's paper summarized the work of his group at Amherst on the effects of temperature and temperature shock on spontaneous mutation. It was Plough's position that there were genuine gene mutations that could be affected by temperature in much the same way that enzyme reactions could and there were chromosome breaks that were not sensitive to temperature but could be affected still by temperature shocks (Plough 1941, 136). This distinction was based on two main lines of reasoning. First, it was argued that if there are genuine gene mutations responding to temperature, they ought to respond as chemical reaction systems and the mutation frequency ought to follow a Van't Hoff curve. Plough's experiments with different temperatures did follow a Van't Hoff curve and as a result the mutations induced were thought to be the result of biochemical reactions. Second, the experiments tested for large translocations resulting from development at high temperature and from temperature shock. A large translocation would require two breaks on two chromosomes. Plough found that large translocations were very rare. Moreover, the creation of these large translocations was not effected by temperature or temperature shock. This justified Plough's distinction between genic and intergenic substances. Mutations were chemical changes in genes or genic substance and were sensitive to temperature and temperature shock, whereas large translocations were created by breaks in intergenic substance. Plough thus concluded that "spontaneous mutability requires the classical gene-chromosome framework involving genes as entities separable from the inter-genic substance" (Plough 1941, 136).

Plough's paper is explicitly directed against Goldschmidt's interpretation of spontaneous mutation, but just to make sure that there was no room for confusion Demerec raised Goldschmidt's views in the discussion period. Demerec asked: "What evidence does the Amherst group have on Goldschmidt's theory that mutation is due to chromosomal rearrangement and not gene changes, and that high mutability lines carry aberrations?" (Plough 1941, 136–137). In response, Plough reaffirmed his opposition to Goldschmidt's interpretation and claimed that while they did in some cases find an inversion on the second chromosome, it may have been there before the experiment began. Demerec continued: "Goldschmidt says that if mutants obtained from our high mutability stock were analyzed cytologically, they would prove to contain aberrations. Actually some of this analysis was done by Slizynski, who studied salivary gland chromosomes of induced and spontaneous lethals. The spontaneous lethals were all from the high mutability stock, and no aberrations except small deficiencies were found" (Plough 1941, 137). Because Plough had not done any cytological work, only analysis based on linkage and recombination, he could only claim that there were no "obvious chromosome rearrangements." The nonobvious rearrangements could still be undetectable deficiencies. The finer details that a cytological analysis could have yielded were missing and would become major issues later in the dispute.

Demerec's own work tried to address this situation, but was still geared toward large scale changes. Demerec's contribution was a review of his work on unstable genes in *Drosophila* and their bearing on questions of spontaneous mutability and the nature of the gene. Demerec made it very clear at the end of his paper that the unstable genes he had been studying were not associated with visible chromosomal aberrations. He had confirmed this with linkage studies and with analysis of the salivary chromosomes by himself and Dr. Eileen Sutton. However, he did note that the "closest known parallel" to unstable genes found in *Drosophila melanogaster* were mottled characters that were "connected with chromosomal aberrations involving heterochromatin" (Demerec 1941a, 149). Perhaps as a result, Demerec refrained from the type of ringing endorsement for the gene that Plough had given earlier.

Immediately following Plough and Demerec at the Cold Spring Harbor Symposium in 1941, was H. J. Muller. Muller had the task of summarizing the current state of work on induced mutations. On this topic, Muller could speak with unrivaled authority and his comments were often at odds with those of Plough and Demerec. Plough's argument was predicated on a simple division between gene mutation and gross rearrangement such that if gross rearrangement was eliminated the best explanation was gene mutation. Muller diversified the range of explanatory options to include the possibility of minute rearrangements detectable by cytological analysis and even minute rearrangements too fine to be cytologically detected. What enabled this diversification was Muller's interpretation of the molecular action of radiation on the chromosome. Muller accepted that a "single atom change" induced by radiation could create, via a chain of reactions, a single break in a chromosome. Two such changes far apart could create two breaks and possibly a gross rearrangement. A minute rearrangement, however, could be produced by a single atom change that initiates "a chain of reactions in two or more directions, so that two or more different but nearby breaks in the chromonema are induced . . . " (Muller 1941, 163). This possibility suggested that mutations may be minute rearrangements. Moreover, it had the consequence that "the concept of the individual gene" may be "only an approximate one, roughly describing, for our convenience, certain chromosome regions having to do with certain functions" (Muller 1941, 161).

Muller's only hesitancy about equating mutations and minute rearrangements had to do with the chemical nature of the chromonema, the fibrous portion of the chromosome capable of coiling and uncoiling. Muller reasoned that the chromonema was a nucleoprotein and so was composed of amino acid units that were "grouped into higher units, in a kind of ascending hierarchy" (Muller 1941, 161). If breaks occur, Muller argued, they were not likely to occur between amino acids since that would require the breaking of a peptide bond with the resulting fragments having opposite charge. In order for a peptide bond to be reformed, fragments with opposite charge must reunite. Yet, Muller argued, this was not what occurred in chromosomes where "two fragments that unite may have the same sign [or charge]" (Muller 1941, 161). Chromosome breaks were, thus, not breaks within protein molecules. Instead they must be breaks between the protein molecules that make up the chromonema. As a consequence, Muller believed that "there must be larger groupings in the chromonema than the amino acid units" (Muller 1941, 161). Muller never came right out and claimed that these larger groupings were in fact genes. Although he did not call for a return to the old corpuscular gene, Muller was not willing to get rid of it entirely. As a result he concluded his survey by claiming that "the underlying facts may be more complicated than we have imagined" (Muller 1941, 162).

By the 1940s, Muller, Dobzhansky, L. C. Dunn, and many others were willing to accept that genes were not the discrete entities that they were represented to be by the classical beads-on-a-string metaphor (Dobzhanksy 1937, 117; Raffel and Muller 1940). They were not willing to give up, however, that there were no mutations, only rearrangements of different sizes. Nor were they willing to give up that there were units that corresponded to specific functions. It was not clear that these functional units corresponded to discrete or discontinuous structures, nor was it clear what kinds of structures were manifest in the chromosome.

GENETIC HIERARCHIES

From 1944 until his death in 1958, Goldschmidt articulated and elaborated a hierarchy of genetic units. Goldschmidt's position was based on the argument that the unit of mutation did not necessarily provide any insight into the wild-type or what he called the hereditary unit. Mutations were detected differences and could be localized; hereditary units were units with structures and functions that need not be localized in the same way mutations were (Goldschmidt 1946, 250–251). Hereditary units were the units necessary for the process of normal development (Goldschmidt 1944a, 197).

Goldschmidt first presented a fully articulated genetic hierarchy in 1944. This hierarchy of hereditary units was summarized in a table based on five levels of visible structures (see Table 5-1). Because this table was constructed as part of an argument against the classical gene, it contrasts the classical interpretation with the cytological data and with Goldschmidt's own interpretation. This table started with the smallest structures and proceeded up to extremely large structural divisions, thereby creating an inclusive structural hierarchy. The higher levels were composed of well-recognized cytological structures. The lowest level was more controversial.

The smallest units in Goldschmidt's 1944 hierarchy corresponded to visible structures that he and his student Masuo Kodani had been able to detect in salivary gland chromosomes after treatment at very high pH (Calvin, Kodani, and Goldschmidt 1940). Structures resembling lamp brushes were produced after treatment in high pH solutions of NaCl and staining with acetocarmine solution. The salivary

Table 5-1 *Goldschmidt's Genetic Hierarchy as of 1944 (after Goldschmidt 1944 a, 206, table 2)*

Level	Lowest	Next Higher	Next Higher	Next Higher	Next Higher
Visible Structure	Goldschmidt Kodani penultimate chromomeres	Salivary band	Belling's ultimate chromomeres	Wenrich's pachytene and diplotene chromomeres	Heitz's chromosomal segments of euchromatin
Approximate Numbers	5,000	1,000	100	610	1–6
Units of Genetic Action	Not yet known rearrangement changes of penultimate chromomeres	Mutants without visible rearrangements	Sections of localized action like y- or sc-segment	Segments of identical action in special circumstances	Situation as found in homeotic mutants
Within Classic Theory	Subgenes	Gene	Muller's redefined gene	Genes influencing each other	Numerous independent genes

gland chromosomes appeared after treatment as irregular cylinders marked by dark staining lateral threads with transverse bristles or loops. These loops were assembled in a rosette structure around a centromere. The interpretation of these structures was controversial (Metz 1941; Goldschmidt 1955, 19). Goldschmidt seemed to favor Koltzoff's interpretation of the loops as individual protein molecules, but in the end left the question open and referred to them as penultimate chromomeres.

What was significant about this structural hierarchy was that Goldschmidt could attribute genetic action to each of the units at the salivary gland chromosome band level and higher. Mutations were regularly attributed to the salivary gland chromosome bands, even though rearrangements were not visible at that level. Localizable genetic action, however, was usually attributed to the next higher level. It was at this level, which Goldschmidt associated with Belling's 1928 description of chromomeres in leptotene chromosomes (Belling 1928), that he put the important results concerning Muller's *scute* mutants and his own *yellow* mutants. In both of these cases, breaks over as much as a five band section of chromosome produced very similar phenotypic effects (Goldschmidt 1944a, 191). To Goldschmidt, this argued strongly for his position that regions of normal action were larger than those associated with mutation and the particulate gene.

The two highest levels of structures were the chromomeres identified by Wenrich in the diplotene stage of meiosis and the alternating blocks of euchromatin and heterochromatin identified by Heitz (Heitz 1929; Wenrich 1916). These structures were associated with position effects and homeotic mutants. It was known that rearrangements with one break in heterochromatin and the other in euchromatin behaved differently from those with both breaks in euchromatin. In particular, rearrangements involving the placement of the heterochromatin next to euchromatin seemed to have position effects that stretched over long distances (Goldschmidt 1946, 255). The grouping of many homeotic mutants within a short segment of the third chromosome of *Drosophila* was similarly suggestive of larger functional units. In general, however, the actions both of heterochromatin and of homeotic mutants were poorly understood in 1944. As a result, Goldschmidt dedicated the final years of his experimental career to researching these two higher levels in his genetic hierarchy. The culmination of these efforts to articulate a genetic hierarchy came in Goldschmidt's 1955 book *Theoretical Genetics.* In many ways this book is Goldschmidt's last word on many of the diverse lines of research he had pursued during his lifetime, including physiological genetics, evolution, and sex determination. For present purposes, what is most significant is the lengthy section on the nature of the genetic material.

As he had years earlier, Goldschmidt's arguments in *Theoretical Genetics* began with position effects and moved to now familiar arguments against the particulate gene. The possible sources of position effects were expanded from the 1940s to include visible and invisible deficiencies as well as inversions ranging from large to so small as to be cytologically invisible (see Figure 5-1). The diversity of position effects and their role in the case against the particulate gene justified the concentration of Goldschmidt's efforts on chromosome segments. Given that there were no discrete genes *identifiable by muta-*

Normal	a	b	c	d	e	f			
Deficiency visible	a	b		d	e	f			
Deficiency invisible	a	b	$\underline{c}$ 2	d	e	f			
Deficiency invisible	a	b	c	$\underline{d}$ 2	e	f			
Large inversion or translocation	a	b	c				x	y	z
or	a	b	c	d	e		x	y	z
Microinversion (visible or not)	a	b	c	d	**e**	f			
or	a	**b**	c	d	e	f			

Figure 5-1   Possible types of position effects. Possible changes in the serial order of the structural elements of a section of a chromosome. c/2 represents that only $c^1 c^2$ is present, instead of a submicroscopic series of $c^1 c^2 c^3 c^4$. **b** (or **e**) represents that the order $b^1 b^3 b^2 b^4$ ($e^1 e^3 e^2 e^4$) is present, instead of submicroscopic series $b^1 b^2 b^3 b^4$ ($e^1 e^2 e^3 e^4$). (Goldschmidt 1955, 161)

tion, Goldschmidt wanted to know if there were nonetheless discrete "well-defined morphological structures." The diagrammatic representation of possible position effects (see Figure 5-1) suggested that there were discrete structures. Indeed Goldschmidt's 1944 hierarchy had been based on well-defined differences in the morphology of the chromosome. In *Theoretical Genetics*, however, Goldschmidt cites two phenomena that bring into question the possibility of morphologically distinct segments and consequently any hierarchy based on sharply defined morphological differences in chromosome structure: the overlap between neighboring segments and position effects with heterochromatic breaks.

Based on his own study of *yellow* mutants, Muller's study of *scute* mutants, and Demerec's study of other mutants, such as *white* and *Notch* in *Drosophila*, Goldschmidt argued that the ability to produce these phenotypes was spread over segments of chromosome and that these segments could overlap. In the case of *yellow* and *scute*, for instance, breaks between the 1B1 and 1B3 bands on the X-chromosome produce *yellow* and *scute*, whereas breaks farther to the left will produce only *yellow* and breaks farther to the right will produce only *scute* (see Figure 5-2). This phenomenon with the *yellow* and *scute* segments led Goldschmidt to conclude that "it cannot be the morphological segment which counts [as the hereditary unit], but a field-like function of the segment which under certain conditions . . . reaches from the center of the segment to different distances" (Goldschmidt 1955, 162). This interpretation in terms of fields required that a segment be thought of in terms of a "definite polarized order on a molecular level" with a specific range of action associated with a specific function.

The same interpretation was offered for position effects produced by breaks in heterochromatin. Rearrangements with one break in heterochromatin (and the other in the euchromatin) were known to have a greater effect than rearrangements with both breaks in euchromatin. In 1941, Demerec had explained this difference in terms of what he called sensitive regions.

According to Demerec, when a rearrangement occurred new parts of the chromosome were brought into contact with each other resulting in a position effect. The sensitive region was the region surrounding the place where this new contact occurred and in which position

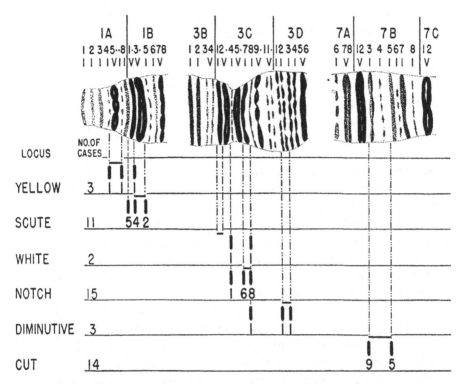

Figure 5-2 Locations of breaks associated with rearrangements in the X-chromosome of *Drosophila*. Vertical bars represent the location of breaks, the numbers below indicate the numbers of breaks per location. According to Demerec's interpretation, the horizontal bars represent the position of loci. Goldschmidt interprets the same data on the location of breaks as evidence that genic segments overlap. (After Demerec 1941b, 4, figure 1.)

effects could be detected. Using his research on *Notch* and *white* mutants, Demerec concluded that rearrangements involving placement of euchromatin next to heterochromatin resulted in a sensitive region that was five to ten times bigger than a similar rearrangement involving only euchromatin (Demerec 1941b, 6). Demerec explained this effect as a change in either the gene or in gene activity. Goldschmidt thought that explanations that appealed to heterochromatin's action on distant genes provided no insight and were "devoid of meaning" (Goldschmidt 1955, 162). Instead, Goldschmidt preferred to think of heterochromatin as stretching the fields of adjacent segments to produce what he thought of as a case of "an extreme type of overlapping." This stretching effect, like overlapping, convinced Goldschmidt that chromosomes had to be understood in terms of segments with associated fields of action.

As he had done in 1944 for genetic structures, Goldschmidt ordered these segments and their fields into a genetic hierarchy. These fields would range from those associated with submicroscopic segments of the chromosome to larger fields covering possibly the entire chromosome (Goldschmidt 1955, 180). The action of genic material at any time could be the result of a field at any one of these levels.[3] Goldschmidt represented these fields in a diagram where a series of numbers represented the molecular structure of a chromosome from the centromere to its end and possible fields at different levels were represented as corresponding to different segments of the number series (see Figure 5-3). Goldschmidt based this type of representation on Kenneth Mather's analysis of what he called fields of coordination or integration (Mather 1946, 1948).

In arguing against the bead-on-a-string view of the gene, Mather claimed that the gene should be delimited as a physiological unit. Using Goldschmidt's results on the *yellow-scute* region of the X-chromosome in *Drosophila* from 1944, Mather reasoned that the chromosome produced a number of major products. Each product was produced by a specific region of the chromosome that was independent of the others, to the extent that its limits could be determined. "Within one such region," Mather continued, "we must sup-

Figure 5-3   Goldschmidt's genetic hierarchy as of 1955. (After Goldschmidt 1955, 180.)

pose that the parts act together to give a single major product in a way depending on their arrangement with relation to each other. Such a region would be the ultimate genetical unit. In this way we can arrive at the idea of a gene as a field of coordinated activity, the property of full activity being conditioned by internal arrangement, but, within limits, independent of external relations" (Mather 1946, 67–68). Moreover, the more complex the system of coordination the greater the likelihood that there were also lesser fields of coordination. Mather's purely physiological interpretation of the gene as fields of co-ordination thus suggested a functional hierarchy. Goldschmidt effectively extended this hierarchy above the level of Mather's gene.

That there could be large fields of action associated with extended segments of the chromosome was a far cry from the localized action associated with a mutant locus. Goldschmidt bolstered his case for higher fields with his analysis of homeotic mutants in *Drosophila*. Homeotic mutants were understood as mutants that produced the substitution of one homologous part for another, e.g., an antenna for a leg. What struck Goldschmidt about homeotic mutants in *Drosophila* was the high concentration of related mutants in a region of the third chromosome. All of these mutants affected the determination of segmental appendages, which led Goldschmidt to interpret this segment of the third chromosome as a field "vitally concerned with the processes of segmental determination" (Goldschmidt 1955, 182). Because homeotic mutants were understood to alter the development of the imaginal discs, Goldschmidt claimed that "the whole intact section [of the chromosome] controls certain parts of the process of normal development of the discs; but a disturbance of this action at individual points inside this field or inside another similar field (in the 2d chromosome) leads to upsets of development in the discs . . . " (Goldschmidt 1955, 182). As he had done earlier with the *yellow* and *scute* regions, Goldschmidt disassociated mutation from normal gene action. The homeotic mutants of the third chromosome, however, were spread over a much larger section of chromosome. This distance between mutants, together with the similarity of their effects and their association with specific developmental processes, allowed Goldschmidt to claim that the homeotic mutants should be understood as affecting a much higher level field than the mutants in the *yellow scute* region. A major result of this difference was the claim that "larger fields act upon more basic (i.e., earlier) developmental processes, while the smaller segments are concerned with the control of late, more or less superficial features of development" (Goldschmidt 1955, 183). Goldschmidt's hierarchy of fields was thus translated into a hierarchy of sequential developmental processes.

CONCLUSION

Richard Goldschmidt valued unification. As a biologist trained in turn-of-the-century Germany, Goldschmidt had what historian Jonathan Harwood calls a comprehensive style of scientific thought. Comprehensive geneticists are characterized by "their broad approach to the problems of genetics, their attitudes toward breadth of biological knowledge, and their cultivation of artistic sensibility, the recurring theme of striving for an all-embracing conceptual synthesis, occasionally manifest in sympathies for holism" (Harwood 1993, 270). Goldschmidt is a clear exemplar of this comprehensive style of thought. Throughout all of his work, Goldschmidt consistently strove to build a unified understanding of vast arrays of biological phenomena. In his work on sex determination, for instance, Goldschmidt produced a vast number of technical articles on the evidence and mechanisms of sex determination in *Lymantria*, but at the same time strove to generalize his findings as laws of nature. As a result, the book marking the culmination of Goldschmidt's research on sex determination, *Die sexuellen Zwischenstufen* (Goldschmidt 1931), attempted to use the regularities discovered for *Lymantria* to explain sex determination in higher orders of animals ranging from amphibians to humans.

Goldschmidt approached genetics in a similar fashion. *Theoretical Genetics* was his last grand synthesis for genetic phenomena, taking into account everything from Weismann to Watson and Crick. In terms of the gene, *Theoretical Genetics* contained Goldschmidt's final repudiation of the corpuscular gene as well as his most complete articulation of its alternative in genetic hierarchies. Rather than try to maintain that there was a new single entity to unify genetics, Goldschmidt articulated a hierarchy of fields of genetic action. This hierarchy of fields allowed Goldschmidt to maintain a unified understanding of genetic structure and function by making crossing-over and mutation secondary considerations when it came to the normal operation of the gene. Because crossing-over and mutation had been traditional tools for delimiting genetic structures, Goldschmidt's fields shift the emphasis in genetics toward function in general and developmental processes in particular. From one of the world's foremost champions of physiological genetics, this emphasis on function should not be surprising.

Richard Goldschmidt's proposal of genetic hierarchies represents an attempt to provide a unified theoretical foundation for genetics without a single unifying object, namely the classical gene. In terms of Petter Portin's (1993) history of gene concepts, Goldschmidt was clearly reacting against the classical gene but was not advocating a neoclassical gene to take its place. Where Portin's neoclassical gene narrowed the meaning of the term "gene" to emphasize structure–function relationships identified by the classical gene, Goldschmidt embraced the diversity of structures and functions known even then. In this sense, Goldschmidt's approach seems to resonate with the sentiment of molecular geneticists who continue to broaden the range of known genetic structures and functions (see review in Portin 1993). As such, the case of Goldschmidt's hierarchies illuminates the possibilities for producing a unified theory of genetics given a vast and diverse array of genetic phenomena. Indeed as the diversity of genetic structures and functions increases proponents of a unifying gene concept are faced with an ever increasing challenge, certainly one that would have continued to reinforce Richard Goldschmidt's belief that the classical gene is dead.

NOTES

1. Research for this paper was supported by a grant from the National Science Foundation (SBER94-12384).
2. See Dietrich 1996 for more on the history of Goldschmidt's early experimental work on *Drosophila*.
3. This genetic hierarchy was meant to be a hierarchy of structures (fields), but was not necessarily a hierarchy of functions since the function of

larger fields was not necessarily a product of the functions of lower level fields. I thank Sara Schwartz for drawing my attention to this ambiguity.

REFERENCES

Allen, G. E. 1974. Opposition to the Mendelian-chromosome theory: The physiological and developmental genetics of Richard Goldschmidt. *Journal of the History of Biology* 7: 49–92.

Belling, J. 1928. The ultimate chromomeres in Lilium and Aloé with regard to the numbers of genes. *University of California Publications in Botany* 14: 307–318.

Burian, R. M. 1985. The 'internal politics' of biology and the justification of biological theories. In *Human Nature and Natural Knowledge: Essays Presented to Marjorie Grene on the Occasion of Her Seventy-Fifth Birthday*, edited by A. Donagan, A. N. Petrovich and M. V. Wedin. Dordrecht: D. Reidel, pp. 3–21.

Calvin, M., M. Kodani, and R. Goldschmidt. 1940. Effects of certain chemical treatments on the morphology of the salivary gland chromosomes and their interpretation. *Proceedings of the National Academy of Sciences USA* 26: 299–301.

Carlson, E. A. 1966. *The Gene: A Critical History*. Philadelphia: Saunders.

Carlson, E. A. 1981. *Genes, Radiation, and Society*. Ithaca: Cornell University Press.

Demerec, M. 1937. Frequency of spontaneous mutations in certain stocks of *Drosophila melanogaster*. *Genetics* 22: 469–478.

Demerec, M. 1941a. Unstable genes in *Drosophila*. *Cold Spring Harbor Symposia on Quantitative Biology* 9: 145–150.

Demerec, M. 1941b. The nature of the gene. In *Cytology, Genetics, and Evolution*. Philadelphia: University of Pennsylvania Press, pp. 1–11.

Dietrich, M. R. 1995. Richard Goldschmidt's "heresies" and the evolutionary synthesis. *Journal of the History of Biology* 28: 431–461.

Dietrich, M. R. 1996. On the mutability of genes and geneticists: The "Americanization" of Richard Goldschmidt and Victor Jollos. *Perspectives on Science* 4: 321–345.

Dobzhansky, T. 1937. *Genetics and the Origin of Species*. New York: Columbia University Press.

Dobzhansky, T. 1941. *Genetics and the Origin of Species*. 2nd ed. New York: Columbia University Press.

Duhem, P. 1981. *The Aim and Structure of Physical Theory*. New York: Atheneum.

Dunn, L. C. 1937. To R. Goldschmidt; November 15: L. C. Dunn Papers, American Philosophical Society Library, Philadelphia, PA.

Dunn, L. C. 1940. To R. Goldschmidt; May 27: L. C. Dunn Papers, American Philosophical Society Library, Philadelphia, PA.

Dunn, L. C. 1965. *A Short History of Genetics*. Ames, Iowa: Iowa State University Press.

Gilbert, S. F. 1991. Cellular politics: Ernest Everett Just, Richard B. Goldschmidt and the attempt to reconcile embryology and genetics. In *The American Development of Biology*, edited by R. Rainger, K. Benson and J. Maienschein. New Brunswick: Rutgers University Press, pp. 311–346.

Goldschmidt, R. 1931. *Die sexuellen Zwischenstufen*. Berlin: Julius Springer.

Goldschmidt, R. 1937a. Spontaneous chromatin rearrangements and the theory of the gene. *Proceedings of the National Academy of Sciences USA* 23: 621–623.

Goldschmidt, R. 1937b. To L. C. Dunn; November 3: L. C. Dunn Papers, American Philosophical Society Library, Philadelphia, PA.

Goldschmidt, R. 1938a. *Physiological Genetics*. New York: McGraw-Hill.

Goldschmidt, R. 1938b. The Theory of the Gene. *Science Monthly* 46: 268–273.

Goldschmidt, R. 1939a. Mass mutation in the Florida stock of *Drosophila melanogaster*: Details of an old experiment reinterpreted. *American Naturalist* 73: 547–559.

Goldschmidt, R. 1939b. To H. J. Muller; May 11: H. J. Muller Papers, Lilly Library, Indiana University, Bloomington, IN.

Goldschmidt, R. 1940a. *The Material Basis of Evolution*. New Haven: Yale University Press.

Goldschmidt, R. 1940b. To L. C. Dunn; May 27: L. C. Dunn Papers, American Philosophical Society Library, Philadelphia, PA.

Goldschmidt, R. 1944a. On some facts pertinent to the theory of the gene. In *Science in the University*. Berkeley: University of California Press, pp. 183–210.

Goldschmidt, R. 1944b. On spontaneous mutation. *Proceedings of the National Academy of Sciences USA* 30: 297–299.

Goldschmidt, R. 1946. Position effect and the theory of the corpuscular gene. *Experientia* 2: 197–230, 250–256.

Goldschmidt, R. 1955. *Theoretical Genetics*. Berkeley, CA: University of California Press.

Goldschmidt, R. 1960. *In and Out of the Ivory Tower*. Seattle: University of Washington Press.

Goldschmidt, R., R. Blanc, W. Braun, M. Eakin, R. Fields, A. Hannah, L. Kellen, M. Kodani, and C. Villee. 1945. A study of spontaneous mutation. *University of California Publications Zoology* 49: 291–500.

Harwood, J. 1993. *Styles of Scientific Thought: The German Genetics Community. 1900–1933*. Chicago: University of Chicago Press.

Heitz, E. 1929. Heterochromatin, Chromocentren, Chromomeren. *Deutsche Botanische Gesellschaft, Berlin* 47: 274–284.

Maienschein, J. 1992. Gene: Historical perspectives. In *Keywords in Evolutionary Biology*, edited by E. F. Keller and E. A. Lloyd. Cambridge, Mass.: Harvard University Press, pp. 122–127.

Mather, K. 1946. Genes. *Scientific Journal of the Royal College of the Sciences* 16: 64–71.

Mather, K. 1948. Nucleus and cytoplasm in differentiation. *Symposia of the Society for Experimental Biology* 2: 196–216.

Metz, C. 1941. The structure of salivary gland chromosomes. *Cold Spring Harbor Symposia on Quantitative Biology* 9: 23–36.

Muller, H. J. 1941. Induced mutations in *Drosophila*. *Cold Spring Harbor Symposia on Quantitative Biology* 9: 151–167.

Muller, H. J., and A. Prokofyeva. 1934. Continuity and discontinuity of the hereditary material. *Dokladi; Academy of Science of the USSR* 4: 74–83.

Muller, H. J., A. Prokofyeva, and D. Raffel. 1935. Minute intergenic rearrangement as a cause of apparent "gene mutation." *Nature* 135: 253–255.

Plough, H. H. 1941. Spontaneous mutability in Drosophila. *Cold Spring Harbor Symposia on Quantitative Biology* 9: 127–137.

Plough, H. H., and C. Holthausen. 1937. A case of high mutational frequency without environmental change. *American Naturalist* 71: 185–187.

Portin, P. 1993. The concept of the gene: Short history and present status. *The Quarterly Review of Biology* 68: 173–223.

Raffel, D., and H. J. Muller. 1940. Position effect and gene divisibility considered in connection with three strikingly similar scute mutations. *Genetics* 25: 541–583.

Rheinberger, H.-J. 1995. Genes: A disunified view from the perspective of molecular biology. In *Gene Concepts and Evolution (Workshop)*. Preprint no. 18. Berlin: Max Planck Institute for the History of Science, pp. 7–13.

Richmond, M. 1986. Richard Goldschmidt and sex determination: The growth of German genetics, 1900–1935. Ph.D. dissertation. Indiana University.

Star, S. L., and J. R. Griesemer. 1988. Institutional ecology, "translations" and boundary objects: Amateurs and professionals in Berkeley's Museum of Vertebrate Zoology 1907–1939. *Social Studies of Science* 19: 387–420.

Stern, C. 1980 (1967). Richard Benedict Goldschmidt (1878–1958): A biographical memoir. In *Richard Goldschmidt: Controversial Geneticist and Creative Biologist*, edited by L. Piternick. *Experientia Supplementum* 35: 68–99.

Wenrich, D. 1916. The spermatogenesis of *Phrynotettix magnus* with special reference to synapsis and the individuality of chromosomes. *Bulletin of the Museum of Comparative Zoology* 60: 57–133.

6

Seymour Benzer and the Definition of the Gene

FREDERIC L. HOLMES

ABSTRACT

The mapping of the fine structure of the gene by Seymour Benzer, between 1954 and 1962, redefined the gene and brought classical genetics into contact with the molecular interpretation of genetic material based on the DNA double helix. The present essay sets Benzer's investigation into its immediate context, follows the development of Benzer's experimental program, and explores the impact of his work on the conceptual structure of genetics as reflected in contemporary textbooks.

INTRODUCTION

In 1954, Seymour Benzer began mapping the fine structure of the rII region of the genome of the bacteriophage T4. By the time he ended this work in 1961 (Benzer 1966, 157), it had become a central support for what Petter Portin has defined retrospectively as the "neoclassical concept of the gene" (Portin 1993, 173–175, 186–187). In his textbook *Molecular Biology of the Gene*, which provided the first pedagogical synthesis of the newly established field of molecular biology, James Watson derived "the geneticist's view of a gene" mainly from Benzer's analysis:

a discrete chromosomal region which (1) is responsible for a specific cellular product and (2) consists of a linear collection of potentially mutable units (mutable sites), each of which can exist in several alternative forms and between which crossing over can occur. (Watson 1965, 233)

This definition was synonymous with what Benzer had called a *cistron*, because he based it on a criterion known as the *cis-trans* test for a functional unit of the genome.

115

In 1957, George Beadle described Benzer's experiments on the rII region as "very beautiful work" (Benzer 1957, 129), a judgment generally shared among contemporaries in the field. Benzer has acquired a prominent place in the epochal events, following the discovery of the double helix, through which classical genetics was brought into direct contact with the identification of DNA as the chemical basis of genetic material (Judson 1979, 271–276).

As several of the papers in this volume emphasize, the concept of the gene, so powerfully advanced by Benzer's analysis, is no longer intact. More recent discoveries have undermined the belief that a gene is a "discrete chromosomal region," and have left a "rather abstract, open, and generalized concept of the gene" (Portin 1993, 173). Where have they left the place of Benzer's experiments in the structure of knowledge that constitutes molecular biology? Was his work bound up with that emerging discipline that Hans-Jörg Rheinberger characterizes as "the heyday of its simplicity"? (Rheinberger, this volume). Was it one of the bulwarks for a concept of the gene that Peter Beurton asserts has been disintegrating (Beurton, this volume)? Or, are Benzer's definitions still represented in the ongoing dialectic that Raphael Falk sees between the instrumental and the material conception of the gene? (Falk 1986, 169–173). Is the work still regarded as beautiful when the conclusions once drawn from it no longer stand in the fullness of their original force? Some recent genetics textbooks still describe Benzer's "classical fine-structure analysis of the gene" (Ayala and Kiger 1984, 157), whereas others mention neither his work nor the word *cistron*. Along with other "classical" investigations of early molecular biology, such as the Meselson-Stahl experiment, which textbook writers treated for nearly three decades as canonical foundations of the field, the recapitulation of Benzer's analysis of the rII region is no longer mandatory to the presentation of molecular genetics.

At the end of the present paper, I shall make a few comments on the legacy of Benzer's mapping project, but my focus will be on its origins and evolution, and on the qualities of his work that made it appear both central and beautiful in its time. During the discussion of a draft of this paper Raphael Falk commented, from the perspective of a post-doctoral fellow in the laboratory of Hermann Muller at the time Benzer's work first appeared, that:

the feeling of the beauty . . . was very strong at the time, it had a very important effect. Benzer's work was, in many aspects, the bridge that helped transmission geneticists to adapt to the molecular age. The elegance of the experiments, the methodological reductionism which his physicist's concept brought into the idea was of great value. . . . I have really experienced . . . the beauty, and the excitement which this beauty brought about. (Falk 1996)

Elof Carlson and others have already described the main features of Benzer's investigation in its historical context (Carlson 1966, 196–209). To retrieve the sense of its beauty, and of the excitement with which it was received, requires a more finely structured historical treatment, toward which I hope here to have taken a first step.

GENE STRUCTURE AND THE GENETIC MAPPING OF BACTERIOPHAGE

Carlson has pointed out that, after a period of intense interest in the structure of the gene during the 1930s, when Hermann Muller and others questioned whether chromosomes were continuous or discontinuous, and whether they were separable material entities or mental isolates, such questions did not again attract sustained attention until the 1950s (Carlson 1966, 156). By then it was commonly imagined that earlier geneticists had simply compared the "genetic structure of an organism to a series of beads threaded along a string." Newer investigations now suggested that the "genetic unit" was complex and "may well be comprised of genetically separable subunits" (Bonner 1956, 163).

Between 1945 and 1950, Edward B. Lewis studied a series of apparent multiple alleles in *Drosophila* that turned out to be separable by crossing over. Adopting a term used earlier by Barbara McClintock and others in a different sense (McClintock 1944, 499), Lewis called them *pseudoalleles*. A pseudoallelic series consisted, he believed, of "component genes," closely linked in function and adjacent to one another on the chromosome. Pseudoalleles exhibited what Lewis called, this time adapting a term originally used by Sturtevant, a *position effect*. If two pseudoallelic mutant genes were located on one of the homologous chromosomes, the other chromosome being wild-type, the phenotype of the heterozygote was wild-type. If one mutant was located on each chromosome, however, the phenotype of the heterozygote was mutant. Lewis represented these two situations, using standard genetic symbols, for two mutant genes *a* and *b*, as *a b*/+ +, and *a* +/+ *b*. Lewis named the first of these situations the *cis-type* and the second the *trans-type* (Lewis 1955). Ordinarily, genes were expected to "complement" each other, so that two separate normal genes, whether located on the same or the homologous chromosome of heterozygotes, produced wild pheno-types. Such a result was, in fact, commonly regarded as the defining criterion for distinct genes. To explain why pseudoallelic genes be-haved differently, Lewis postulated a scheme in which they were supposed to control separate but closely related metabolic reactions (Lewis 1951).

In a review entitled "Genetic Formulation of Gene Structure and Gene Action," published in 1952 in *Advances in Enzymology*, Guido Pontecorvo incorporated Lewis's positional effect of pseudoalleles into a broader discussion of the nature of the gene. There are, Pon-tecorvo pointed out, "various ways in which a gene can be defined; they are consistent with one another at certain levels of genetic anal-ysis, but not at others." The three definitions of a gene he sum-marized were:

(1) as a part of a chromosome which is the ultimate unit of mutation; (2) as the ultimate factor of inheritable differences, *i.e.*, as unit of physiological action; and (3) as the ultimate unit of hereditary recombination.

In most genetic analyses, these three definitions were interchangea-ble, because any two genes were distinguishable according to all three criteria. The definitions expressed "different properties of one and the same thing." In "extreme cases," however, inconsistencies had arisen, "particularly in cases of very close linkage" (Pontecorvo 1952, 123–129).

In normal breeding experiments "we say that we are dealing with differences in *one* gene" (one pair of alleles) when only two kinds of nuclei are produced through meiosis in a diploid nucleus – that is, when there are no recombinants. If recombinants are found, then more than one gene is involved. We can define nonallelism in this way, "but we cannot define allelism: between two very closely linked genes . . . and one gene . . . the distinction becomes a technical one,

118

limited by our ability to analyze large numbers of progeny" (Pontecorvo 1952, 128).

The resolving power of crossing-over was, at present, limited to the order of frequencies of 10^{-5}, the lowest measurable proportion of recombinants to nonrecombinants, because of the technical difficulty of obtaining and counting huge numbers of progeny. "By the use of appropriate selective techniques it should be possible, however, to push the resolution one or two orders of magnitude further." By comparing the lowest crossover frequencies observable with *Drosophila* to the length of their giant salivary chromosomes, Pontecorvo arrived at an estimate of the maximum length of a gene when defined as a unit of recombination. From studies of the target size of ionizing radiation necessary to produce a mutation he inferred a minimum diameter of the "ultimate unit of mutation." These units could, he asserted, be "at least in some cases one or two orders of magnitude shorter than the chromosome segment which forms the basis for the unit of physiological action." The traditional picture of a gene as a "sharply delimited portion of a chromosome – the 'corpuscular gene'" had, Pontecorvo believed, lost its heuristic value. Genes based on more than one crossover and mutation unit can be arranged in various ways: "neighboring genes may share some units." He preferred to "keep the term gene only for the unit of physiological action." Influenced by the views of Richard Goldschmidt (see Dietrich, this volume), Pontecorvo adopted the radical position that "genes as units of physiological action . . . are obviously not megamolecules. They are processes or functions, not atomic edifices" (Pontecorvo 1952, 129–134).

Discussing the "positional aspects of gene action" in support of this view, Pontecorvo invoked the example of pseudoallelism in *Drosophila*:

In individuals heterozygous for *two different* mutant alleles the effect is quite different according to whether the mutant alleles are both on the same chromosome and the normal allele on its homologue, or the mutant and the normal alleles are distributed between the two homologues. In the former case the effect is normal, in the latter it is mutant. (Pontecorvo 1952, 135)

This effect, he proposed afterward, should be named the *Lewis effect* (Pontecorvo 1955, 173). In contrast to Lewis, however, Pontecorvo

treated the pseudoalleles not as closely linked genes, but as subunits
of a single gene. The most plausible interpretation of such cases,
Pontecorvo thought, was that "we are dealing with one gene and the
. . . alleles are due to mutation at . . . different mutation sites of that
gene, in . . . [both] case[s] the result of the mutation being that of
inactivating the gene" (Pontecorvo 1952, 134–137; see also Carlson
1966, 189).

The organisms that Pontecorvo mentioned in the course of his
discussion ranged from the classical fruit fly to microorganisms. At
the time his paper appeared in 1952, some biologists had come to the
conclusion that deeper understanding of the nature of the gene
would most likely come from experiments on the simplest known
organisms, in particular the bacteriophage viruses. The extreme
rapidity of their reproduction and very large numbers of progeny
made bacteriophage especially favorable subjects for crossing-over
experiments. The nature of genetic recombination in phage was,
however, less clear than in such long-studied organisms as
Drosophila. The first well-defined mutations, affecting the range of
the bacterial hosts that the phage can infect, were detected in 1946 by
Salvador Luria. In that year Alfred Hershey discovered a second
class of mutants that differed from the parent type "in causing
prompt rather than delayed lysis in undiluted culture." Hershey
designated the mutant *r* for *rapid lysis*. It soon appeared that there
was a family of independent mutations having the common effect of
rapid lysis. He designated these by giving them numbers following
the *r* (Hershey 1946, 620–640).

In 1948 Hershey showed that when bacterial cells were infected
simultaneously with any two of the independent *r* mutants, they
gave rise to wild-type and double mutant progeny phage. Evidently
a process of genetic recombination, analogous to that produced by
crossing-over in higher organisms, took place during the reproduction of phage. Testing various combinations of *r* mutants and host
range (*h*) mutants, Hershey found that the "linkage data support
fairly well the idea of linear structure, but independent evidence for
crossing over is meager" (Hershey and Rotman 1948). By 1951, he
had gathered more convincing numerical evidence that the loci for *r*
and *h* mutants, and a third class that produced minute plaques, could
be arranged in linear order. He represented the relation between the *r*

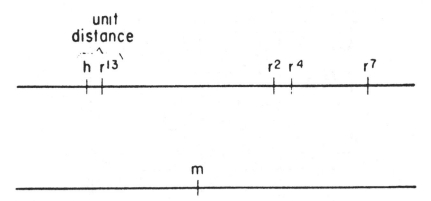

Figure 6-1   Hershey's genetic map of *r* mutants in T2 bacteriophage. (From Hershey and Chase 1951. Copyright © by Cold Spring Harbor Symposia on Quantitative Biology. Reprinted by permission.)

mutants in a genetic map (Hershey and Chase 1951, 471–472) (see Figure 6-1).

In 1952 A. H. (Gus) Doermann added six more *r* mutants to those found by Hershey (which he gave numbers beginning with r_{41}), and discovered also a new class of mutants that produced turbid plaques. He isolated twenty-six of these *tu* mutants of independent origin. The recombination data he was able to gather on these mutants was "compatible" with a linear arrangement of the mutant loci, and showed that the loci fit into two linkage groups, that Doermann thought were linked to Hershey's linkage groups (Doermann and Hill 1952) (see Figure 6-2).

By this time it was clear, therefore, that recombinant phage ratios could be used to produce phage genetic maps analogous to those of classical genetics; but the physical meaning of recombination in phage was not clear. The cross-over model that explained how recombinations took place during meiosis in diploid organisms was not necessarily applicable to phage, which were mainly haploid, and whose reproductive cycle within bacterial cells remained beyond direct observation. Hershey described the situation in 1951 with caution. Analysis of mutations and recombination data agree, he said,

in showing that mutations occur in localized genes. Recombination tests reveal that the genes are organized in linkage groups. For one of these

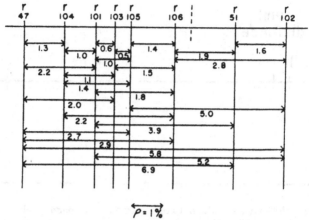

Figure 6-2   Doerman's genetic map of *r* mutants in T4 bacteriophage. (From Hershey and Rotman 1948. Copyright © by Genetics Society of America. Reprinted by permission.)

groups, it appears that the arrangement of genes is linear. Inheritance in bacteriophage is therefore amenable to the same kind of genetic analysis that has served to elucidate nuclear organization in other organisms. The limitations peculiar to viral genetics should not be overlooked. It is not possible to recover the immediate products of recombination, . . . the mechanism of recombination is unknown; and cytogenetic techniques are inapplicable. (Hershey and Chase 1951, 477)

ORIGIN AND EARLY DEVELOPMENT OF BENZER'S INVESTIGATION

Immediately after finishing a Ph.D. thesis on germanium semiconductors at Purdue University in 1947, Seymour Benzer was appointed Assistant Professor in the Department of Physics. Having attended the famous phage course at Cold Spring Harbor while still a graduate student, however, Benzer had already become an enthusiastic phage biologist, and it was understood that he would participate in the formation of a biophysics program at Purdue. To prepare himself for that role, he spent four years on leave of absence. After one year at the Division of Biology of the Oak Ridge National Laboratory, he took a two year postdoctoral fellowship in the laboratory of Max Delbrück at the California Institute of Technology. There

Benzer investigated the replication of bacteriophage by examining the changes in the sensitivity of the bacterial host cells to ultraviolet radiation. Although his results did not reveal what he had hoped to learn about the mode of phage replication, his work was, in Delbrück's opinion, "of very high quality" (Benzer 1966, 152–160; Delbrück 1957a).

During one of the camping trips into the desert, on which Delbrück frequently took the members of his lab, Benzer met André Lwoff, who invited Benzer to spend a year in Lwoff's laboratory at the Pasteur Institute in Paris. There in 1951 Benzer attempted to use *r* mutant T2 phages to lyse a K12 (λ) strain of *E. coli* bacteria during research on the variation in the amount of induced galactosidase enzyme produced in bacterial cells. When the phage failed to lyse the bacteria, he turned to a tougher strain that had been obtained by Jacques Monod and Elie Wollman from sewer water. That phage, together with a different bacterial strain, worked perfectly (Benzer 1966, 159–160; 1999).

Returning to Purdue in September, 1952, Benzer participated in the establishment of a Biophysical Laboratory supported jointly by the Departments of Biological Sciences and Physics. By January 1953 he had equipped his lab to continue experiments with phage, and had identified the "local phagophiles" with whom he could discuss his work. He intended, he reported to Delbrück, to "take up the problem of the number of enzyme-synthesizing centers per cell for galactosidase (i.e., Are the genes the thing?)." In June of that year, Benzer was present at the Cold Spring Harbor Symposium at which James Watson presented the double helix model of DNA that he and Francis Crick had proposed two months earlier in *Nature* – an event that transformed for the phage group the context in which "the genes" were discussed (Benzer 1966, 152–160; Benzer 1952, 1953; Delbrück 1957a).

Asked to give a seminar in genetics at Purdue, Benzer took as his subject, "The Size of the Gene," a choice directly inspired by reading Pontecorvo's review article on that topic. His interest in the paper had been motivated at first by a discussion Pontecorvo included on the estimation of gene size by means of radiation target theory. Referring to the classic text on this subject by Douglas Lea, Pontecorvo had summarized the way in which the diameter of a gene can be estimated from the "reactive volume" associated with a dose of ionizing radiation, assuming that one ionization within the target produces one mutation. As he studied the paper further, however, Benzer became increasingly attracted to the genetic approach and the possibility that the problem of gene size could be solved by recombination experiments. Pontecorvo's argument as a whole, in fact, anticipated with remarkable clarity the strategy of the investigation Benzer soon afterward began (Benzer 1966, 160–161; Benzer 1999; Pontecorvo 1952, 131–132).

All of the work that Benzer had done since acquiring his Ph.D. was concerned with bacteriophage (Delbrück 1957a). When he contemplated how one might pursue Pontecorvo's suggestion that crossing-over experiments could be pushed beyond the present limits of resolution, he inevitably assumed that the work would be done with phages.

Pontecorvo had published his review just before the appearance of the Watson-Crick double helix. By the time Benzer read it, the new model of DNA had already altered discourse about the question of whether genes were megamolecules. Pontecorvo's assertion that the resolution of chromosomes into their "linear array of genes" by crossing-over experiments could be pushed "down to interatomic distances," (Pontecorvo 1952, 129) thereby took on a significance quite different from what its author had intended.

At Purdue Benzer had again used an r mutant stock, but for a different purpose than in Paris. Then an unexpected observation set him off in another direction. The story he has told about the beginning of his gene-mapping project is a classic example of the serendipitous events with which scientists illustrate the adage of Louis Pasteur, "chance favors the prepared mind." It is best reproduced in Benzer's own words:

I had started out to attempt the Hershey-Chase experiment with genetic markers, to show sequential injection of the various parts of the phage genome, as Jacob and Wollman had done with bacterial conjugation. For that experiment, a stock of an r mutant of phage T2 was needed. Now stocks of r mutants grown on strain B of E. coli usually have titers much lower than the wild type r^+ stocks, since r^+ phages induce lysis inhibition and hold the cells together for a longer period of intracellular phage multiplication. But I had just read in George Streisinger's thesis that on certain strains of E. coli other

124

than strain B, r mutants of T2 do yield titers as high as r^+. Could that mean that the r mutant can produce lysis inhibition on those strains? To test this possibility, I plated out some T2r and T2r^+ on the strains I had on hand in my laboratory. If r produced lysis inhibition, it should make small, fuzzy-edged plaques similar to r^+, rather than the large, sharp-edged r-type plaques seen on strain B. On that day, I happened to be preparing an experiment on lysogeny for my phage class and was growing cultures of K12(λ) and its non-lysogenic derivative K12S (obtained via Luria from Esther Lederberg). Plating T2r and T2r^+ on these strains certainly gave different results from plating them on strain B. On K12S, r and r^+ both gave small, fuzzy plaques. On K12(λ), r^+ made small fuzzy plaques, but the plate to which r was supposed to have been added had no plaques at all. I was sure that in the rush to prepare for class, I had neglected to add phage to that plate. But repetition confirmed the result.

To me, the significance of this result was now obvious at once; here was a system with the features needed for high genetic resolution. (Benzer 1966, 161; Benzer 1999)

What Benzer saw at once was that if crosses between two r mutants were grown on strain B and plated on strain K12(λ), they would not grow on the latter, but any wild-type T2 arising from recombination between the mutants would grow there. The test would be so sensitive that mutants arising in frequencies far lower than those detectable in other organisms would be made visible. He immediately dropped the problem of adapting the Hershey-Chase experiment with a Waring Blender to the Jacob-Wollman experiment, and "embarked on this project" (Benzer 1966, 161; Benzer 1999).

Assuming that his recollection twelve years afterward was accurate, Benzer's account constitutes one of the richest "flash of insight" experiences in the scientific autobiographical literature. There was not just one, but a combination of four chance occurrences required to produce the observation under circumstances that would give it the significance it took on for him. There was an experiment he had in mind to do that caused him to use r mutant phage; the reading of a thesis which highlighted a particular property of r mutants; the convenient presence of a particular strain of *E. coli* that had reached him through a chain of personal connections; the necessity to prepare a class demonstration also involving r mutants; and finally, the background interest in the size of the gene triggered by reading Pontecorvo's essay for a genetics seminar. None of these juxtaposed events were causally related to the others.

Benzer's account leaves unanswered the question of whether he had already intended to pursue recombination experiments with phage, and was prompted by this chance observation only to see the way for which he had been searching to detect very rare recombinants, or whether it was this flash of insight that gave him the initial motivation to do such experiments.

These events probably took place in January 1954. Because his wife and two children were away at the time, Benzer had a week of "unrestricted time" to begin work on the problem (Benzer 1999). A few months later, while doing summer research at Cold Spring Harbor, he detected recombination frequencies so small as to convince him that it was "technically possible to study, by genetic means, the structure of the . . . 'gene' down to the individual nucleotide" (Benzer 1954). He wrote a paper, which he showed to Delbrück in Amsterdam. Delbrück criticized his manuscript so strongly (Benzer 1966, 162) that Benzer took six months to prepare and send him a "more subdued second draft of the rII story." He was, Benzer wrote on February 3, 1955,

so grateful for your advice against publishing the first one that I will be almost delighted if you find fault with this. . . . It suffers from not quite answering any of the $64 questions, but may nevertheless be acceptable as a preliminary report. (Benzer 1955a)

Delbrück made further suggestions for changes in the discussion section of the paper. Benzer must not have agreed with some of them, because they do not appear in the published version. The points Delbrück made throw an interesting light on the doubts that a leading member of the phage group held about the genetic implications of the Watson-Crick model of DNA three years after its announcement, but these can best be discussed after a summary of Benzer's paper, "The Fine Structure of a Genetic Region in Bacteriophage" (Benzer 1955b).

"This paper," Benzer began, "describes a functionally related region in the genetic material of bacteriophage that is finely subdivisible by mutation and genetic recombination." The group of mutants he had found resembled what had been described in other organisms as "pseudo-alleles" – that is, as mutants that arose independently at different sites, but were so close together as to show no recombination. Echoing Pontecorvo's general discussion of genetic recombination, Benzer asserted that that phenomenon provides a powerful tool for separating mutations and discerning their positions along a chromosome. When it comes to very closely neighboring mutations, a difficulty arises, since the closer two mutations lie to one another, the smaller is the probability that recombination between them will occur. Therefore, failure to observe recombinant types among a finite number of progeny ordinarily does not justify the conclusion that the two mutations are inseparable but can only place an upper limit on the linkage distance between them. A high degree of resolution requires the examination of very many progeny. This can best be observed if there is a selective feature for the detection of small proportions of recombinants. (Benzer 1955b, 344–345)

Briefly Benzer introduced the selective feature that he had discovered, based on the inability of *r*II type mutants of T4 bacteriophage to produce plaques on one of two bacterial hosts on which the wild-type does so. By making a cross between two *r*II mutants "any wild-type recombinants which arise, even in proportions as low as 10^{-8}, can be detected by plating" on the strain on which the mutants cannot produce plaques (Benzer 1955b, 345). Pontecorvo had predicted that selective techniques could push the resolution "one or two orders of magnitude" beyond the 10^{-5} frequencies previously attainable (Pontecorvo 1952, 130). Now Benzer had already reached three orders of magnitude further.

Benzer also followed Pontecorvo's approach in attempting to define the physical distance between the closest mutations observable by recombination by equating the length of the entire genetic map with that of the "chromosome." Where Pontecorvo had used the microscopic length of a *Drosophila* giant salivary gland and come out with "say, 100 Angstroms" as the upper limit of the distance (Pontecorvo 1952, 130–131), Benzer used the "total length of DNA per T4 particle of 2×10^5 nucleotide pairs." The recombination "per nucleotide pair," he concluded from this "exceedingly rough" calculation, "is 10^{-3} per cent" (Benzer 1955b, 345).

Even while emulating his strategy, Benzer subverted the aim of Pontecorvo's analysis. Where Pontecorvo had inferred that "it is clear that the (maximum) length of a gene estimated from calculations" based on crossing-over "is such that the value of picturing the gene as of megamolecular size is disputable" (Pontecorvo 1952, 131), Benzer wrote, "We wish to translate linkage distances, as derived from genetic recombination experiments, into molecular units" (Benzer 1955b, 345). Where Pontecorvo had questioned the continued value of the "picture of the gene as a sharply delineated portion of a chromosome" (Pontecorvo 1952, 134), Benzer presented a "program designed to extend genetic studies to the molecular (nucleotide) level." Aside from personal differences in outlook, the contrast in their approaches reflected that the intervening years had, as Benzer later put it, "brought the Watson-Crick model, and now DNA was really *in*." The calculation he introduced in 1955 was only meaningful "If we accept the model of DNA structure proposed by Watson and Crick" (Benzer 1955b, 345; Benzer 1966, 160).

Phage biologists formed a small, closely knit group in 1955. Benzer had begun his experiments with stocks of genetically mapped T2 mutants supplied by Hershey, and T4 mutants from Doermann. Of the various *r* mutants mapped by Doermann and Hershey, Benzer selected only those that shared the property he had discovered by chance with one of them, the failure to grow on bacterial strain K12S(λ). He called these mutants the "*r*II group." These mutants could be isolated from wild-type T4 by plating on another strain, *E. coli* B, where they produced large, sharp-edged plaques, in contrast to the small, fuzzy ones of the wild-type. In addition to two of Doermann's mutants, Benzer isolated six new ones that fit his criteria. When he determined their recombination frequencies, all of the *r*II mutants fell within a small region of the T4 map located near the two *r*II mutants of Doermann, *r*47 and *r*51. Benzer gave his new *r*II mutants the numbers 101–106. His map of the region is shown below (Benzer 1955b, 346–349) (see Figure 6-3). "While all *r*II mutants of this set fall into a small portion of the phage linkage map," he commented, "it is possible to seriate them unambiguously, and their positions *within* the region are well scattered" (Benzer 1955b, 349).

These *r*II mutants were so closely linked (as can be seen from the map, their recombination frequencies ranged only from 1.0 to 6.9 percent), that the question arose whether they belonged to the same, or to separate functional units. To find out, Benzer "simulated" the test for pseudoallelism described by Lewis and by Pontecorvo. In diploid higher organisms one could decide whether two mutations were in the same functional unit by constructing heterozygotes in two configurations. When both mutations were on one chromosome, the heterozygote was expected to be that of a wild-type, because the second chromosome supplied the complete functioning unit. Follow-

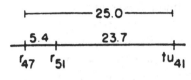

LINKAGE GROUP II

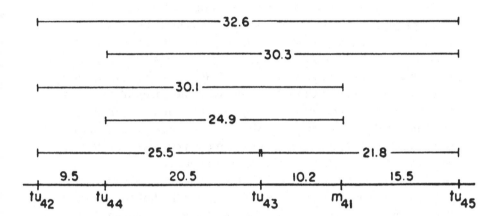

LINKAGE GROUP III

Figure 6-3   Benzer's genetic map of the *r*II region of T4 bacteriophage in 1955. (From Benzer 1955b. Copyright © by S. Benzer. Reprinted by permission.)

ing Lewis's terminology, Benzer called the two configurations the *cis* and *trans* forms. When one mutation was on each chromosome, (the *trans* form) then the heterozygote was that of a wild-type if the mutations were not part of the same unit, because each could supply what was missing on the other; but that of a mutant type if they were part of the same functional unit, because neither chromosome would contain an intact unit. (According to the standard contemporary view of the gene, the functional unit in question in this "complementation test" was the gene itself, but Pontecorvo had placed the simple identification of genes with functional units in doubt.) For phage, the equivalent to constructing the heterozygotes was to infect bacteria simultaneously with two kinds of phage. The *cis* configuration would result when both mutations were in one of the phages, the *trans* form when one mutation was in each. Although he had not yet tested the cis form, Benzer was able through the production of the trans forms of every pair of *r*II mutants to divide the *r*II region into two segments. Benzer indicated the point of division on his map of the region by a short vertical dotted line. These two portions of the region "correspond," he concluded, "to independent functional units" (Benzer 1955b, 349–350).

Thus far, Benzer had worked at a level of resolution only a little greater than that of Hershey or Doermann. His next step took him to the level that justified the title of his paper, the "fine structure of a genetic region." Through what he called a "rough mapping spot test," in which he could test one mutant "against several others on a single plate by first seeding the plate" with the bacterial strain on which *r* mutants do not grow, then spotting it with drops containing the other mutants, he could quickly tell which mutants were in the same functional segment of his map, and which were in different segments. Those in different segments constituted a "mixed infection" and gave extensive lysis. For those in the same segments, only wild recombinants gave plaques. "The greater the linkage distance between them the larger the number of plaques that appear in the spot." With this method he could quickly establish the order of a group of mutants along a segment of the map. The rapidity of the test enabled him to isolate a much larger number of *r*II mutants than had been known, and to give them "preliminary locations" on the map. From this array he picked out four "microclusters" of "very closely neighboring mutations" to map more accurately. Now the extreme sensitivity of his method for selecting wild-type recombinants showed its full power. He was able to produce maps with "reasonably good additivity properties" for mutants whose recombination frequencies ranged between 0.015 and 0.29 percent (Benzer 1955b, 350–352).

Not all of the mutants showed this property. Some gave "violently anomalous results." They produced no wild recombinants at all with several of the other mutants that recombined with each other. To explain this situation, Benzer suggested that such mutations "extend over a certain length of the chromosome." This observation, he said, "raises the question whether there exist true 'point' mutations (i.e., involving an alteration of only one nucleotide pair) or whether all mutations involve more or less long pieces of the chromosome." He had encountered some mutations that were "leaky," – that is, that were not fully expressed in the phenotype, or that reverted easily to the wild-type by back mutations. Since such effects interfered with the recombinant results, he had favored those mutants that were extremely stable against reversion. In doing so, he realized, he may also have selected for mutants containing larger alterations of the chromosome. "In continuing these experiments," he wrote, "it would seem well advised to employ only mutants for which some reversion is observed" (Benzer 1955b, 351).

In the "Discussion" section of his paper, Benzer speculated about the meaning of the two "functionally distinguishable units" into which the region he had studied was divided. He did so within the context of the prevailing one-gene – one-enzyme concept. The segments could be

imagined to affect two necessary sequential events or could go to make up a single substance the two parts of which must be unblemished in order for the substance to be fully active. For example, each segment might control the production of a specific polypeptide chain, the two chains later being combined to form an enzyme. (Benzer 1955b, 353)

Returning to the theme of his introduction, Benzer reinforced his earlier characterization of his results as preliminary, by indicating where the investigation was headed:

By extension of these experiments to still more closely linked mutations, one may hope to characterize, in molecular terms, the sizes of the ultimate units of genetic recombination, mutation, and "function." Our preliminary results suggest that the chromosomal elements separable by recombination are not larger than the order of a dozen nucleotide pairs (as calculated from the smallest non-zero recombination value) and that mutations involve variable lengths which may extend over hundreds of nucleotide pairs.

The unit of function, defined by the *cis-trans* test, he estimated to "contain of the order of 4×10^3 nucleotide pairs" (Benzer 1955b, 353).

Thus, Benzer gave striking confirmation to Pontecorvo's view that the three definitions of the gene were not interchangeable and were of different dimensions, the unit of function being one or two orders of magnitude larger than either of the other two units; yet he reversed Pontecorvo's inference that genes are "not atomic edifices" (Pontecorvo 1952, 134). That he did so seems to be the convergent outcome of his own ability to carry the resolution of the genetic map much closer to atomic dimensions, and the advent of a specific atomic edifice to which the dimensions of the map could be matched.

BENZER AND MAX DELBRÜCK

The discussion section in the "second draft" of his paper that Benzer had sent to Delbrück in February may not have been identical with the published one, but was probably less different from it than the suggestions for a discussion and summary that Delbrück sent him in response. In his covering letter, Delbrück wrote:

I have a few questions about the experimental part, but hardly any suggestions for alterations. Your present discussion section seems to me to stand too much under the influence of the pre-conceived notion of an ideal WC [Watson and Crick] chromosome. Since the principal result of your experiments show up a gross incompatibility between this ideal and the facts of life it seems to me more profitable to let the discussion be dominated by a rational account of the facts, and to point out just marginally that the ideal WC chromosome, if some people should be so deluded as to have it in the back of their minds, is a far cry from your findings. (Delbrück 1955)

Coming from this charismatic leader of the phage group – from the mentor to whom James Watson had written the first description of the double helix before its publication and from one who had regarded the Watson-Crick model for DNA as a momentous event – the opinion of that model that Delbrück expressed here two years later seems startling. It is less so when we recall that he had recently questioned in print the mode of replication of the double helix proposed by Watson and Crick. Ever iconoclastic, Delbrück was prepared both to support the importance of, and severely criticize, the views of the colleagues whom he most valued.

The "Outline for a revised discussion and summary" that Delbrück included with his letter makes clear the basis for his position. The first five points stated the "rational account of the facts" that he thought could be drawn from the experiments. The sixth point dealt with the "WC model":

6) Our principal interest attaches to a group of questions which could be given the heading "the topological structure of the map in relation to the topological structure of DNA."

The WC model and the EM pictures do not suggest the existence of any structural units in the DNA larger than those of the single nucleotide link, or perhaps that of a turn of the helix. The resolving power of our method would be ample to detect recombination between mutants which are only one nucleotide link apart, if map distance is simply proportional to DNA distance and if the whole map (about 100 units) corresponds to the whole DNA (about 200,000 links). It is obvious at a glance that our random sample of mutants does not correspond to a random sample from an ideal map, the ideal map being defined as one in which

a) every mutation is a change in a single nucleotide pair, b) every nucleotide pair mutates with equal probability, c) recombination can occur at every nucleotide pair with equal probability.

Indeed the real sample differs vastly from an ideal sample, and most grossly in the following features:

a) we find double and even triple repeats in a relatively small sample from a region of map length 8%, corresponding to 16,000 links,
b) we find more than two classes of reversion rates,
c) we find a smallest non-zero map distance of .1%

These deviations from the ideal sample may have one or more of three obvious causes

a) recombination cannot occur at every link, but only in selected points which lie about 200 links apart,
b) the mutations are not point mutations, i.e., alterations affecting a single nucleotide pair only, but they affect a stretch of about 200 links,
c) mutation rates at different points in the region differ enormously and what we pick out are the few points of relatively enormously great mutability. (Delbrück 1955)

Benzer replied on February 11, "Your comments and suggestions on the manuscript are apt and much appreciated, and I am working on the revisions." He arranged to fly to Pasadena at the end of March to discuss the manuscript and to give a seminar (Benzer 1955a). Delbrück communicated the final paper on April 6 to *PNAS*. Benzer did incorporate several of Delbrück's suggestions verbatim into his summary, but did not take up any of Delbrück's discussion of the differences between the real map and an ideal map. In contrast to Delbrück's assertion that the experiments showed that recombination can occur only at points "about 200 DNA links apart," Benzer concluded that the elements separable by recombination "are not larger than the order of a dozen nucleotide pairs." Was Delbrück's criticism of an "ideal WC chromosome" a serious commentary on contemporary fashions, or was this ideal model one he conjured up only in his own mind? Did he see patterns in Benzer's results that had escaped their author, or did he misunderstand those results? Delbrück was not persuaded by Benzer during their reunion at Caltech to abandon his objections, but the younger man had the self-confidence to reject some of the views of the very person about whom he acknowledged in his publication his indebtedness "for his invaluable moderating influence" (Benzer 1955b, 354; Benzer 1999).

PUBLICATION AND EARLY RESPONSE

Benzer was not alone at this time in his quest to subdivide the "genetic material" into its finer subunits. At Cold Spring Harbor, Milislav Demerec was at the same time using autotrophic mutants of the bacterium *Salmonella typhimurium* to examine the "structure and organization of gene loci." Taking advantage of "transduction" – a process in which bacteriophage grown in one bacterial culture can carry a small fragment of the bacterial chromosome into the bacteria they subsequently infect – to create heterozygote regions of bacterial chromosomes, Demerec, too, was able to gather evidence "that a gene can be subdivided by crossing over into small, lineally arranged units," that he called "sites" (Demerec 1956; Carlson 1966, 200–206). At Yale, David Bonner used *Neurospora* to explore the possibility that "the genetic unit as described in physiological terms may well be comprised of specific genetically separable subunits" (Bonner 1956). In Glasgow, Pontecorvo, with his associates J. A. Roper and R. H. Pritchard, was pursuing study of the "organization of the chromosome over minute regions" in the plant mould *Aspergillus*. In linkage experiments with this organism, Pontecorvo and Roper were able by 1955 to achieve recombination fractions of "the order of 10^{-6}" (Pontecorvo and Roper 1956, 83).

In this field of activity, Benzer's publication was received with enthusiasm. Among the first to respond was Pontecorvo, who commented in June 1956, at the Cold Spring Harbor symposium on Genetic Mechanisms, that "the remarkable work of Benzer (1955) on bacteriophage and Demerec (this Symposium) on *Salmonella*" provided support for the "model of gene structure" Pontecorvo had suggested in 1952 (Pontecorvo 1956). In a paper coauthored with Roper, Pontecorvo focused specifically on Benzer's work:

> We may . . . attempt to give a chemical meaning to what genetic analysis resolves, and perhaps even to what is the basis of the process on which genetic analysis is based, that is, crossing over. In this direction a remarkable paper by Benzer on bacteriophage has already given a lead. In the present paper, Benzer's attempt will be used in a comparative approach.

From a comparison of the number of mutational sites so far identified with the recombination fractions observed in the experiments of Lewis on *Drosophila,* Benzer on phage T4, and Roper on *Aspergillus,* Pontecorvo calculated "what part of the genetic map the smallest recombination fractions" in each of these organisms represented. Then, giving "chemical meaning" to these results by following "Benzer's argument," he calculated that the distances were equivalent to "eight nucleotide pairs for *Aspergillus,* 216 for *Drosophila,* and 12 for phage" (Pontecorvo and Roper 1956).

It was not only the particular argument of Benzer, but also the spreading impact of the Watson-Crick structure that had driven Pontecorvo to consider this chemical meaning of the results of genetical analysis. But the classical geneticist who had so recently maintained that genes are not atomic edifices made this concession only guardedly. All these "crude estimates," he wrote, "are amusing exercises in numerology. They may even have some heuristic value if they are taken for nothing more than what they are." The analysis of the genetic map by crossing over possessed an "amazing sensitivity, a sensitivity which biochemistry still lacks" (Pontecorvo and Roper 1956, 84).

According to Carlson's account, it was the theoretical "daring" with which Benzer translated "genetic concepts . . . into molecular terms," rather than his experimental results themselves, that distinguished his work from that of the more conservative Demerec (Carlson 1966, 202–203). Pontecorvo's reaction supports the view that it was Benzer's attempt to give "chemical meaning" to "genetical analysis" that forced classical geneticists, as Falk puts it, to "enter

F. L. HOLMES

the era of DNA." But as Falk's previously cited recollections suggest, it was also the elegance of the experiments themselves that excited and persuaded geneticists of the importance of the work. Falk remembers particularly the effect of the beauty of the work when Benzer visited Hermann Muller in Muller's laboratory in Indiana (Falk 1996).

Meanwhile, Benzer pressed forward in his own investigation. The discussion section of his first publication suggested the two directions that his research now took. His questions about the definition of the "function" of a functional unit led toward identification of the polypeptide chain or enzyme whose production it might control. Shortly afterward, Alan Garen came to Purdue and attempted to isolate "the rII protein." He and a succession of other associates of Benzer were unsuccessful in this quest (Benzer 1966, 162–163).

The other direction was to extend the mapping experiments to "still more closely linked mutations." Perhaps it was during their conversations at Caltech that Delbrück told Benzer that, to determine the minimal size of a unit of recombination he would have to "run the map into the ground" (Benzer 1957, 74). Only that way could it be settled whether the distance was 200 nucleotide links as he maintained, or no more than a dozen, as Benzer maintained.

The anomalous stable mutants that Benzer had planned, when he wrote his first paper, to avoid in future experiments, turned out instead to provide the "trick" that enabled him to extend his recombination experiments to much larger numbers of mutants than he or anyone else had previously uncovered. By the usual methods, the magnitude of the task increased with the square of the number of mutants tested. But if the anomalous mutants were, as Benzer assumed, caused by deletions extending over a segment of the map, then any other mutant that could produce with it a wild recombinant must be located outside of the region of a deletion. By crossing each new mutant with several of these stable mutants that represented overlapping segments of the map, he could quickly establish which region of the map it occupied. Then it was necessary only to cross the mutant with all possible pairs within the group of mutants found to occupy the same region (Benzer 1957, 76–77).

By applying this and several other time-saving methods to make preliminary classifications, Benzer mapped 241 r mutants of T4. Of these, 33 "species" fell in one of the two "functional units" of the *r*II region defined by the cis-trans test, and 18 species in the other. This was enough, he believed, to "define reasonably well the limits" of each functional unit. It did not sufficiently saturate the map to determine the size of the minimum units of recombination and mutation. Therefore, he isolated and located on the map 923 more *r* mutants. The smallest recombination distance he found in this way was equivalent to 0.02 percent, but since only one such interval was found, still smaller values were "not ruled out." To measure the size of the smallest unit of mutation, he compared the long distance between three closely linked mutants with the sum of the two shorter distances. The difference between them gave the size of the center mutation (Benzer 1957, 83–87).

Benzer presented the results of this second stage in his gene mapping project at a symposium on "The Chemical Basis of Heredity," held at the Johns Hopkins University in June 1956, at which many of the leaders in the emerging field of molecular biology were present. In the title, the introduction, and discussion of his paper, he focused more sharply than in his initial paper on the refinement that genetic fine structure studies could bring to the definition of "the elementary units of heredity." Within a lucid discussion of the principles of genetic mapping, he gave definitions and new names for the units of recombination, mutation, and function:

The classical "gene," which served at once as the unit of genetic recombination, of mutation, and of function, is no longer adequate. These units require separate definition. A lucid discussion of this problem has been given by Pontecorvo.

The unit of recombination will be defined as the smallest element in the one-dimensional array that is interchangeable (but not divisible) by genetic recombination. One such element will be referred to as a "recon." The unit of mutation, the "muton," will be defined as the smallest element that, when altered, can give rise to a mutant form of the organism. A unit of function is more difficult to define. It depends on what level of function is meant.

After discussing, similarly to his first paper, whether such a unit may specify an ensemble of enzymatic steps, the synthesis of an enzyme, one peptide chain, or even one critical amino acid, Benzer bypassed the difficulty by stating that "A functional unit can be defined genet-

137

ically, independent of biochemical information, by means of the elegant *cis-trans* comparison." Briefly describing the test, he added that "It turns out that a group of non-complementary mutants falls within a limited segment of the genetic map. Such a map segment . . . will be referred to as a 'cistron'" (Benzer 1957, 70–71).

At the end of his paper, Benzer gave estimates of the molecular sizes of each of these units. A recalculation of the percent recombination corresponding to a single nucleotide pair that took into account recent evaluations of the portion of the T4 DNA particle that contains genetic information, and of the deviations from additivity in the genetic map due to "negative interference" yielded a figure, (0.01 percent) ten times larger than what he had estimated in the previous paper. The smallest observed nonzero-recombination value now led to the conclusion that "the size of the recon would be limited to no more than two nucleotide pairs" (compared to his previous inference of "not larger than the order of one dozen nucleotides"). So far no mutation size larger than "around 0.05 percent recombination" had been demonstrated, indicating that "alteration of very few nucleotides . . . is capable of causing a mutation" (Benzer 1957, 91–93).

"A cistron," Benzer wrote, "turns out to be a very sophisticated structure." He expected that, by the time he completed his analysis of the 923 *r* mutants, there would turn out to be about 60 species of mutants in the A cistron. Since there were probably "many more yet to be found," he surmised that there were probably "over a hundred . . . locations at which a mutational event leads to an observable phenotypic effect" (Benzer 1957, 93).

In the discussion of the papers presented at the session in which Benzer defined the recon, the muton, and the cistron, George Beadle commented: "I would like to say that I don't want to get into any argument with Benzer about terminology. If he wants to call a functional unit a cistron that's all right with me. I would like to say, however, that I agree that his is a very beautiful work" (Benzer 1957, 129). Whether other geneticists would begin calling a functional unit a cistron remained an unanswered question in 1956, but there was little question that some of them already shared Beadle's estimate of the work. A year later, in a letter of recommendation for Benzer, Delbrück wrote:

Benzer picked up a casual observation regarding the behaviour of certain mutants of phage T4 and developed from this a powerful method for the study of genetic fine structure, with sensational results. . . . He worked out this method in minute detail and the results obtained with it then strongly influenced our whole thinking (and terminology) concerning the organization and method of functioning of genetic material. (Delbrück 1957a)

"Our thinking" probably referred mainly to the still small group of phage biologists and the international circle of enthusiasts for the Watson-Crick DNA structure. The fact that Benzer was both well-connected and well-respected within these groups undoubtedly helped to give his work rapid prominence. Delbrück's statement suggests that, already in 1957, Benzer's redefinition of the elementary units of heredity was winning consensus support within these circles.

At a deeper, more personal level, Delbrück himself remained unpersuaded – not only of the meaning that others gave to Benzer's work, but of the basic assumptions of genetic mapping on which it rested. He gave voice to his doubts in a lecture on "Atomic Physics in 1910 and Molecular Biology in 1957" that he presented at MIT in November of the latter year. "*If*" the linear genetic map whose fine structure Benzer studied "is a direct image of its material carrier, the DNA molecule," Delbrück acknowledged, "then it can be estimated from Benzer's studies that a point on the genetic map cannot be larger than a few unit links of the molecule." From the fact that points on the genetic map can be equated with points on the molecule, however, Delbrück did not believe that it necessarily followed that they were direct images of one another. "Certain odd features," such as the "positive correlation" of markers very close together in three factor crosses, and the small heterozygote regions in phage genomes, had so far not been accounted for in this scheme. Possibly, he inferred, the topology of the material carrier is not really a "simple line," but involves branches or ladders. Or, "the non-simplicity results from the fact that we are confronted with a true complementarity" in the sense of Niels Bohr's argument that "experimental situations used to define observations may be mutually exclusive." In this case, "A rational account which includes both sides of the picture, chemical structure *and* genetic map, may turn out to be an abstract one, not a visualizable one" (Delbrück 1957b, 6, 16–19).

Delbrück's doubts about the identification of the genetic map with
the material carrier were expressions of his personal quest for paradoxes in biology that would force a reexamination of "the scientific
method itself" (Delbrück 1957b, 22). Few, even of those colleagues
deeply influenced by the power of his personality, shared this quest.
For them, Benzer's results seemed to show exactly that the genetic
map *was* a direct image of the topology of the material carrier, an
expression of the sequence of the nucleotides comprising a DNA
double helix. In 1958, when Francis Crick formulated the "Sequence
Hypothesis," according to which "the specificity of a piece of nucleic
acid is expressed solely by the sequence of its bases, and that this
sequence is a (simple) code for the amino acid sequence of a particular protein," the primary evidence he cited for the "genetic linearity
within the functional gene," was "the work of Benzer (1957)" (Crick
1958, 152).

FURTHER REFINEMENTS OF THE FINE STRUCTURE

Ever since he had met Sydney Brenner at Cold Spring Harbor in
1954, Benzer had been hoping that his map of the rII region could
somehow be exploited to establish the "colinearity of alterations in
amino acid sequence [of the protein whose formation was the expected function of the region] with the locations of the mutations in
the genetic map." In a year spent at Cambridge in 1957, he and
Brenner tried without success to find a structural difference in a
protein of T2 or T4 that could be related to a single mutation in their
genomes (Benzer 1966, 162–164).

As Benzer resumed his genetic mapping of the rII region back at
Purdue in 1958, Delbrück continued to exert "his usual moderating
influence" (Benzer 1959, 1619). Expanding his study of the r mutants
still further, Benzer had by 1959 screened "some 2000 spontaneous
mutants." With associates such as Ernst Freese he also examined
induced mutations, such as those caused by substituting 5-Brom-
ouracil for the thymine in their DNA molecules (Benzer and Freese
1958). During this period, however, Benzer shifted both his aims and
his method. In place of his earlier goal to establish the size of the
elementary hereditary units, he now focused his attention on the
"topology of the genetic fine structure." His central objective now

was to "make a rigorous test of the notion that the structure is linear" (Benzer 1959, 1610). It is tempting to attribute his motivation for doing so to Delbrück's resistance to the notion that the genetic map was a "direct image of the sequence in the DNA molecule" (Delbrück 1957b, 17).

The question arose only at the level of the "finest structural details," involving the "molecular subunits of the hereditary material." The linearity of the genetic map at higher levels had long been settled by classical genetics. As Benzer noted,

> From the classical researches of Morgan and his school, the chromosome is known as a linear arrangement of hereditary elements, the "genes." These elements must have an internal structure of their own. At this finer level, within the "gene" the question arises again: what is the arrangement of the *sub*-elements? Specifically are *they* linked together in a linear order analogous to the higher level of integration of the genes in the chromosome?

In his earlier experiments, Benzer had assumed this analogy and focused on the distances between elements linearly ordered. But, as Delbrück had argued in his lecture, it was just at the shortest distances that the results of recombination frequencies became least reliable. The lack of strict additivity at this range had been attributed to phenomena such as "negative interference," but now Delbrück was suggesting more radical interpretations of such discrepancies. Consequently, Benzer decided to explore the topology by a method independent of recombination frequencies. Such experiments, he asserted, "should ask *qualitative* questions (e.g., do two parts of the structure touch each other or not?) rather than *quantitative* ones (how far apart are they?)" (Benzer 1959, 1607). It should be obvious to us that he had previously been most interested in just those quantitative questions that he now put aside.

For his new purpose, Benzer again selected nonreverting *r*II mutants, taken to represent deletions of various lengths rather than point mutations. Again he established whether the mutants occupied parts of the same segment or not by determining whether or not in mixed infections they produced recombinants; but the role of this test was now elevated from a preliminary mapping procedure to the main source of the data he would test for linearity. By screening all the spontaneous *r* mutants, he chose 145 nonreverting *r*II mutants, that "were crossed in many pairs, in each case testing for the appearance of standard type progeny." He then arranged the results in a matrix, in which each cross that gave such progeny was represented with a figure 1, each which did not, with a figure 0. If the overlaps of the mutants, defined by the cases in which no recombination resulted, formed a linear topology, he showed, then this matrix could be rearranged in such a way that when one moved from a 0 on the diagonal downward or to the right, any series of 0s encountered was unbroken by a 1.

In the paper in which he presented these results in 1959, he showed first such a matrix for a family of 19 of the rII mutants (see Figure 6-4). The order in which the mutants must be arranged to

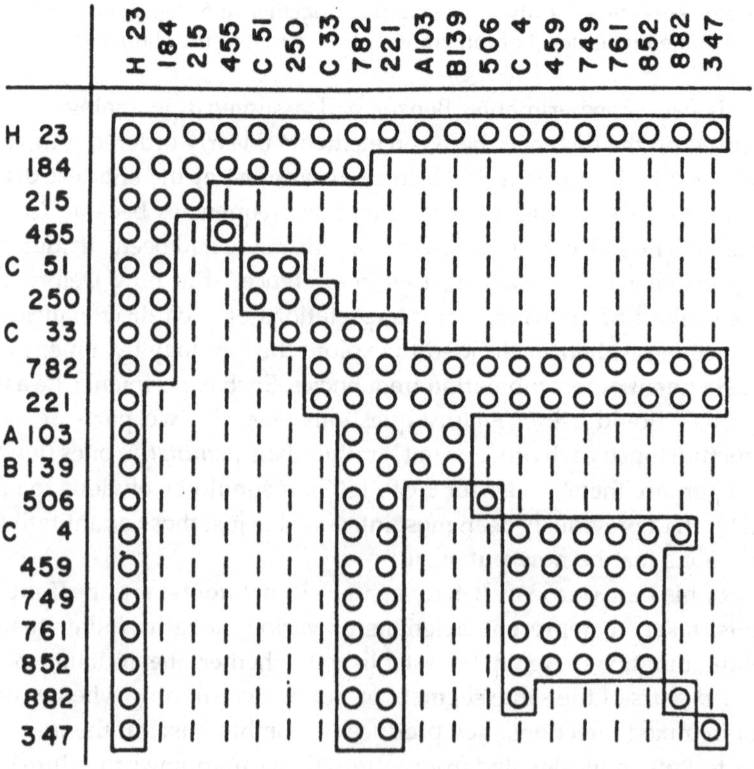

Figure 6-4 Benzer's matrix for 19 rII mutants in 1959. (From Benzer 1959. Copyright © by S. Benzer. Reprinted by permission.)

142

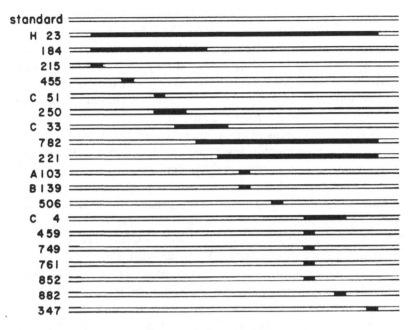

Figure 6-5   Linear order of 19 *r*II mutants, 1959. (From Benzer 1961. Copyright © by S. Benzer. Reprinted by permission.)

provide such a matrix defined the relative positions of the mutations along a linear structure. He next deduced an array of relative positions for the same 19 mutants that was consistent with this matrix (see Figure 6-5). Benzer then presented a corresponding, more complicated matrix and linear ordering for all 145 *r*II mutants (not reproduced here). The resulting genetic map fell into two separate segments consistent with the two *r*II cistrons he had previously defined through the *cis-trans* test (Benzer 1959, 1608–1617).

The chances that one could produce such a matrix from random data were improbably low, Benzer showed. He could not rule out that certain types of branched maps might produce similar matrixes. The "present analysis" could not, therefore, disprove the "existence of complex situations. However," he concluded, "the fact of the matter is that a simple linear model suffices to account for the data" (Benzer 1959, 1618–1619).

In 1956 Benzer had given a "taxonomy" of *r*II mutants, in which he defined as members of the same "species," those which when crossed yielded no wild recombinants. The 145 mutants belonging to one subclass defined by the segment of one particular deletion mutant separated into 11 such species, but they were far from equally distributed. Five of the species included only one independent mutant and four of them contained from two to four mutants, but "one of the species accounted for 123 of the mutants!" (Benzer 1957, 84–85). Despite this striking evidence that the mutation rates were very uneven along a segment of the genetic map, Benzer mentioned the phenomenon only in passing in the paper in which he first reported it. In 1959, he turned his attention more fully to this feature of what he now called the "topography of the genetic fine structure." Topography, as distinct from the "topology" that he had just studied, meant to Benzer the delineation, "in minute detail . . . of the physical features . . . of a region" (Benzer 1961, 403, 415).

For his latest purpose Benzer developed a further refinement of the method he had earlier devised to assign point mutations preliminary locations on the map by crossing them with deletion mutations to find out which segment of the map they fit into. By mapping "many more deletions than were described in the previous paper," he had now identified seven long deletion mutations that divided the two cistrons of the rII region into seven segments. Twenty-five additional deletion mutations further divided each of these segments into a total of 47 segments. By crossing a point mutation first with each of the long mutations, to find out which of the larger segments it belonged in, and then with each of the deletion mutants dividing that particular segment, he could, "in two steps," map the point mutation "into one of the 47 segments." With this method he analyzed "some thousands" of spontaneous and induced mutations and displayed their locations in a "topographic map" (Figure 6-6): (The map defined the order of sites from one of the 47 segments to another, but not within a segment.) The result not only amplified Benzer's earlier finding that the number of mutation sites within a cistron "is very large," but that the "spontaneous mutability" of the sites varied greatly. There were several "hotspots" separated by regions of much lower mutability. The topography of maps constructed from induced mutations also gave different "spectra" from that for spontaneous mutations. In discussing these results in 1960,

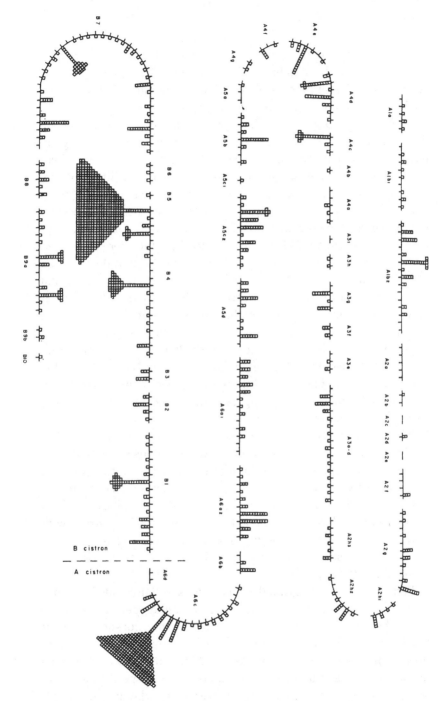

Figure 6-6 Benzer's topographic map of *r*II mutants, 1960. (From Benzer 1961. Copyright © by S. Benzer. Reprinted by permission.)

Benzer suggested that the differences in mutability might be related to differences in the stability of the AT and GC base pairs, and proposed that it might be possible to "translate the genetic map into a nucleotide sequence" (Benzer 1961).

At the beginning of 1961, Benzer was still aiming to use newly discovered mutagens to saturate further the map of the *r*II region. Irwin Tessmann at Purdue was also isolating *r*II mutants using Benzer's technique for isolating them through their inability to reproduce on *E. coli* K12(λ). Benzer asked him for samples of any of them that filled previously unoccupied sites. Some of these mutants did not, however, "behave like *r*II mutants at all – they multiplied happily on my strain of K12(λ)": that is, on the strain Benzer used to distinguish *r*II mutants by their failure to grow! When Benzer tried some of his own "good" *r*II mutants on the bacterial strain that Tessmann used for the same purpose, a few of them also grew. It occurred to Benzer immediately that his and Tessman's strains of *E. coli* might differ in some mutation that permitted certain subsets of the *r*II mutants to become active respectively in the one or the other (Benzer 1966, 164). Even though the two strains had derived from the same source, they had "passed through various treatments in different hands," and had had "opportunities to accumulate changes" (Benzer and Champe 1961, 1034).

With Sewell Champe, Benzer investigated this effect further, eliminating the possibility that it was due to extraneous suppressor mutations. By screening several thousand *r*II mutants with spot tests, they found two distinct subsets of mutants that were active in one but not the other of the two strains, and a third subset active in another bacterial variant obtained by treating Benzer's strain with ethyl methane sulfonate. By the usual techniques, they were able to map the three subsets onto the A and B cistrons of the *r*II region (Benzer and Champe 1961).

Benzer and Champe called the new subsets "ambivalent *r*II mutants of phage T4." Impressed by the "extreme specificity" of the effects he had found, Benzer was drawn to the question of what "modifications in the cellular mechanism for translating genetic information" could account for the fact that a mutation in the bacterium could "lead to an altered response with respect to an entire subset of phage mutations." The answer, he felt, lay in the "specificities of the system for activating amino acids and attaching them to acceptor RNA." Consequently, he was finally "jarred out" of his long project to run the map into the ground and became "embroiled" instead in the biochemistry of RNA and the amino acid-activating enzymes (Benzer and Champe 1961, 1036–1038; Benzer 1966, 164).

EXPERIMENTAL SYSTEMS AND INVESTIGATIVE PATHWAYS

While recounting in 1966 his research on the mapping of the *r*II region, Benzer called it an example of Alfred Hershey's idea of scientific happiness: "to have one experiment that works, and keep doing it all the time" (Benzer 1966, 163). We might qualify and extend this insight by applying to Benzer's research trajectory the views about the role of experimental systems recently put forth by Hans-Jörg Rheinberger. Drawing on ideas earlier expressed by Ludwig Fleck, Rheinberger argues that an experimentalist does not deal with single experiments, but with "a whole experimental arrangement designed to produce knowledge that we do not yet have." The "experimental *system*" is "the smallest functional unit of research, designed to give answers to questions which we are not yet able clearly to ask" (Rheinberger 1992, 309).

Although Benzer did repeat many times what reasonably can be called a single experiment, in the sense that he carried out an identical sequence of procedures on many *r* mutants, these experiments did not remain constant, but evolved over the decade in which he pursued his project. The experiments remained embedded, however, within a recognizable system built around a small region in the genome of a particular bacteriophage, and the specific set of bacterial strains that provided his extremely sensitive test for a wild-type recombinant.

Within this system one source of change in the individual experiments was the continually growing number of isolated mutants. Their increase not only carried Benzer forward toward his goal of saturating the map, but enabled him to introduce increasingly powerful modifications of his initial methods. There were also qualitative changes in the types and uses for the mutants already identified in the mapping of further mutants. The nonreverting deletion mutants evolved from an obstacle to be avoided, to the basis for a preliminary test of the location of point mutations on the map, to the primary means of dividing the map up into segments small enough to examine the topology and topography of the map. The objective of Benzer's experiments evolved from the determination of the sizes of the elementary units of heredity, to tests for linearity, to exploration of the variability of mutation rates along the linear structure. All this occurred within the limits of an experimental system that remained stable in its overall outlines throughout these years.

In his own treatment of experimental systems as units of historical analysis, Rheinberger emphasizes the autonomy and the momentum they acquire, the ways in which they lead the investigators who wield them toward situations the investigators cannot anticipate. "Once the system can be handled reproducibly," he asserts, "its own momentum takes over. It forces researcher and research into a kind of *internal exclusion*. The more the experimenter learns to manipulate the system, the better the system realizes its internal capacities" (Rheinberger 1993, 465). Benzer's experience mapping the fine structure of the *r*II region illustrates both the force and the limitations of Rheinberger's viewpoint.

That Benzer's program acquired a momentum that kept him "occupied for ten years," as Delbrück predicted to him at the beginning that it would (Benzer 1966, 157), is self-evident. The more fully the region was mapped, the more powerful the incentive to continue using the system became. "Experimental systems have a life of their own" (Rheinberger 1993, 444), according to Rheinberger, and the T4 *r*II system did acquire such a life – a life that eventually extended beyond Benzer himself. Just as the exploitation of *Drosophila* by the Morgan school early in the century made it the organism of choice for many other geneticists, so the unique knowledge Benzer gained of the fine structure of the *r*II region invited other molecular geneticists at the end of the 1950s to choose this system for their own work. As Frank Stahl noted in his treatise *The Mechanics of Inheritance* in 1964, "Seymour Benzer recognized the virtues of the T4 *r*II mutants and set an extraordinary example of vigorous exploitation." Stahl then enumerated the uniqueness of the *r*II mutants that made them so important for testing mechanisms of mutation. Later Stahl returned to the subject with the comment that Benzer "led the way into the area of fine-structure recombination analysis in phage. Hundreds of the thousands of rII mutants isolated have been mapped" (Stahl 1964, 52, 128). Had Benzer found the unique way into this area, or did his originally contingent choice of an experimental system give rise to "the thinking machinery" (Rheinberger 1992, 307, footnote 9) that eventually overshadowed the intentions of its inventor and dictated the direction of movement?

In the case history that Rheinberger used to work out his conception of the experimental system, the objects of inquiry changed several times in ways that the investigators could not anticipate. The momentum of their experimental system led them in directions they could not foresee, caused them to drop the problems for which they had designed it, and to take up problems that had not been among their original intentions. In this instance it appears plausible to ascribe to "the dynamic body of knowledge, the network of practices structured by laboratories, instruments and experimental arrangements" the driving force behind the ongoing research trajectory. Despite the intrinsic robustness of Benzer's experimental system, its performance appears more clearly subordinate to his own primary intentions. What he achieved in the end was, by and large, what he had set out to do.

That is not to say that his experimental system led Benzer nowhere except where he had originally intended to go. The shifts from the size of the elementary units to the linearity and topography of the map were changes of direction made along the way. Were they the products of potentialities that unfolded from the nature of the system itself (realizations of the "internal capacities" of the system), or were they changes of objective that the investigator imposed on his system because of influences arising from outside it? Did he and his system together constitute the "thinking machinery" that carried them forward, or were his interactions with other scientists, such as Delbrück, and his own evolving perspective the directing forces?

Benzer, too, encountered unexpected phenomena that forced him to change course. The ambivalent mutants finally induced him to alter, more drastically than any prior events had, the direction of his investigation. This instance, however, illustrates more the limits on than the priority of experimental systems as directing forces. These mutants raised new questions for Benzer, but they were questions that his system was powerless to address. The result of the encounter was that Benzer left his experimental system for other systems rooted in another discipline – biochemistry – that he had until then been able to circumvent in his study of the genetics of bacteriophage. If systems have lives of their own, so do scientists. They may sometimes become trapped within their systems, but they retain the flexibility to escape from them when opportunity or need presses them to exert their independence from what they, themselves, have created.

BENZER AND THE DEFINITION OF THE GENE

By the time Benzer moved on from his mapping project in 1962, his approach was already being extended beyond phage genetics. Using a selective procedure analogous to that of Benzer for detecting rare recombinations among large populations, Arthur Chovnick and his group at Cold Spring Harbor were able to identify, in *Drosophila*, mutant markers that lay clearly "within a single cistron" (Chovnick et al. 1962). Benzer's results were at the same time being added to the canonical structure of genetics. Discussing the "Topology and Topography of the Phage Genetic Map" in his authoritative review of the "Genetics of Bacteriophage" in 1962, Salvador Luria wrote that "The key system for this study has been Benzer's *r*II region in phage T4," and proceeded to describe the "essential features" of this "powerful system" (Luria 1962, 207). The implications of the study were quickly generalized. The introductory textbook *Genetics*, published by Robert King in 1962, asserted that "The subdivision by Benzer of the *r*II region of the T_4 bacteriophage chromosome into two multirecon cistrons constitutes one of the major advances in genetics of recent years." He devoted three pages to a summary of the experiments, including a reproduction of Benzer's schematic drawing of the topography of the *r*II region. Adopting Benzer's definitions of the *recon, muton,* and *cistron,* King adduced evidence that in *Drosophila,* too, "Many instances are now known where a cistron is composed of several recons" (King 1962, 233–240).

A preliminary sampling of genetics textbooks of the following years suggests that Benzer's definitions occasioned a widespread reassessment of the concept of the gene. George Burns's *The Science of*

Genetics introduced a chapter on the "Molecular Structure of the Gene" with the question "How much of the nucleotide sequence comprises a gene?" Working up from theories of mutation derived from the structure of the DNA base pairs, Burns then analyzed the "Fine Structure of the Gene" in terms of the cistron, the muton, and the recon. He discussed the relation between Benzer's categories, functional definitions based on the sequence of amino acids and hemoglobin, and other considerations external to Benzer's own experimental system, before turning to a description of the A and B cistrons of the *r*II region (Burns 1969, 291–304).

In the *Biology of the Gene,* published (in second edition) in 1973, Louis Levine introduced a chapter on "Gene concepts," only after eight chapters on classical genetics. It seems evident, Levine wrote, "that we may examine the gene from three points of view: (1) as a mutational unit, (2) as a functional unit, and (3) as a recombination unit." He, too, discussed the mutational unit in terms of the theory of base pairing, then invoked the "studies on *r*II mutants . . . of phage T4" as "the experimental evidence" for these transitions (Levine 1973, 185–201).

Benzer's work on the "T4 *r*II locus" also provided "much of the knowledge" on which Levine based his discussion of the gene as a functional unit, including the evidence "for our present ideas on its spatial limits." After reviewing Benzer's experiments, Levine concluded that "a cistron is that section of the DNA molecule which specifies the composition of a particular polypeptide chain" (Levine 1973, 197–199).

Benzer's definitions of the units comprising the fine structure of the gene thus became firmly embedded into the conceptual structure of genetics during the 1960s; but the terminology with which he introduced these concepts attained only a tenuous hold in the discipline. Even those authors who repeated his names seemed to question their utility. In 1962, for example, King ended his discussion of the subject with the comment that "The current tendency is to equate the terms gene and cistron" (King 1962, 240). Even though Burns headed his descriptions of the three units in 1969 with Benzer's terms, he remarked that "crossing-over occurs within a functional gene or, if we use Benzer's concept, a cistron" (Burns 1969, 301). Levine mentioned the cistron without adopting it, stating that

"A cistron is, in effect, the gene when it is viewed as a functional unit" (Levine 1973, 199). In *Molecular Biology: An Introduction to Chemical Genetics,* J. M. Barry wrote in 1973 that:

Benzer . . . tried to introduce a term less ambiguous than gene. He called the two regions *cistrons.* . . . However, *cistron,* like most scientific terms, is not entirely without ambiguity, and *gene* has tended to persist and usually refers to what Benzer meant by cistron. (Barry 1973, 66)

Intending to introduce more sharply defined concepts with new names to replace a name and a concept that seemed too diffuse to withstand the analysis of its fine structure, Benzer succeeded instead in capturing the old term for his new concepts. Although the word *cistron* continues to appear on occasion in scientific texts when authors feel the need to specify the particular aspect of the gene for which he had proposed that term (Benzer 1999), the continued dominance of the all-purpose "gene" in the basic language of genetics has left the lasting effect of his achievement less than fully visible.

Definitions of the gene and its subunits based on and illustrated by the genetic mapping of a small region of a bacteriophage genome could only appear paradigmatic for genetics as a whole so long as the simplified genetics of phage and bacteria were taken as the models from which general principles, applicable also to the genetics of higher organisms, could be inferred. Since the 1970s, genetic research has focused increasingly on eukaryotic organisms, revealing ever more complexity differentiating them from those earlier models. It is not coincidental that Maxine Singer and Paul Berg's magisterial text *Genes and Genomes,* which is strongly oriented to eukaryotic genes and gene expression, bears little explicit trace of the great impact that Benzer's work exerted on the molecular genetics of thirty years ago (Singer and Berg 1991). Had his terminology been more generally accepted by those who adopted his concepts then, the situation today might be different.

We have seen that Benzer borrowed three definitions of the gene from a classical geneticist who did not believe that the gene was a molecular edifice, and attached those definitions precisely to contiguous lengths of the subunits of a macromolecule. A quarter-century later, the gene is again viewed as a "general, open, and abstract" concept resembling more closely in some respects Pontecorvo's than Benzer's view (Portin 1993, 207). Although not in the same way that Delbrück intended, it turned out in some sense, as he predicted, that the relation between the two "sides of the picture, chemical structure *and* genetic map," is not clearly visualizable. (Delbrück 1957b). Does that mean, as Portin suggests, that Benzer's work belonged to an era of the "neoclassical view of the gene" that has ended (Portin 1993, 186–191)?

The matter is more subtle. The recent developments that have complicated the "modern view" of the gene have demonstrated that not all genes can be contiguous stretches of DNA transcribed as single units into messenger RNA and coding for a single polypeptide. They have not undermined the validity of Benzer's identification of the A and B cistrons of the *r*II region of T4, as functional genetic units, but only the scope of the generalizations that were once extrapolated from this simple case. The meanings that Benzer incorporated into three new terms – the *cistron, recon,* and *muton* – altered instead the meaning of the older, more general concept of the gene. These alterations still did not incorporate complexities later discovered, which have continued to modify what is meant by a gene. That is true also of other basic concepts in biology. Any given cell is more complex than the idealized cell used to define the cell theory, and every general postulate about the character of cells is violated by some cells. Repeatedly in the past, biologists have argued that the cell theory had lost its usefulness, yet the cell continues to be one of the principal organizing frameworks for biology. So, too, will the gene.

REFERENCES

Ayala, F. J., and J. A. Kiger. 1984. *Modern Genetics.* 2nd ed. Menlo Park, CA: Benjamin Cummings, p. 157.

Barry, J. M. 1973. *Molecular Biology: An Introduction to Chemical Genetics.* Englewood Cliffs, NJ: Prentice-Hall.

Benzer, S. 1952–1953. To M. Delbrück; September 26, 30, 1952, October 27, 1952, January 23, 1953: Max Delbrück Collection, Caltech Archive.

Benzer, S. 1954. To Gunther Stent; July 17, 1954: Stent Collection, University of California Archive, Berkeley.

Benzer, S. 1955a. To Delbrück, M.; February 3, 1955, March 5, 1955: Max Delbrück Collection, Caltech Archive.

Benzer, S. 1955b. Fine structure of a genetic region in bacteriophage. *Proceedings of the National Academy of Sciences USA* 41: 344–354.

Benzer, S. 1957. The elementary units of heredity. In *A Symposium on the Chemical Basis of Heredity*, edited by W. D. McElroy and B. Glass. Baltimore: The Johns Hopkins Press, pp. 70–133.

Benzer, S. 1959. On the topology of the genetic fine structure. *Proceedings of the National Academy of Sciences USA* 45: 1607–1620.

Benzer, S. 1961. On the topography of the genetic fine structure. *Proceedings of the National Academy of Sciences USA* 47: 403–415.

Benzer, S. 1966. Adventures in the rII region. In *Phage and the Origins of Molecular Biology*, edited by J. Cairns, G. S. Stent, and J. D. Watson. Cold Spring Harbor: Cold Spring Harbor Laboratory of Quantitative Biology, pp. 157–165.

Benzer, S. 1999. To F. L. Holmes, September 14.

Benzer, S., and S. P. Champe. 1961. Ambivalent rII mutants of phage T4. *Proceedings of the National Academy of Sciences USA* 47: 1025–1038.

Benzer, S., and E. Freese. 1958. Induction of specific mutations with 5-bromouracil. *Proceedings of the National Academy of Sciences USA* 44: 112–119.

Bonner, D. M. 1956. The genetic unit. *Cold Spring Harbor Symposia on Quantitative Biology* 21: 163–170.

Burns, G. W. 1969. *The Science of Genetics: An Introduction to Heredity*. New York: The Macmillan Company.

Carlson, E. A. 1966. *The Gene: A Critical History*. Philadelphia: W. B. Saunders.

Chovnick, A., A. Schalet, R. P. Kernaghan, and J. Talsma. 1962. The resolving power of genetic fine structure analysis in higher organisms as exemplified by Drosophila. *American Naturalist* 46: 281–296.

Crick, F. H. C. 1958. On protein synthesis. *Symposia of the Society for Experimental Biology* 12: 138–163.

Delbrück, M. 1955. To Seymour Benzer; February 8: Max Delbrück Collection, Caltech Archive.

Delbrück, M. 1957a. To B. Davis; April 12: Max Delbrück Collection, Caltech Archive.

Delbrück, M. 1957b. Atomic physics in 1910 and molecular biology in 1957 (unpublished manuscript). Max Delbrück Collection, Caltech Archive 35.2.

Demerec, M. 1956. A comparative study of certain gene loci in *Salmonella*. *Cold Spring Harbor Symposia on Quantitative Biology* 21: 113–121.

Doermann, A. H., and M. B. Hill. 1952. Genetic structure of bacteriophage T4 as described by recombination studies of factors influencing plaque morphology. *Genetics* 38: 79–90.

Falk, R. 1986. What is a gene? *Studies in the History and Philosophy of Science* 17: 133–173.

Falk, R. 1996. Taped recording of discussions at workshop "Gene concepts in development and evolution, II." Berlin: Max Planck Institute for the History of Science, October 17–19, 1996.

Hershey, A. D. 1946. Mutation of bacteriophage with respect to type of plaque. *Genetics* 31: 620–640.

Hershey, A. D., and M. Chase. 1951. Genetic recombination and heterozygosis in bacteriophage. *Cold Spring Harbor Symposia on Quantitative Biology* 16: 471–479.

Hershey, A. D., and R. Rotman. 1948. Genetic recombination between host-range and plaque-type mutants of bacteriophage in single bacterial cells. *Genetics* 34: 44–71.

Judson, H. F. 1979. *The Eighth Day of Creation: Makers of the Revolution in Biology.* New York: Simon and Schuster.

King, R. C. 1962. *Genetics.* New York: Oxford University Press.

Levine, L. 1973. *Biology of the Gene.* Saint Louis: C.V. Mosby.

Lewis, E. B. 1951. Pseudoallelism and gene evolution. *Cold Spring Harbor Symposia on Quantitative Biology* 16: 159–174.

Lewis, E. B. 1955. Some aspects of position pseudoallelism. *American Naturalist* 89: 73–89.

Luria, S. E. 1962. Genetics of bacteriophage. *Annual Review of Microbiology* 16: 205–240.

McClintock, B. 1944. The relation of homozygous deficiencies to mutations and allelic series in maize. *Genetics* 29: 478–502.

Pontecorvo, G. 1952. Genetic formulation of gene structure and gene action. *Advances in Enzymology* 13: 121–149.

Pontecorvo, G. 1955. Gene structure and action in relation to heterosis. *Proceedings of the Royal Society of London. Series B.* 144: 171–177.

Pontecorvo, G. 1956. Allelism. *Cold Spring Harbor Symposia on Quantitative Biolgy* 21: 172–174.

Pontecorvo, G., and J. A. Roper. 1956. Resolving power of genetic analysis. *Nature* 178: 83–84.

Portin, P. 1993. The concept of the gene: Short history and present status. *The Quarterly Review of Biology* 68: 173–223.

Rheinberger, H.-J. 1992. Experiment, difference, and writing: I. Tracing protein synthesis. *Studies in the History and Philosophy of Science* 23: 305–331.

Rheinberger, H.-J. 1993. Experimentation and orientation: Early systems of in vitro protein synthesis. *Journal of the History of Biology* 26: 443–471.

Singer, M., and P. Berg. 1991. *Genes and Genomes: A Changing Perspective.* Mill Valley, CA: University Science Books.

Stahl, F. W. 1964. *The Mechanics of Inheritance.* Englewood Cliffs, NJ: Prentice-Hall.

Watson, J. D. 1965. *Molecular Biology of the Gene.* New York: W.A. Benjamin.

3

Genetic Programs and Developmental Genes

7

Decoding the Genetic Program

Or, Some Circular Logic in the Logic of Circularity

EVELYN FOX KELLER

ABSTRACT

Taken as a composite, the meaning of the composite term *genetic program* – offered (or taken for granted) as the name of an explanatory theory of developmental biology – simultaneously depends upon and underwrites the particular presumption that a "plan of procedure" for development is itself written in the sequence of nucleotide bases. Is this presumption correct? Certainly, it is almost universally taken for granted, but I want to argue that, at best, it is misleading, and at worst, simply false: To the extent that we may speak at all of a developmental program, or of a set of instructions for development, in contradistinction to the data or resources for such a program, current research obliges us to acknowledge that these instructions are not written into the DNA itself (or at least, are not all written in the DNA), but rather are distributed throughout the fertilized egg. Indeed, if the distinction between program and data is to have any meaning in biology, it has become abundantly clear that it does not align (as had earlier been assumed) either with a distinction between "genetic" and "epigenetic," or with the precursor distinction between nucleus and cytoplasm. I want to suggest that the notion of genetic program depends upon, and sustains, a fundamental category error in which two independent distinctions, one between genetic and epigenetic, and the other, between program and data, are pulled into mistaken alignment. The net effect of such alignment is to reinforce two outmoded associations: on the one hand, between genetic and active, and, on the other, between epigenetic and passive.

INTRODUCTION

Not long ago, in a lecture on the continuing conceptual gaps between genetics and development, I suggested that we still have no adequate theory to explain the emergence of phenotype from genotype, i.e., no adequate theory for developmental biology. Whereupon a prominent biologist in the audience, unmistakably angry, stood up to say, "We most certainly do – it is called 'Development'!" We laugh, of course. But if he had said "it is called the genetic program," we wouldn't regard his response as quite so much of a joke. Quite simply, the concept of a developmental program written in the genome – i.e., of the genetic program – has come to be widely regarded as a fundamental explanatory concept in developmental biology. Indeed, it is a mainstay of the molecular biology of development, referred to ubiquitously in both the popular and the scientific literature. My fear, however, is that should we ask "What exactly is a genetic program?" we would soon find ourselves in much the same boat as if we had taken "development" as our explanation, confusing explanation with that which is to be explained.

So far as I have been able to determine, the term *genetic program* appears neither in indexes, in dictionaries, nor as a keyword in library data bases. Each of its components has a present-day definition in biology, and along with that definition, at least a tacit present-day contrast: What I want to explore here is how the histories of these two contrasts contribute to the apparent efficacy of the composite today. Since 1953, the term *genetic* has come to be widely understood as referring to the sequence of nucleotide bases in the DNA, with its contrast term, epigenetic, referring to everything else in the cell, organism, or even environment. "Program," even without the help of computer science, is defined by Webster as "a plan of procedure" (in contrast with both its execution and the resources for that execution). The advent of computers adds the more specific meaning of "a sequence of coded instructions" that can be fed into a computer, and the contrast it invokes is similarly dual – depending on context, either machine or data.[1] The notion of a biological program is avowedly metaphorical, for what could be more teleological than a "plan" or "set of instructions"? But metaphors, like word histories, have consequences, and the more commonplace the metaphor, the more covert its effects. Thus, it is one of the aims of this paper to show how the behavior of the composite term *genetic program* is colored by the memory of past debates, and to argue that those effects have now become scientifically counterproductive.

Development results from the temporally and spatially specific activation of particular genes, which, in turn, depends on a vastly complex network of interacting components including not only the "hereditary codescript" of the DNA, but also a densely interconnected cellular machinery made up of proteins and RNA molecules. Necessarily, each of these systems functions in relation to the others alternatively as data and as program. If development cannot proceed without the "blueprint" of genetic memory, neither can it proceed without the "machinery" embodied in cellular structures. To be sure, the elements of these structures are fixed by genetic memory, but their assembly is dictated by cellular memory.[2] As Richard Lewontin states,

The linear sequence of nucleotides in DNA is used by the machinery of the cell to determine what sequence of amino acids is to be built into a protein, and to determine when and where the protein is to be made. But the proteins of the cell are made by other proteins, and without that protein-forming machinery *nothing* can be made. There is an appearance here of infinite regress . . . but this appearance is an artifact of another error of vulgar biology, that it is only the genes that are passed from parent to offspring. In fact, an egg, before fertilization, contains a complete apparatus of production deposited there in the course of its cellular development. We inherit not only genes made of DNA but an intricate structure of cellular machinery made up of proteins. (Lewontin 1992, 33)

Assuming one is not misled by Lewontin's colloquial use of the term *inherit* to refer to transmission over a single generation (as distinct from multigenerational transmission), none of this is either controversial or news, nor does it depend on the extraordinary techniques now available for molecular analysis. Yet, however surprisingly, it is only within the last decade or two that the developmental and evolutionary implications of so-called *maternal effects* have begun to be appreciated.[3] Current research is now providing us with the kind of detail about the mechanisms involved in the processing of genetic data that make the errors of what Lewontin calls "vulgar biology" manifest. Yet, even when elaborated by the kind of detail we now have available, such facts are still not sufficient to dislodge the confidence that many distinguished biologists continue to have in both the meaning and explanatory force of the genetic program. The question I want, therefore, to ask is twofold: First, when and how did the presumption built into the very term *genetic program* come to seem so self-evident? And second, what grants it its apparent explanatory force, even in the face of such obvious caveats as those above? I begin with the first question, namely, that of origins.

"PROGRAMS" IN THE BIOLOGICAL LITERATURE OF THE 1960s

The metaphor of a *program*, borrowed directly from computer science, entered the biological literature in the 1960s not once, but several times, and in at least two distinctly different registers. In its first introduction, simultaneously by Mayr (1961) and by Monod and Jacob (1961), the locus of the program was explicitly identified as the genome, but, over the course of that decade, another notion of program, a *developmental program*, also surfaced, and repeatedly so. This program was not located in the genome, but instead, distributed throughout the fertilized egg (see, e.g., Apter 1966). By the 1970s, however, the program for development had effectively collapsed into a genetic program, with the alternative, distributed sense of a developmental program all but forgotten.

François Jacob, one of the earliest to use the concept of a genetic program, contributed crucially to its popularization. In *The Logic of Life*, first published in 1970, Jacob describes the organism as "the realization of a programme prescribed by its heredity" (Jacob 1976, 2), claiming that, "when heredity is described as a coded programme in a sequence of chemical radicals, the paradox [of development] disappears" (Jacob, 1976, 4). For Jacob, the genetic program, written in the alphabet of nucleotides, is what is responsible for the apparent purposiveness of biological development; it and it alone gives rise to "the order of biological order" (Jacob 1976, 8). He refers to the oft-quoted characterization of teleology as a "mistress" whom biologists "could not do without, but did not care to be seen with in public," and writes, "The concept of programme has made an honest woman of teleology" (Jacob 1976, 8–9). Although Jacob does not exactly define the term, he notes that "[t]he programme is a model borrowed from electronic computers. It equates the genetic material of an egg with the magnetic tape of a computer" (Jacob 1976, 9).

However, equating the genetic material of an egg with the magnetic tape of a computer does not imply that that material encodes a program; it might just as well be thought of as encoding data to be processed by a cellular program. Or by a program residing in the machinery of transcription and translation complexes. Or by extranucleic chromatin structures in the nucleus. Computers have provided a rich source of metaphors for molecular biology, but they cannot by themselves be held responsible for the notion of genetic program. To be sure, the informational content of the DNA is essential – without it development (life itself) cannot proceed. But for many developmental processes, it is far more appropriate to refer to this informational content as data rather than as program (Atlan and Koppel 1990). Indeed, as already indicated, other, quite different, uses of the program metaphor for biological development were already in use. One such use was in the notion of a *developmental program* – a term that surfaced repeatedly through the 1960s, and that stood in notable contrast to that of a genetic program.

Let me give an example of this alternative use. In 1965, a young graduate student, Michael Apter, steeped in information theory and cybernetics, teamed up with the developmental biologist Lewis Wolpert to argue for a direct analogy not between computer programes and the genome, but between computer programes and the egg:

if the genes are analogous with the sub-routine, by specifying how particular proteins are to be made . . . then the cytoplasm might be analogous to the main programme specifying the nature and sequence of operations, combined with the numbers specifying the particular form in which these events are to manifest themselves . . . In this kind of system, instructions do not exist at particular localized sites, but the system acts as a dynamic whole. (Apter and Wolpert 1965, 257)

The following year, Apter (1966) published a book-length elaboration of the argument (based on his doctoral dissertation in psychology) titled *Cybernetics and Development*. This work may well have been the first comprehensive application of automata theory to biological development, and, ironically perhaps, it can be seen as an important precursor to current work on genetic algorithms and Artificial Life (about which more to be discussed later). But it was not unique; during the 1960s, a number of developmental biologists attempted to employ ideas from cybernetics to illuminate development, and almost all shared Apter's starting assumptions (see Keller

1995, chapter 3 for examples) – i.e., they located the program (or instructions) for development in the cell as a whole.

The difference in where the program is said to be located is crucial, for it bears precisely on the controversy over the adequacy of genes to account for development that had been raging among biologists since the beginning of the century (see Keller 1995, chapter 1). By the beginning of the 1960s, this debate had subsided, largely as a result of the eclipse of embryology as a discipline during the 1940s and 1950s. Genetics had triumphed, and after the identification of DNA as the genetic material, the successes of molecular biology had vastly consolidated that triumph. Yet the problems of development, still unresolved, lay dormant. Molecular biology had revealed a stunningly simple mechanism for the transmission and translation of genetic information, but, at least until 1960, it had been able to offer no account of developmental regulation, of how different genes come to be activated at different times and different places in the developing embryo.

James Bonner, professor of biology at Cal Tech, in an early attempt to bring molecular biology to bear on development, put the problem well. Granting that "the picture of life given to us by molecular biology . . . applies to cells of all creatures," he goes on to observe that this picture

is a description of the manner in which all cells are similar. But higher creatures, such as people and pea plants, possess different kinds of cells. The time has come for us to find out what molecular biology can tell us about why different cells in the same body are different from one another, and how such differences arise. (Bonner 1965, v)

Bonner's own work was on the biochemistry and physiology of regulation in plants, in an institution well known for its importance in the birth of molecular biology (see, e.g., Kay 1992). In this work published in 1965, like Apter and a number of others of that period, Bonner employs the conceptual apparatus of automata theory to deal with the problem of developmental regulation. But unlike them, he does not locate the program in the cell as a whole, but rather, in the chromosomes, and more specifically in the genome. Indeed, he begins with the by then standard credo of molecular biology, asserting that "[w]e know that . . . the directions for all cell life [are] written in the DNA of their chromosomes" (Bonner 1965, v). Why? An obvious answer is suggested by his location. Unlike Apter and other developmental biologists of the time, Bonner was situated at a major thoroughfare for molecular biologists, and it is hard to imagine that he was uninfluenced by the enthusiasm of his colleagues at Cal Tech. In any case, Bonner's struggle to reconcile the conceptual demands posed by the problems of developmental regulation with the received wisdom among molecular biologists is at the very least instructive, especially given its location in time, and I suggest it is worth examining in some detail for the insight it has to offer on our question of how the presumption of a genetic program came – in fact, over the course of that very decade – to seem self-evident. In short, I want to take Bonner as representative of a generation of careful thinkers about an extremely difficult problem who opted for this (in retrospect, inadequate) conceptual shortcut.

EXPLANATORY LOGIC OF THE GENETIC PROGRAM

From molecular biology, Bonner inherited a language encoding a number of critical if tacit presuppositions. That language shapes his efforts in decisive ways. Summarizing the then current understanding of transcription and translation, he writes:

Enzyme synthesis is therefore an information-requiring task and . . . the essential information-containing component is the long punched tape which contains, in coded form, the instructions concerning which amino acid molecule to put next to which in order to produce a particular enzyme. (Bonner, 1965, p. 3)

The principal keywords – prior to the term *genetic program* – are *in formation, instruction,* and *code.* To his credit, Bonner recognizes that premolecular genetics had already provided language to much the same effect:

We have in a sense known this since 1941 when Beadle and Tatum showed that for each enzyme . . . there is a gene . . . which *specifies* that enzyme. Or, to put it the other way round, that the function of each gene is to *supervise* the production of a particular kind of enzyme. (Bonner 1965, 4, my italics)

Although Bonner does not seem to take note of the conceptual work performed in the transition from "information" to "instruction," or from "specify" to "supervise," he does clearly recognize that, to date, only the composition of the protein had been accounted for, and not the regulation of its production required for the formation of specialized cells, i.e., cell differentiation remained unexplained. As he wrote,

Each kind of specialized cell of the higher organism contains its characteristic enzymes but each produces only a portion of all the enzymes for which its genomal DNA contains information. (Bonner 1965, 6)

But he continues:

Clearly then, the nucleus contains some further mechanism which determines in which cells and at which times during development each gene is to be active and produce its characteristic messenger RNA, and in which cells each gene is to be inactive, to be repressed. (Bonner 1965, 6)

Two important moves have been made here. Bonner argues that something other than the information for protein synthesis encoded in the DNA is required to explain cell differentiation (and this is his main point), but on the way to making this point, he has placed this "further mechanism" in the nucleus, with nothing more by way of argument or evidence than his "clearly then." Why does such an inference follow? And why does it follow "clearly"? Perhaps the next paragraph will help:

The egg is activated by fertilization . . . As division proceeds cells begin to differ from one another and to acquire the characteristics of specialized cells of the adult creature. There is then within the nucleus some kind of programme which determines the property [sic] sequenced repression and derepression of genes and which brings about orderly development. (Bonner 1965, 6)

Here, the required further mechanism is explicitly called a program and once again, it is located in the nucleus. But this time around, a clue to the reasoning behind the inference has been provided in the first sentence, "The egg is activated by fertilization." This is how I believe the (largely tacit) reasoning goes: If the egg is "activated by

166

fertilization," the implication is that it is entirely inactive prior to fertilization. What does fertilization provide? The entrance of the sperm, of course, and unlike the egg, the sperm has almost no cytoplasm: it can be thought of as pure nucleus. Ergo, the active component must reside in the nucleus and not in the cytoplasm. Today, the supposition of an inactive cytoplasm would be challenged, but in Bonner's time, it would have been taken for granted as a carryover from what I have called the discourse of gene action of classical genetics (Keller 1995). And even then, it might have been challenged had it been made explicit, but as an implicit assumption encoded in the language of activation, it was likely to go unnoticed by Bonner's readers as by Bonner himself.

Bonner then goes on to ask the obvious questions:

What is the mechanism of gene repression and derepression which makes possible development? Of what does the programme consist and where does it live? (Bonner 1965, 6)

And answers them as best he can:

We can say that the programme which sequences gene activity must itself be a part of the genetic information since the course of development and the final form are heritable. Further than this we cannot go by classical approaches to differentiation. (Bonner 1965, 6)

In these few sentences, Bonner has completed the line of argument leading him to the conclusion that the program *must* be part of the genetic information, i.e., to the genetic program. And again, we can try to unpack his reasoning. Why does the heritability of the course of development and the final form imply that the program must be part of the genetic information? Because – and only because – of the unspoken assumption that it is only the genetic material that is inherited. The obvious fact – that the reproductive process passes on (or transmits) not only the genes but also the cytoplasm (the latter through the egg for sexually reproducing organisms) – is not mentioned. But even if it were, this fact would almost certainly be regarded as irrelevant, simply because of the prior assumption that the cytoplasm contains no active components. The conviction that the cytoplasm could neither carry nor transmit effective traces of intergenerational memory had been a mainstay of genetics for so long that it had become part of the memory of that discipline, working silently but effectively to shape the very logic of inference employed by geneticists.

Yet another ellipsis becomes evident (now, even to Bonner himself) as he attempts to integrate his own work on the role of histones in genetic regulation. Not all copies of a gene (or a genome) are in fact the same: Because of the presence of proteins in the nucleus, capable of binding to the DNA, "in the higher creature, if it is to be a proper higher creature, one and the same gene must possess different attributes, different attitudes, in different cells" (Bonner 1965, 102). The difference is a function of the histones. How can we reconcile this fact with the notion of a genetic program? There is one simple way, and Bonner takes it – namely, to elide the distinction between genome and chromosome. The genetic program is saved (for this discussion) by just a slight shift in reference: Now it refers to a program built into the chromosomal structure – i.e., into the complex of genes and his tones, where that complex is itself referred to as the genome.

But the most conspicuous inadequacy of the location of the developmental program in the genetic information becomes evident in the chapter in which Bonner attempts to sketch out an actual computer program for development; it is called "Switching Networks for Developmental Processes." Here, the author reframes what is known about the induction of developmental pathways in terms of a master program, proposing to "consider the concept of the life cycle as made up of a master programme constituted in turn of a set of subprogrammes or subroutines" (Bonner 1965, 134). Each subroutine specifies a specific task to be performed – for a plant, his list includes: cell life, embryonic development, how to be a seed, bud development, leaf development, stem development, root development, reproductive development. Within each of these subroutines is a list of cellular instructions or commands, such as, e.g., "divide tangentially with growth"; "divide transversely with growth"; "grow without dividing"; "test for size or cell number"; etc. (Bonner 1965, 137). He then asks the next obvious question: "[H]ow might these subroutines be related to one another? Exactly how are they to be wired together to constitute a whole programme?" (Bonner 1965, 135). Yet nowhere in the text is this question answered. We might say, fortunately not. For if it had been, the answer would have necessarily undermined Bonner's core assumption. To see this, two points emerging from his discussion need to be underscored: First, the list of subroutines, although laid out in a linear sequence – as if following from an initial master program, actually constitutes a circle, as indeed they must if they are to describe a life cycle. The master program is in fact nothing but this composite set of programs, wired together in a structure exhibiting the characteristic cybernetic logic of *circular causality.*

The second point bears on Bonner's earlier question, "Of what does the programme consist and where does it live?" The first physical structures that were built to embody the logic of computer programs were built out of electrical networks[4] (hence the term *switching networks*), and this is Bonner's frame of reference. As he writes, "[t]hat the logic of development is based upon [a developmental switching] network, there can be no doubt" (Bonner 1965, 148). But what would serve as the biological analogue of an electric (or electronic) switching network? How are the instructions specified in the subroutines that comprise the life cycle actually embodied? Given the dependence of development on the regulating activation of particular genes, Bonner reasonably enough calls the developmental switching network a *genetic switching network.* But this does not quite answer our question, rather it obfuscates it. The clear implication is that such a network is constituted of nothing but genes, whereas in fact, many other kinds of entities also figure in this network, all playing critical roles in the control of genetic activity. Bonner himself writes of the roles played by histones, hormones, and RNA molecules; today, the list has expanded considerably to include enzymatic networks, metabolic networks, transcription complexes, signal transduction pathways, etc., with many of these additional factors embodying their own switches. We could of course still refer to this extraordinarily complex set of interacting, controlling factors as a genetic switching network – insofar that the regulation of gene activation remains central to development – but only if we can manage to avoid the implication (an implication tantamount to a category error) that that network is embodied in and by the genes themselves.

Indeed, it is this category error that confounds the very notion of a genetic program. If we were now to ask Bonner's question, "Of what does the programme consist and where does it live?" we would have to say, just as Apter saw long ago, that it consists not of particular gene entities, and lives not in the genome itself, but of and in the cellular machinery integrated into a dynamic whole. In current jargon, it would seem more reasonable to describe the fertilized egg as a massively parallel processor in which programs (or networks) are distributed throughout the cell.[5] The roles of "data" and "program" here are relative, for what counts as data for one program is often the output of a second program, and the output of the first is 'data' for yet another program, or even for the very program that provided its own initial data. Thus, for some developmental stages, the DNA might be seen as encoding programs or switches that process the data provided by gradients of transcription activators, or alternatively, one might say that DNA sequences provide data for the machinery of transcription activation (some of which is acquired directly from the cytoplasm of the egg). In later developmental stages, the products of transcription serve as data for splicing machines, translation machines, etc. In turn, the output of these processes make up the very machinery or programs needed to process the data in the first place. Sometimes, this exchange of data and programs can be represented sequentially, sometimes simultaneously.

When, in the mid 1960s, Bonner, Apter, and others were attempting to represent development in the language of computer programs, automata theory was in its infancy, and cybernetics was at the height of its popularity. During the 1970s and 1980s, these efforts lay forgotten: Cybernetics had lost its appeal to computer scientists and biologists alike, and molecular biologists found they had no need of such models. The mere notion of a genetic program sufficed by itself to guide their research. Today, however, provoked in large part by the construction of hard-wired parallel processors, the project to simulate biological development on the computer has returned in full force, and in some places has become a flourishing industry. It goes by various names – Artificial Life, adaptive complexity, or genetic algorithms. But what is a genetic algorithm? Like Bonner's subroutines, it is "a sequence of computational operations needed to solve a problem" (see, e.g., Emmeche 1994). And once again, we need to ask, why "genetic"? Furthermore, not only are the individual algorithms referred to as genetic, but "In the fields of genetic algorithms and artificial evolution, the [full] representation scheme is often called a 'genome' or 'genotype'" (Fleischer 1995, 1). And, in an account of the sciences in *Complexity* written for the lay reader, Mitchell Waldrop quotes Chris Langton, the founder of *Artificial Life*, as saying:

> [Y]ou can think of the genotype as a collection of little computer programs executing in parallel, one program per gene. When activated, each of these programs enters into the logical fray by competing and cooperating with all the other active programs. And collectively, these interacting programs carry out an overall computation that is the phenotype: the structure that unfolds during an organism's development. (Waldrop 1992, 194)

Workers in the field well understand, and when pressed, readily acknowledge, that the biological analogues of these computer programs are not in fact *genes* (at least as the term is used in biology), but complex biochemical structures or networks comprised of proteins, RNA molecules, and metabolites that often, although certainly not always, execute their tasks in interaction with particular stretches of DNA.[6] Artificial Life's "genome" typically consists of instructions such as "reproduce," "edit," "transport," or "metabolize," and the biological instantiation of these algorithms is found not in the nucleotide sequences of DNA, but in specific kinds of cellular machinery such as transcription complexes, spliceosomes, and metabolic networks. Why then are they called "genetic," and why is the full representation called a genome? I suggest that the primary justification for such terminology is that it readily follows usage of the term *genetic program* already acquired in genetics.

In short, words have a history, and their usage depends on this history, as does their meaning. But history does not fix meaning; rather, it builds into words a kind of memory. In the field of genetic programming, "genes" have come to refer, not to particular sequences of DNA, but to the computer programs required to execute particular tasks (as Langton puts it, "one program per gene"); yet, at the same time, the history of the term ensures that the word *gene*, even as adapted by computer scientists, continues to carry its original meaning. And perhaps most importantly, that earlier meaning remains available for deployment whenever it is convenient to do so.

Much the same can be said for the use of the terms *gene* and *genetic programs* by geneticists. I have taken some time in examining Bonner's argument for 'genetic programs', not because his book played a major role in establishing the centrality of this notion in biological discourse, but rather because of the critical moment in time it was written and because of the relative accessibility of the kinds of slippage on which his argument depends. The very first use of the term *program* that I have been able to find in the molecular biology literature appeared only four years earlier.[7] In 1961, Jacob and Monod published a review of their immensely influential work on a genetic mechanism for enzymatic adaptation in *E. coli*. This was the first mechanism for regulation to be identified, and even though pertaining only to the regulation of protein synthesis in a single-celled organism that does not undergo developmental differentiation, it was (with the explicit encouragement of the authors) widely acclaimed as a "resolution" of the paradox of differentiation that had for so long divided embryology from genetics (Monod and Jacob 1961, 397). In this work, Monod and Jacob found a genetic structure that could be characterized as a molecular switch, triggered by the presence or absence of the product of a particular gene, and they rapidly assimilated all possible regulatory mechanisms to this model. The introduction of the term *program* appears in their concluding sentence:

The discovery of regulator and operator genes, and of repressive regulation of the activity of structural genes, reveals that the genome contains not only a series of blueprints, but a coordinated program of protein synthesis and the means of controlling its execution. (Jacob and Monod 1961, 354)

Three decades later, Sydney Brenner refers to the belief "that all development could be reduced to [the operon] paradigm" – that "[i]t was simply a matter of turning on the right genes in the right places at the right times" – in rather scathing terms. As he puts it, "[o]f course, while absolutely true this is also absolutely vacuous. The paradigm does not tell us how to make a mouse but only how to make a switch" (Brenner et al. 1990, 485).[8] And even in the first flush of enthusiasm, not everyone was persuaded of the adequacy of this particular regulatory mechanism to explain development.[9] Lewis Wolpert, e.g., wrote in 1969, "Dealing as it does with intracellular regulatory phenomena, it is not directly relevant to problems where the cellular bases of the phenomena are far from clear" (Wolpert 1969, 2–3). In those days, Wolpert seemed certain that an understanding of development required a focus not simply on genetic information, but also on cellular mechanisms. But by the mid 1970s, even Wolpert had been converted to the notion of a genetic program (see, Wolpert and Lewis 1975).

What carried the day? Certainly not more information about actual developmental processes. Rather, I suggest, it was the consonance of this formulation with the prior history of genetic discourse, fortified both by the rhetorical links forged with the new science of computers and by frequent reassertion by figures of authority. Jacob's *Logic of Life* was of key importance. To provide support for the concept of genetic program, Jacob invokes the authority of both Schrödinger and Wiener, managing the transition from past to future metaphors with finesse. "According to Norbert Wiener, there is no obstacle to using a metaphor 'in which the organism is seen as a message'" (Jacob 1976, 251–252). And two pages later,

According to Schrödinger, the chromosomes contain in some kind of code-script the entire pattern of the individual's future development and of its functioning in the mature state . . . The chromosome structures are at the same time instrumental in bringing about the development they foreshadow. They are law-code and executive power – or, to use another simile, they are architect's plan and builder's craft all in one. (Jacob 1976, 254)

His direct inference is that "[t]he order of a living organism therefore is based on the structure of a large molecule. . . . Heredity functions like the memory of a computer" (Jacob 1976, 254). For Jacob, as indeed, for anyone steeped in the metaphors of premolecular genetics, it is a short step from Schrödinger's metaphor of *law-code and executive power* to the new metaphors of *information and instruction* (or of *codes and programs*). He concludes:

Everything then leads one to regard the sequence contained in genetic material as a series of instructions specifying molecular structures, and hence the properties of the cell; to consider the plan of an organism as a message transmitted from generation to generation; to see the combinative system of the four chemical radicals as a system of numeration to the base four. In short, everything urges one to compare the logic of heredity to that of a computer. Rarely has a model suggested by a particular epoch proved to be more faithful. (Jacob 1976, 264–265)

Yet Jacob himself recognizes a problem with the model: Toward
the end of this treatise that has been designed to demonstrate that the
DNA contains the program needed "to direct the synthesis of proteins and guide their organization" (Jacob 1976, 275), he confronts
the problem head-on. The genetic program is, he acknowledges, a
rather peculiar kind of program: It is "a program which needs the
products of its reading and execution to be read and executed."
Indeed, he admits that "the genetic message can do nothing by itself
... Only the bacterium, the intact cell, can grow and reproduce,
because only the cell possesses both the programme and the directions for use, the plans and the means of carrying them out" (Jacob
1976, 278). And with even greater clarity, he writes in the final
chapter,

The only elements that can interpret the genetic message are the products of
the message itself. The genetic text makes sense only for the structures it has
itself determined. There is thus no longer a cause for reproduction, simply a
cycle of events in which the role of each constituent is dependent on the
others. (Jacob 1976, 297)

Jacob's integrity, it would seem, has brought him to an impasse. Must
he – with this recognition – give up on the picture of development as
a linear process, beginning with fertilization and culminating in the
mature organism, with the program for that process encoded in the
genome? Well, no. His explicit and seemingly conclusive acknowledgment of the circularity of the developmental process, here, at the
end of his text, segues immediately into a reassertion of the primacy
of the genetic program – a program made special by the very fact
that it is written in the linear text of the genome:

Organization could reproduce itself and living organisms emerge, *because*
the complexity of structures in space happened to be generated by the simplicity of a linear combinative system . . . (Jacob 1976, 297, my italics)

In the end, by anticipating the inevitable objection, Jacob's acknowledgment serves only to strengthen the force of the genetic program
as the fundamental explanatory concept; indeed, so strong is it that
(at least by implication) it can even explain its own rebuttal. By
turning the very logic into a circle, a linear causal scheme can be
invoked to explain a cycle of events that has no causal starting point.

CODA

Let me close with a recent quote on the future prospects of molecular biology from Harvey Lodish at MIT:

It will [soon] be possible, by sequencing important regions of the mother's DNA, to infer important properties of the egg from which the person develops. . . . [The information that results] will be transferred to a super-computer, together with information about the environment. . . . The output will be a color movie in which the embryo develops into a fetus, is born, and then grows into an adult, explicitly depicting body size and shape and hair, skin, and eye color. Eventually the DNA sequence base will be expanded to cover genes important for traits such as speech and musical ability; the mother will be able to hear the embryo – as an adult – speak or sing. (Lodish 1995)

I have chosen this quote to help illustrate my main point, namely, that words matter. They shape the way we think, and how we think inevitably shapes the ways we act. In particular, the use of the term *genetic* to describe developmental instructions (or programs) encourages the belief, even in the most careful of readers (as well as writers), that it is only the DNA that matters. To lose sight of the fact that, if that term is to have any applicability at all, it is primarily to refer to the *entities upon which instructions directly or indirectly act* and *not of which they are constituted*. The necessary dependency of genes on their cellular context, not simply as nutrient but for causal agency, is all too easily forgotten – in laboratory practice, in medical counseling, and in popular culture.

NOTES

1. For a related but more general critique of the very concept of a program for development, Stent (1985), Newman (1988), Oyama (1989), Moss (1992), and de Chadarevian (1998); a good overview of the literature on the biological usage of the term *genetic* in the twentieth century may be found in Sapp (1987).
2. A vivid demonstration of this interdependency was provided in the 1950s and 1960s with the development of techniques for interspecific nuclear transplantation. Such hybrids almost always fail to develop past gastrulation, and in the rare cases when they do, the resultant embryo exhibits characteristics intermediate between the two parental species. This dependency of genomic function on cytoplasmic structure follows as well

from the asymmetric outcomes of reciprocal crosses demonstrated in earlier studies of interspecific hybrids (Markert and Ursprung 1971, 135–137).

3. *Maternal* (or *cytoplasmic*) *effects* refers only to the effective agency of maternal (or cytoplasmic) contributions (such as gradients). Because such effects need not be (and usually are not) associated with the existence of permanent structures that are transmitted through the generations, they should not be confused with maternal inheritance.

4. In modern computers such networks are electronic.

5. Supplementing Lenny Moss's observation that a genetic program is "an object nowhere to be found" (Moss 1992, 335), I would propose the developmental program as an entity that is everywhere to be found.

6. Executing a task means processing data provided both by the DNA and by the products of other programs i.e., by information given in nucleotide sequences, chromosomal structure, gradients of proteins and RNA molecules, the structure of protein complexes, and so on.

7. Simultaneously, and probably independently, Ernst Mayr introduced the notion of "program" in his 1961 article on "Cause and Effect in Biology," adapted from a lecture given at MIT on February 1, 1961. There he wrote, "The complete individualistic and yet also species-specific DNA code of every zygote (fertilized egg cell), which controls the development of the central and peripheral nervous system . . . is the *program* for the behavior computer of this individual" (Mayr 1961, 1504).

8. As Soraya de Chadarevian points out (1998), Brenner had taken a critical stance toward the use of the operon model for development as early as 1974 (see his comments in Brenner 1974).

9. Or even of the appropriateness of the nomenclature. Waddington noted not only that it "seems too early to decide whether all systems controlling gene-action systems have as their last link an influence which impinges on the gene itself," but also redescribed this system as genotropic rather than genetic in order "to indicate the site of action of the substances they are interested in" (Waddington 1962, 23).

REFERENCES

Apter, M. J. 1966. *Cybernetics and Development*. Oxford: Pergamon Press.

Apter, M. J., and L. Wolpert. 1965. Cybernetics and development. *Journal of Theoretical Biology* 8: 244–257.

Atlan, H., and M. Koppel. 1990. The cellular computer DNA: Program or data. *Bulletin of Mathematical Biology* 52: 335–348.

Bonner, J. 1965. *The Molecular Biology of Development*. Oxford: Open University Press.

Brenner, S. 1974. New directions in molecular biology. *Nature* 248: 785–787.

Brenner, S., W. Dove, I. Herskowitz, and R. Thomas. 1990. Genes and development: Molecular and logical themes. *Genetics* 126: 479–486.

de Chadarevian, S. 1998. Development, programs and computers: Work on the worm (1963–1988). *Studies in History and Philosophy of Biological and Biomedical Sciences* 29: 81–105.

Emmeche, C. 1994. *The Garden in the Machine.* Princeton: Princeton University Press.

Fleischer, K. 1995. A multiple-mechanism developmental model for defining self-organizing geometric structures. Ph.D. Thesis, California Institute of Technology.

Jacob, F. 1976. *The Logic of Life.* New York: Vanguard.

Jacob, F., and J. Monod. 1961. Genetic regulatory mechanisms in the synthesis of proteins. *Journal of Molecular Biology* 3: 318–356.

Kay, L. 1992. *The Molecular Vision of Life.* Oxford: Oxford University Press.

Keller, E. F. 1995. *Refiguring Life: Metaphors of Twentieth-Century Biology.* New York: Columbia University Press.

Lewontin, R. C. 1992. The dream of the human genome. *New York Review of Books,* May 28, 31–40.

Lodish, H. 1995. Through a glass darkly. *Science* 267: 1617.

Markert, C. L., and H. Ursprung. 1971. *Developmental Genetics.* Englewood Cliffs, NJ: Prentice-Hall.

Mayr, E. 1961. Cause and effect in biology. *Science* 134: 1501–1506.

Monod, J., and F. Jacob. 1961. General conclusions: Teleonomic mechanisms in cellular metabolism, growth, and differentiation. *Cold Spring Harbor Symposia on Quantitative Biology* 26: 389–401.

Moss, L. 1992. A kernel of truth? On the reality of the genetic program. *Proceedings of the Philosophical Society of America 1992* 1: 335–348.

Newman, S. A. 1988. Idealist biology. *Perspectives in Biology and Medicine* 31: 353–368.

Oyama, S. 1989. Ontogeny and the central dogma: Do we need the concept of genetic programming in order to have an evolutionary perspective? In *Systems and Development,* edited by M. R. Gunar and E. Thelen. Hillside, NJ: Lawrence Erlbaum, pp. 1–34.

Sapp, J. 1987. *Beyond the Gene: Cytoplasmic Inheritance and the Struggle for Authority in Genetics.* New York: Oxford University Press.

Stent, G. S. 1985. Hermeneutics and the analysis of complex biological systems. In *Evolution at a Crossroads,* edited by D. J. Depew and B. Weber. Cambridge, Mass.: MIT Press, pp. 209–225.

Waddington, C. H. 1962. *New Patterns in Genetics and Development.* New York: Columbia University Press.

Waldrop, J. M. 1992. *Complexity.* New York: Simon and Schuster.

Wolpert, L. 1969. Positional information and the spatial pattern of cellular differentiation. *Journal of Theoretical Biology* 25: 1–48.

Wolpert, L., and J. H. Lewis. 1975. Towards a theory of development. *Federation Proceedings* 34: 14–20.

8

Genes Classical and Genes Developmental

The Different Use of Genes in Evolutionary Syntheses

SCOTT F. GILBERT

ABSTRACT

Dobzhansky (1964) stated that "Nothing in biology makes sense except in the light of evolution," and the function of the gene is no exception. The use of *genes* in population genetics and developmental genetics differs significantly. This is reflected in the roles that genes are postulated to play in evolution. In the Modern Synthesis of population genetics and evolution, genes become manifest by differences in alleles that are active in conferring differential reproductive success in adult individuals. The gene is thought to act as a particulate, atomic unit. In current syntheses of evolution and developmental genetics, important genes are manifest by their similarities across distantly related phyla, and they are active in the construction of embryos. These developmental genes are thought to act in a context-dependent network. In the population genetics model of evolution, mutations in genes provide insights into the mechanisms for natural selection and microevolution. Different individuals will be selected and their genes will be represented in higher proportions in the next generation. For the developmental geneticist, mutations in the genes provide insights into the mechanisms of phylogeny and macroevolution. Different modes of regulation may enable the production of new types of structures or the modification of existing ones. The importance of developmental approaches to the role of genes is exemplified by the discovery and subsequent analysis of the developmental gene *MyoD*.

The concept of the gene has had its own radiation once it entered into the territory of developmental biology. As Morange has shown (1996 and this volume), the concept of the *developmental gene* was a major insight, and it changed the way development was discussed. It also changed the ways the gene was discussed with regard to evolution. Developmental biology and evolutionary biology are converging on a new synthesis for macroevolution, and this synthesis is very different from the *Modern Synthesis* of population genetics and evolutionary biology that accounted for microevolutionary processes (see Carroll 1997; Gerhart and Kirschner 1997; Gilbert 1997; Gilbert, Opitz, and Raff 1996; Hall 1992, 1996; Raff 1996). Moreover, the concept of the gene is very different between the two types of synthesis.

GENES IN THE MODERN SYNTHESIS AND IN DEVELOPMENTAL SYNTHESES

There are several differences that distinguish the gene of population genetics from the gene of developmental genetics. The main difference concerns the levels of events being explained. The gene of the Modern Synthesis of the 1940s was an abstraction. It was not sequenced, its structure was unknown (and generally thought to be protein), and the mechanisms accounting for genetic change (mutation, recombination) were unexplained. Moreover, given the abstract, mathematical, nature of the gene, none of this mattered. The gene could be anything that had the properties of transmittal with infrequent change. Alleles of *A* and *a* did not even need to be DNA. The genes of the Developmental Synthesis (to use a convenient shorthand for these new syntheses) are specific sequences of DNA containing not only protein-encoding regions, but regulatory sequences such as promoters, enhancers, silencers, insulators, introns, 5' untranslated regions, and 3' untranslated regions (see Gilbert, 1997). Thus, for a population geneticist, the problem of *Drosophila* sex determination was solved as early as 1905 (Stevens 1905; Wilson 1905) when it was discovered that the females have two X chromosomes (XX) while males have but one (XY or XO). This information was necessary and sufficient for modeling populations. However, for a developmental geneticist, this is but the starting point. The mechanisms of sex determination involve the binding of specific proteins to specific bases of DNA and RNA, and they can differ widely between phyla. What is mechanism to the population geneticist is correlation to the developmental geneticist.

A second difference concerns the tension between constancy and divergence. The genes of the Modern Synthesis are manifest by the differences they cause (Dobzhansky 1937, 19–49; Goldschmidt 1952;

see Dietrich, Gifford, and Schwartz, this volume). These differences could be selected. The genes of the Developmental Synthesis are manifest by their similarities. That the *Pax6* genes (encoding a particular transcription factor) are expressed in photoreceptive cells throughout the kingdom indicates that they may have an important role in photoreceptor development and evolution (Quiring et al. 1994). More importantly, there are conserved *pathways* of conserved genes. For example, the BMP4-chordin pathway (by which chordin blocks the epidermal induction of the ectoderm and permits the ectoderm to develop neurons) is critical for neural specification in both arthropods and vertebrates (see De Robertis and Sasai 1996). PCR allows these genes to be discovered through their similarities.

A third, related difference concerns what aspects of evolution these genes attempt to explain. The gene of the Modern Synthesis is a gene that could explain the mechanisms of natural and sexual selection (hence, the allelic genes are manifest by differences in adult phenotypes; Dobzhansky 1937; Lewontin 1974). The gene of the Developmental Synthesis attempts to explain phylogeny. This macroevolutionary program harks back to what Bowler (1996) calls the "first evolutionary biology" – the attempt to discover the origins of the different phyla and classes. Thus, differential *Hox* gene expression is postulated to have brought about (a) the transformation from fins to limbs in vertebrates (Shubin, Tabin, and Carrol 1997; Sordino, van der Hoeven, and Duboule 1995) and (b) the transformation of limbs into maxillipedes during crustacean development (Averof and Patel, 1997). Developmental syntheses look at the possibilities and constraints for the arrival of the fittest, while population genetics can model their survival. Both approaches are obviously needed to understand evolution.

This leads to a fourth difference between the genes of the Modern Synthesis and those of the Developmental Synthesis. In the Modern Synthesis, evolutionary change was conceived to originate from alterations in the *coding* region that altered the performance of enzymes or structural proteins. For example, did the gene make a functional or nonfunctional protein? Did the gene encode a slower or faster variant of the enzyme? The genes that are important in the Developmental Synthesis are not those necessarily encoding metabolic enzymes or structural proteins. Rather, they are genes encoding signal transduction components or gene expression proteins (such as transcription factors and splicing factors). Moreover, the important portions of these genes are not so much the protein encoding exons as they are the regulatory regions of these genes or the portion of the protein that binds to these regions.

A fifth difference involves when and where these genes are expressed. The genes of the Modern Synthesis are expressed in adults competing for reproductive advantage. The genes of the Developmental Synthesis are expressed during the construction organs within the embryo (see Raff 1992; Waddington 1953).

The sixth difference between the gene of the Modern Synthesis and the gene of the Developmental Synthesis concerns atomicity. The gene of the Modern Synthesis was independent from all other genes. It might be physically close to other genes on the chromosome, and this might cause the physically linked genes to be inherited together, but gene action was an individual phenomenon. The gene acted as an autonomous unit. The genes involved in the Developmental Synthesis are not autonomous actors. First, many of them are linked in physical aggregates, such as the *Hox* genes. Not only are these *Hox* genes, themselves, conserved throughout evolution, but their linkage is conserved between arthropods and vertebrates. The reason for this close linkage appears to be that these genes share regulatory elements in their promoters and enhancers (Duboule 1994; Morange, this volume). Therefore, the entire entity – consisting of many linked genes – is a developmentally functional unit. Another example of genes linked together in a developmental sequence are the mammalian globin genes. Here, they are each regulated by a Locus Control Region. Whereas the *Hox* genes are ordered in the genome according to their spatial expression patterns, the globin genes are present in the genome according to their temporal expression patterns (see Martin, Fiering, and Groudine 1996). Second, the developmental genes are linked together into networks of interacting genes and gene products (Gilbert, Opitz, and Raff 1996). For instance, the deletion of the muscle-forming *MyoD* gene in mice does not lead to marked deficiencies of muscle development, since the *MyoD* protein suppresses the activation of the muscle-forming *Myf-5* gene. In the absence of *MyoD*, this suppression is lifted and *myf-5* can be expressed and transforms the cells into muscle (Rudnicki et al. 1993).

Studies of the classical limb fields and the imaginal discs have dissected them into numerous pathways of reciprocal activation and suppression by genes and their protein products (see Gilbert 1997, for review). In development the gene is not an independent entity, but is part of a pathway.

This lack of autonomy has important consequences. First, "what" a gene does depends upon its context. In the liver, enolase is a glycolytic enzyme, while in the lens cell, it's a structural crystallin (Piatigorsky and Wistow 1991). The GSK-3β gene (He et al. 1995) can play a role in the Wnt signal transduction pathway for fly segmentation or frog neural axis formation, or it can help regulate glycolysis. Beta-catenin can hold cells together as part of the desmosome or it can be a developmentally critical transcription factor, depending on the cell in which it is expressed (Schneider et al. 1996).

A second consequence of the context dependency of genes is that a gene's effects can differ when it is placed in a pathway containing different alleles of the other genes in that pathway. This constitutes the *background effect* well known to developmental biologists, immunologists, and clinical geneticists (Wolf 1995, 1997). For instance, in the formation of the limb, a gene deficiency in one individual may cause an absent limb; in a different individual the same genetic mutation may cause an absent thumb (Freire-Maia 1975). Certain histocompatibility alleles predispose mice to some disease, but only in particular strains. This has practical consequences for agriculture and pharmaceutical manufacture. The rationale for cloning both transgenic sheep and cattle is that the transgene does not function the same way when sexual reproduction places it in different backgrounds (Meade 1997; Schnieke et al. 1997).

The differences between the developmental and population approaches to evolution (and their different views of genes) were appreciated as early as 1953 by Conrad Hal Waddington. He claimed that in addition to "normative selection" (the elimination of less favorable phenotypes by natural selection), there must also be "stabilizing selection" within the embryo. At the same meeting where Waddington presented this view, J. B. S. Haldane (1953) concluded, "The current instar of evolutionary theory may be defined by such books as those of Huxley, Simpson, Dobzhansky, Mayr, and Stebbins. We are certainly not ready for a new molt, but signs of new organs are perhaps visible." He pointed to "a broader synthesis in the future." This is what we are embarking upon now. The data of developmental genetics is complementing that of population genetics to provide a broader evolutionary synthesis that can explain macroevolutionary, as well as microevolutionary, phenomena.

THE DISCOVERY OF A DEVELOPMENTAL REGULATORY GENE: *MYOD*

The discovery of developmental genes has been accomplished by methods far removed from the ways that "classical" genes have been discovered. Classical genes were identified by looking for differences in individuals within populations; developmental genes have often been identified by looking at differences in the expression of these genes in a normal embryo. A muscle cell should be expressing a set of genes that differs from that of a fibroblast.

Michel Morange (this volume) has focused on one set of developmental genes – the homeotic (*Hom-C/Hox*) genes. As Morange and others have correctly noted, this work originally fell within the research programs of classical genetics, especially that of pseudoallelism. However, just as *Drosophila* is a very derived insect, so was research on the homeotic genes a very special case of developmental genetics. Most research in developmental genetics sought the causes of cell commitment and differentiation. This meant that one looked to see why a given cell became a muscle cell and not a fat cell, a skin cell and not a neuron, a blood cell and not a lymphocyte. The research on *Drosophila* maternal effect, segmentation, and homeotic genes was not in this category. It sought the means by which parasegments, segments, and compartments were specified. Each of these parasegments had nerves, blood cells, integument, etc., but arranged differently. The homeotic genes did not concern cell differentiation; rather, they regulated segment identity. So *Drosophila* research was studying a higher plane of development than most developmental geneticists, who were studying cell differentiation[1] (see Emmons 1996).

I would like to trace the history that led to the identification of the first developmental gene involved in cell differentiation, the *MyoD* gene of vertebrates. This gene encodes a transcription factor protein that binds to a specific region of DNA, and the activation of this gene in any particular cell will transform that cell into a muscle precursor.

The beginnings of this history are rooted in the search to find the eukaryotic equivalent of the operon. Sol Spiegelman and others had convinced biologists that differentiation was nothing but changes in protein synthesis, and the operon gave the first testable model concerning how different cells made different proteins (see Gilbert 1996). Britten and Davidson's (1969) developmental operon hypothesis became one of the most quoted papers in all biology, and it predicted regulatory elements (sensors) in the DNA and diffusible regulators that would bind to them. This model would explain not only differential protein synthesis but also coordinated protein synthesis.[2] The program to find the eukaryotic operon can be divided into two main fronts – first, the search for the eukaryotic promoter, and second, the search for eukaryotic regulatory proteins. Both branches would be remarkably successful, and their first success was the discovery of MyoD – the "master regulatory gene" of muscle development.

The discovery of MyoD, unlike the discovery of the homeotic genes, was not a surprise. However, it did have its origins from a relatively unappreciated source – somatic cell genetics. Somatic cell genetics was one of Boris Ephrussi's brainchildren, and it flourished during the late 1960s and early 1970s, declining precipitously by the 1980s (see Burian, Gayon, and Zallen 1991). Its technique was to fuse different types of cells together and look at the resulting state of differentiation. Most cell types (liver, neurons, melanocytes) lost their differentiated phenotypes when fused with other cells, so some intranuclear diffusible negative regulator was hypothesized for the disappearance of their specific differentiated functions (see Davidson, Ephrussi, and Yamamoto 1968). The nuclear constitutions of these hybrids were often unstable, and chromosome loss occurred. If particular lost chromosomes were correlated to the retention of a differentiation cell enzyme, then the genes for the negative regulatory protein (that would have repressed that "luxury" enzyme) could be postulated to reside on the absent chromosome. This was proposed to be the case for kidney-specific esterase (Klebe, Chen, and Ruddle 1970) and hepatic aminotransferase inducibility (Weiss and Chaplain 1971).

The exception to this rule was the skeletal muscle cell. First, myocytes retained their differentiated state in pure cell culture better than other cells. Second, proliferating myoblasts that had not yet made contractile protein retained this commitment in culture (Konigsberg 1963). Third, several laboratories (Blau, Chiu, and Webster 1983; Ringertz, Krondal, and Coleman 1978; Wright 1981; see Pinney, de la Brousse, and Emerson 1990) found that when differentiating myoblasts were fused with other cells, not only was the muscle-specific phenotype retained, but the myoblasts could cause the nucleus of the *other* cell type to make muscle-specific proteins. There appeared, then, to be a *positive* regulator of muscle gene transcription. This agreed well with other experiments that showed coordinate transcriptional control of muscle gene expression (Devlin and Emerson 1978, 1979).

In 1984, Konieczny and Emerson showed that the mouse embryonic cell line C3H10T1/2 (generally called the *T-one halfs*) could be converted into stable proliferating myogenic, chondrogenic, or adipogenic cell lines following treatments that inhibited DNA methylation. They predicted that the high rate of myogenic phenotypes resulted from the activation of one or a very few regulatory loci. In 1986, Emerson's laboratory and Weintraub's laboratory both reported that the transfection of C3H10T1/2 cells with cDNA from either cultured muscle cells or from 5-azaC-treated C3H10T1/2 cells would transform the cells into myocytes (Konieczny, Baldwin, and Emerson 1986; Lassar, Paterson, and Weintraub 1986). The Weintraub group (Davis, Weintraub, and Lassar 1987; Weintraub et al. 1989) made cDNA copies of the mRNA, cloned them, and transfected the clones individually into the C3H10T1/2 cells. One of these clones, *MyoD*, was found to convert the C3HT101/2 cells solely into myoblasts, and at high frequency. Moreover, it converted freshly cultured endodermal gut cells, ectodermal neurons, and other cells as well, into skeletal muscle.

MyoD turned out to be a muscle-cell-specific transcription factor. It controls cell determination and differentiation by binding to regions of the DNA that precede several muscle-specific protein-encoding genes. It also binds to its own promoter to retain its own transcription, and it binds to the promoters or enhancers of other muscle-specific transcription factors to activate them, as well. (This MyoD-binding DNA sequence was discovered through a collaboration between Weintraub's laboratory and David Baltimore's group; Murre et al. 1989.)

This muscle differentiation research program was occurring at the same time as the *Drosophila* research program, yet along a fundamentally different line of approach. Charles Emerson is a well-known muscle developmental biologist; Hal Weintraub – until his recent death – was a major investigator of chromatin structure and transcription factors. Nowhere in this research program was a mutant used.[3] This research program was strictly epigenetic. It was based on phenotype analysis – the appearance of muscle contractile proteins. When recombinant DNA became available, it was used to see if the cloned gene encoded a protein capable of changing the "phenotype" of the cell. Also, although other organisms were found to make *MyoD* as a muscle-specific transcription factor, that data did not play any major role in forming the research program or (at least initially) strengthening the program.

DEVELOPMENTAL GENES AND THE REGULATORY NETWORK

Homeotic genes and *MyoD* have been called *master regulators*. They are seen as being at the top of the developmental hierarchy, controlling the genes below them. Are they master regulators? Yes – the products of homeotic genes can convert a haltere segment into a wing or an antenna into a leg; *MyoD* can convert a neuron into a muscle if activated there. However, are they also "slaves," genes that are themselves told what to do? Yes – homeotic genes such as *abd-A* are regulated to make the parapods in the abdominal segments of caterpillars (Carroll 1995; Carroll, Weatherbee, and Langeland 1995) and genes such as *Ubx* are regulated temporally to distinguish the third thoracic from the first abdominal segment (Castelli-Gair and Akam 1995). Some homeotic genes, such as the *Abd-B*-like *mab-5* in *C. elegans*, are regulated by the cell lineage in which the gene resides. In this last case, cell lineage plays a greater role than cell region in determining the gene's expression (Harris et al. 1996; Salser and Kenyon 1996). Similarly, recent research suggests the paradoxical view that *MyoD* is both a master control gene and also one of the most tightly controlled genes in the genome.

So it appears that these master control genes are themselves under masterful regulation. There can be no top of the hierarchy in a life cycle. The hierarchy has become a network of interactions. *MyoD* is such a powerful protein that the cell must control it at all ectopic times and places so that it is not expressed in the wrong cell or at the wrong time. If even a small amount of MyoD is made, that cell will become muscle. So *MyoD* is regulated at transcription, RNA processing, and by at least two post-translational regulators[4] (see Gilbert 1997). Governors govern the governor. Regulators must be regulated by factors that are themselves both regulated and regulators. Moreover, *MyoD* regulation works within a field – the limb field or the somite field – because the regulators are soluble proteins coming from outside the cell: BMP-4, Wnt-4, FGFs (Kopan, Nye, and Weintraub 1994; Li et al. 1992; Vaidya et al. 1989). The basic state of *MyoD* and the *Hox* genes is to be inhibited. Like so much in developmental biology, activation consists of inhibiting the inhibitor; suppression is the inhibition of the inhibitor of the inhibitor.

So these developmental genes have to be both regulators and regulatees. The things they regulate and the things that regulate them are part of a pathway. In the end, it is not the conservation of the gene that is important, but the conservation of these developmental pathways that include them. The inheritance is not of a gene but of a regulated network of genes and the binding regions for their products. The genes encoding GSK-3β and β-catenin are particularly instructive cases. They can be considered structural genes or developmental genes depending upon which tissue is being considered. This is to be expected from our knowledge of evolution. As Jacob (1977) noted, nature should use what it has before inventing something new. Proteins have multiple sites. The fact that a gene can be used for different purposes within the body should not be troubling except by those people trying to name the gene. The interesting questions of evolutionary biology will involve how these pathways were modified to bring about the formation of new cell types and new body plans during the development of life on earth.

CODA

The genes of the Modern Synthesis and those of the Developmental Synthesis are quite different. They were invoked to explain different aspects of evolution, and they emphasize different aspects of genetic structure and function. The discovery of the first developmental gene associated with cell commitment and differentiation, *MyoD*, was accomplished in a manner very distinct from the methods used to identify classical genes. Furthermore, the analysis of the *MyoD* and homeotic genes has given rise to a principle of developmental biology – that the major regulatory genes are themselves highly regulated in a complex network. It would not be surprising if different "alleles" of regulatory genes were to be found between closely related species and that morphological changes may involve changes in the dissociation constants between ligands and their receptors or between components of the chromatin around the promoter. While the molecular bases of macroevolution may reside in the changes in these networks of gene regulation, both the classical and developmental "natures" of the gene are required to account for evolutionary processes.

NOTES

1. The aptness of Christiane Nüsslein-Volhard's address is worthy of noting: The Friedrich-Miescher Laboratorium on Spemannstraße. It combines the institutional authority of DNA (the lab named for the discoverer of DNA is part of the Max Planck Institut) and the epigenesis of Spemann. The perfect place for the molecular biology of development.
2. E. B. Lewis (1963) also used the operon model, but he used it very differently than developmental biologists. He saw it as a mechanism for sequential gene activation, not for the differentiation of particular cell types.
3. It would have been extremely difficult to discover MyoD by mutational analysis. There is overlapping redundancy in the myogenic bHLH transcription factors and Myf-5 can compensate for the absence of MyoD (Rudnicki et al. 1993; Wang et al. 1996). Mutants were eventually constructed to test the binding site of the MyoD protein (Murre et al. 1989; Davis et al. 1990).
4. The wisdom of the control genes appears to be Sophrosyne. This Apollonian principle of Greek ethics is characterized by the disciplined self-restraint of great power, or as Helen North (1966) defined it, "the harmonious product of intense passion under perfect control."

REFERENCES

Averof, M., and N. H. Patel. 1997. Crustacean appendage evolution associated with changes in Hox gene expression. *Nature* 388: 682–686.

Blau, H. M., C.-P. Chiu, and C. Webster. 1983. Cytoplasmic activation of human nuclear genes in stable heterokaryons. *Cell* 32: 1171–1180.

Bowler, P. J. 1996. *Life's Splendid Drama.* Chicago: University of Chicago Press.

Britten, R. J., and E. H. Davidson. 1969. Gene regulation for higher cells: A theory. *Science* 165: 349–357.

Burian, R. M., J. Gayon, and D. T. Zallen. 1991. Boris Ephrussi and the synthesis of genetics and embryology. In *A Conceptual History of Modern Embryology,* edited by S. F. Gilbert. New York: Plenum. pp. 207–227.

Carroll, R. L. 1997. *Patterns and Processes of Vertebrate Evolution.* New York: Cambridge University Press.

Carroll, S. B. 1995. Homeotic genes and the evolution of arthropods and chordates. *Nature* 376: 479–485.

Carroll, S. B., S. D. Weatherbee, and J. A. Langeland. 1995. Homeotic genes and the regulation and evolution of insect wing number. *Nature* 375: 58–61.

Castelli-Gair, J., and M. Akam. 1995. How the *Hox* gene *Ultrabithorax* specifies two different segments: The significance of spatial and temporal regulation within metameres. *Development* 121: 2973–2982.

Davidson, R. L., B. Ephrussi, and K. Yamamoto. 1968. Regulation of melanin synthesis in mammalian cells as studied by somatic cell hybridization I. Evidence for negative control. *Journal of Cell Physiology* 72: 115–127.

Davis, R. L., P. F. Cheng, A. B. Lassar, and H. Weintraub. 1990. The MyoD binding domain contains a recognition code for muscle-specific gene activation. *Cell* 60: 733–746.

Davis, R. L., H. Weintraub, and A. B. Lassar. 1987. Expression of a single transfected cDNA converts fibroblasts into myoblasts. *Cell* 51: 987–1000.

De Robertis, E. M., and Y. Sasai. 1996. A common plan for dorsoventral patterning in the Bilateria. *Nature* 380: 37–40.

Devlin, R. B., and C. P. Emerson Jr. 1978. Coordinate regulation of contractile protein synthesis during myoblast differentiation. *Cell* 13: 599–611.

Devlin, R. B., and C. P. Emerson Jr. 1979. Coordinate regulation of contractile protein mRNAs during myoblast differentiation. *Developmental Biology* 69: 202–216.

Dobzhansky, T. 1937. *Genetics and the Origin of Species.* New York: Columbia University Press.

Dobzhansky, T. 1964. Biology: Molecular and organismic. *American Zoologist* 4: 443–452.

Duboule, D. 1994. Temporal colinearity and the phylotypic progression: A basis for the stability of the vertebrate Bauplan and the evolution of morphologies through heterochrony. *Development* 1994 (Supplement): 135–142.

Emmons, S. W. 1996. Simple worms, complex genes. *Nature* 382: 301–302.

Freire-Maia, N. 1975. A heterozygote expression of a "recessive" gene. *Human Heredity* 25: 302–304.

Gerhart, J., and M. Kirschner. 1997. *Cells, Embryos, and Evolution*. London: Blackwell Science.

Gilbert, S. F. 1996. Enzyme adaptation and the entrance of molecular biology into embryology. In *The Philosophy and History of Molecular Biology: New Perspectives*, edited by S. Sarkar. Dordrecht: Kluwer Academic Publishers, pp. 125–151.

Gilbert, S. F. 1997. *Developmental Biology*. 5th ed. Sunderland: Sinauer Associates.

Gilbert, S. F., J. M. Opitz, and R. A. Raff. 1996. Resynthesizing evolutionary and developmental biology. *Developmental Biology* 173: 357–372.

Goldschmidt, R. B. 1952. The theory of the gene. *Cold Spring Harbor Symposia on Quantitative Biology* 16: 1–11.

Haldane, J. B. S. 1953. Foreword. In *Symposium of the Society for Experimental Biology*, 7: ix–xix.

Hall, B. K. 1992. *Evolutionary Developmental Biology*. London: Chapman and Hall.

Hall, B. K. 1996. Evolutionary Developmental Biology. *McGraw-Hill Yearbook of Science and Technology:* 110–112.

Harris, J., L. Honigberg, N. Robinson, and C. Kenyon. 1996. Neuronal cell migration in *C. elegans:* regulation of *Hox* gene expression and cell position. *Development* 122: 3117–3131.

He, X., J.-P. Saint-Jeannet, J. R. Woodgett, H. E. Varmus, and I. B. Dawid. 1995. Glycogen synthase kinase-3 and dorsoventral patterning in *Xenopus* embryos. *Nature* 374: 617–622.

Jacob, F. 1977. Evolution and tinkering. *Science* 196: 1161–1166.

Klebe, R. J., T. R. Chen, and F. H. Ruddle. 1970. Mapping of a human genetic regulator element by somatic cell genetic analysis. *Proceedings of the National Academy of Sciences USA* 66: 1220–1227.

Konieczny, S. F., A. S. Baldwin, and C. P. Emerson Jr. 1986. Myogenic determination and differentiation of 10T1/2 cells lineages: Evidence for a single genetic regulatory system. *Molecular Cell Biology* 29: 21–34.

Konieczny, S. F., and C. P. Emerson Jr. 1984. 5-Azacytidine induction of stable mesodermal stem cell lineages from 10T1/2 cells: Evidence for regulatory genes controlling determination. *Cell* 38: 791–800.

Konigsberg, I. R. 1963. Clonal analysis of myogenesis. *Science* 140: 1273–1284.

Kopan, R., J. S. Nye, and H. Weintraub. 1994. The intracellular domain of mouse *Notch:* A constitutively activated repressor of myogenesis directed at the basic helix-loop-helix region of MyoD. *Development* 120: 2421–2430.

Lassar, A. B., B. M. Paterson, and H. Weintraub. 1986. Transfection of a DNA locus that mediates the conversion of 10T1/2 fibroblasts into myoblasts. *Cell* 47: 649–656.

Lewis, E. B. 1963. Genes and developmental pathways. *American Zoologist* 3: 33–56.

Lewontin, R. C. 1974. *The Genetic Basis of Evolutionary Change.* New York: Columbia University Press.

Li, L., J. Zhou, J. Guy, R. Heller-Harrison, M. P. Czech, and E. N. Olson. 1992. FGF inactivates myogenic helix-loop-helix proteins through phosphorylation of a conserved protein kinase C site in their DNA-binding domains. *Cell* 71: 1181–1194.

Martin, D. I. K., S. Fiering, and M. Groudine. 1996. Regulation of β-globin gene expression: Straightening out the locus. *Current Opinions in Genetics and Development* 6: 488–495.

Meade, H. M. 1997. The dairy gene. *The Sciences* (Sept./Oct.): 20–25.

Morange, M. 1996. Construction of the developmental gene concept. The crucial years: 1960–1980. *Biologisches Zentralblatt* 115: 132–138.

Murre, C., P. S. McCaw, H. Vaessin, M. Caudy, L. Y. Jan, Y. N. Jan, C. V. Cabrera, J. N. Buskin, S. D. Hauschka, A. B. Lassar, H. Weintraub, and D. Baltimore. 1989. Interactions between heterologous helix-loop-helix proteins generate complexes that bind specifically to a common DNA sequence. *Cell* 58: 537–544.

North, H. 1966. *Sophrosyne: Self-knowledge and Self-restraint in Greek Literature.* Ithaca: Cornell University Press.

Piatigorsky, J., and G. Wistow. 1991. The recruitment of crystallins: New functions precede gene duplication. *Science* 252: 1078–1079.

Pinney, D. F., F. C. de la Brousse, and C. P. Emerson Jr. 1990. Molecular genetic basis of skeletal myogenic lineage determination and differentiation. In *Genetics of Pattern and Growth Control,* edited by A. Mahowald. New York: Wiley-Liss, pp. 65–89.

Quiring, R., U. Walldorf, U. Kloter, and W. J. Gehring. 1994. Homology of the *eyeless* gene of *Drosophila* to the *Small eye* gene in mice and *Aniridia* in humans. *Science* 265: 785–789.

Raff, R. A. 1992. Evolution of developmental decisions and morphogenesis: The view from two camps. In *Gastrulation,* edited by C. D. Stern and P. W. Ingham. Cambridge: Company of Biologists, pp. 15–22.

Raff, R. A. 1996. *The Shape of Life: Genes, Development, and the Evolution of Animal Form.* Chicago: University of Chicago Press.

Ringertz, N., U. Krondahl, and J. R. Coleman. 1978. Reconstruction of cells by fusion of cell fragments I. Myogenic expression after fusion of minicells from rat myoblasts (L6) with mouse fibroblast (A9) cytoplasm. *Experimental Cell Research* 113: 233–246.

Rudnicki, M. A., P. N. J. Schnegelsberg, R. H. Stead, T. Braun, H.-H. Arnold, and R. Jaenisch. 1993. MyoD or Myf-5 is required in a functionally redundant manner for the formation of skeletal muscle. *Cell* 75: 1351–1359.

Salser, S. J., and C. Kenyon. 1996. A *C. elegans Hox* gene switches on, off, and on again to regulate proliferation, differentiation, and morphogenesis. *Development* 122: 1651–1661.

Schneider, S., H. Steinbeisser, R. M. Warga, and P. Hausen. 1996. β-catenin translocation into nuclei demarcates the dorsalizing centers in frog and fish embryos. *Mechanisms of Development* 57: 191–198.

Schnieke, A., A. J. Kind, W. A. Ritchie, K. Mycock, A. R. Scott, M. Ritchie, I. Wilmut, A. Colman, and K. H. S. Campbell. 1997. Human factor IX transgenic sheep produced by transfer of nuclei from transfected fetal fibroblasts. *Science* 278: 2130–2133.

Shubin, N., C. Tabin, and S. Carroll. 1997. Fossils, genes and the evolution of animal limbs. *Nature* 388: 639–648.

Sordino, P., F. van der Hoeven, and D. Duboule. 1995. *Hox* gene expression in teleost fins and the origin of vertebrate digits. *Nature* 375: 678–681.

Stevens, N. M. 1905. Studies in spermatogenesis with especial reference to the "accessory chromosome." *Carnegie Institute of Washington Report* 36: 3–30.

Vaidya, T. B., S. J. Rhodes, E. J. Taparowsky, and S. F. Konieczny. 1989. Fibroblast growth factor and transforming growth factor-β repress transcription of the myogenic regulatory gene MyoD1. *Molecular Cell Biology* 9: 3576–3579.

Waddington, C. H. 1953. Epigenetics and evolution. *Symposia of the Society for Experimental Biology* 7: 186–199.

Wang, Y., P. N. J. Schnegelsberg, J. Dausman, and R. Jaenisch. 1996. Functional redundancy of the muscle-specific transcription factors Myf5 and myogenin. *Nature* 379: 823–826.

Weintraub, H., S. J. Tapscott, R. L. Davis, M. J. Thayer, M. A. Adam, A. B. Lassar, and D. Miller. 1989. Activation of muscle-specific genes in pigment, nerve, fat, liver, and fibroblast cell lines by forced expression of MyoD. *Proceedings of the National Academy of Science USA* 86: 5434–5438.

Weiss, M. C., and M. Chaplain. 1971. Expression of differentiated function in hepatoma cell hybrids III. Reappearance of tyrosine aminotransferase inducibility after loss of chromosomes. *Proceedings of the National Academy of Sciences USA* 68: 3026–3031.

Wilson, E. B. 1905. The chromosomes in relation to the determination of sex in insects. *Science* 22: 500–502.

Wolf, U. 1995. The genetic contribution to the phenotype. *Human Genetics* 95: 127–148.

Wolf, U. 1997. Identical mutations and phenotypic variation. *Human Genetics* 100: 305–321.

Wright, W. E. 1981. The synthesis of rat myosin light chains in heterokaryons formed between undifferentiated rat myoblasts and chick skeletal myocytes. *Journal of Cell Biology* 91: 11–16.

9

The Developmental Gene Concept

History and Limits[1]

MICHEL MORANGE

ABSTRACT

Today, developmental genes are at the focus of attention of many developmental biologists as well as evolutionary biologists and paleontologists. The formation of the concept of developmental genes progressed stepwise during the twentieth century. However, the significance of these genes was not fully recognized until the end of the 1980s when, thanks to new genetic engineering techniques, homologous genes were isolated from very different organisms, ranging from *Drosophila* or nematodes to mammals and humans. Such genes are highly conserved. The proteins encoded by developmental genes are transcription factors that regulate the expression of other genes and components of the cellular signaling pathways that allow cells to communicate with their neighbors. Frequently, the conservation is not limited to one gene but extends to the entire pathway.

The significance of these genes remains puzzling: Their functions during development are difficult to state precisely, as well as the role they play in the evolutionary shifts such as the Cambrian explosion. Are they the master genes that guide development and constrain evolution or only the toolbox with which evolution tinkers? The very notation of *developmental genes* appears increasingly problematic.

The concept of the developmental gene occupies a central position in contemporary biological research at the crossroads between developmental and evolutionary biology. In this brief text, I would like to outline the major steps that led to the formation of this concept and emphasize some of the difficulties encountered by biologists in its present-day use. Today, all biologists agree that a limited number of genes, highly conserved during evolution, play a major role in the control of development. However, the precise terminology has not yet stabilized: Biologists speak of master genes, switch genes, selector genes, control genes, master control genes, key regulatory genes, genes controlling development, developmental genes, developmental regulatory genes, developmental control genes, or even morphology sculptors. The history of these different names and their usage is worthy of an independent study. The use of a particular expression is a subtle way to recognize a debt to one of the traditions that led to the formation of the developmental gene concept.

PROGRESSIVE HISTORICAL FORMATION OF THE CONCEPT

The Early Genetic Works

The separation that occurred between the science of heredity, genetics, and embryology at the beginning of the twentieth century has been well documented (Allen 1978). The scientific and sociological reasons behind this separation have been analyzed (Sapp 1983, 1987). This separation was, however, specific to American genetics. In Germany, the two disciplines remained tightly linked (Harwood 1993). The first attempts to reduce this rift were made during the 1930s after publication of *Embryology and Genetics* (Morgan 1934) in which Thomas H. Morgan advocated that these two disciplines join efforts. Although Morgan did not really show the way to do so, studies were undertaken to bridge the gap (Gilbert 1988, 1991a). Some of these studies were intended to characterize the role of genes in development as a way to uncover the normal physiological function of genes. Such were the studies of Richard Goldschmidt (1938; see Allen 1974), Boris Ephrussi, and George Beadle (Kohler 1991; Burian, Gayon, and Zallen 1991). The existence of a specific developmental function of genes, distinct from their normal physiological function, was absent from the writings of these authors. In fact, these works led to the characterization of the one-gene – one-enzyme relationship. However, this major step in early molecular biology did not provide any immediate clue to the role of genes in development.

The studies performed by Paul Chesley (1935) and Salomé Gluecksohn-Schoenheimer (1938, 1940; see Gilbert 1991a) on the brachyury mutation (and later by Gluecksohn-Schoenheimer on the kinky mutation [1949]) attempted to correlate the alterations of development observed in these mutants to the mechanisms revealed by the German school of embryology headed by Hans Spemann (1938; see Saha 1991). Calvin B. Bridges, followed by many other geneticists (Balkaschina 1929; Bridges 1917; Bridges and Dobzhansky 1933; Li 1927; Waddington 1939), isolated numerous mutations that deeply affected the structure of *Drosophila* and its development. Whereas the first descriptions of these mutants by American geneticists were very brief, the later ones, primarily under the impact of the Russian school of genetics, were more precise and the mechanisms leading to their formation were discussed at length. Walter Landauer described very early a mutation affecting fowl development – the creeper factor – although the embryological characteristics were given only later (Landauer 1944). D. F. Poulson (1937) systematically correlated chromosomal modifications and alterations in the development of *Drosophila melanogaster*, demonstrating the early involvement of genes in development and defining the number of genes involved (the significance of D. F. Poulson's work is discussed in Keller 1996). These studies were fundamental in showing that some genes actively participate in the first steps of embryogenesis. They contributed to the reversal of the ideas of the most conservative embryologists who thought that genes were responsible only for the determination of the most superficial characteristics of the organisms. During these years (1915–1950), the mutations that would play a major role in the future history of developmental biology, and that affect what would later become known as developmental genes, were described. However, the characterization of these genes did not immediately lead to the idea that a specific class of genes was involved in development. Some genes affected by mutations were taken to play definite roles in the embryonic developmental process, whereas others were thought to affect more superficial characteristics. Yet there was a continuum between the two categories of genes and their mechanism of action. The physiology of the different genes was thought to follow similar rules.

The Homeotic Complex

The work initiated by C. B. Bridges and pursued by Edward B. Lewis in 1940 on the bithorax mutations led geneticists a step further

(Lewis 1992), though without bringing about the formation of the developmental gene concept. The careful description and analysis of these different mutations showed that some of them were pseudoallelic and did not obey the rules of genes as independent entities of recombination. To explain this pseudoallelism, and to correlate it with the macroscopic modifications affecting what became the bithorax complex, Ed Lewis (1951) suggested the existence of different but related genes, resulting from the duplication of an ancestral gene: The action of these genes controlled the nature of the segments of the larva or adult. It was, however, falsely deduced from these observations that the different pseudoallelic genes controlled a series of sequentially related biochemical reactions. The association of the bithorax complex with a metabolic pathway was to be abandoned in the 1960s for an interpretation of the data within the framework of the operon model (Lewis 1963). Only in his model of 1978 did Lewis establish a clear correlation between the organization of the genetic material and the construction of the organism. The future successful development of the studies on the homeotic genes, with the characterization of the homeobox and the discovery of similarly organized genes in higher organisms, does not legitimize a retrospective interpretation of these earlier investigations. The article published in *Nature* in 1978 was difficult to read, and contained an incorrect estimate of the number of genes involved. But most of all, this article did not make any allusion whatsoever to a possible generalization of the results: The existence of the bithorax complex was linked with the segmented nature of insects (and arthropods); it provided no general solution for problems in developmental biology.

The Regulatory Gene Concept and its Importance for Developmental Biology

It is obvious, however, that the studies on the bithorax complex emphasized the importance of some genes in controlling the formation of living organisms. The idea that development is controlled by a subfamily of genes can be traced back to the work of François Jacob and Jacques Monod on the regulatory mechanisms operating in micro-organisms (Brenner et al. 1990; Jacob and Monod 1961; Monod and Jacob 1961). In fact, by the 1960s it was rapidly recognized that the models of the Pasteurian molecular biologists (Jacob and Monod 1963), when extended to higher organisms, provided a solution to the problem raised as early as 1934 by T. H. Morgan, namely the need for control of the activities of genes during development and in different tissues of the adult organism. As early as 1962, the British embryologist Conrad H. Waddington recognized the power of these models to explain the control of the activity of the nuclear genes by signals coming from the cytoplasm. This appeared to provide a unique way to reconcile the role of cytoplasm (as emphasized by embryologists) and nucleus (as favored by geneticists) in the development of the organism (Gilbert 1991b; Waddington 1962).

The most important contribution of the models of F. Jacob and J. Monod, however, was elsewhere: These models distinguished in the genome two different kinds of genes, the structural genes coding for proteins and enzymes, and the regulatory genes coding for repressors, the sole function of which was to control the activity of structural genes (Jacob and Monod 1959). Jacob and Monod introduced a hierarchy in the genome (Morange 1990; Thieffry 1996): They suggested that, to understand development, one had to characterize the network of regulatory genes controlling this development, not the details of the functioning of structural genes. It was also implicit in the models elaborated by Jacob and Monod that modifications of regulatory genes, and regulatory networks, played central roles in the evolution of living beings. This idea was made explicit only later by F. Jacob (1977, 1981) and was reminiscent of the distinction between micromutations and macromutations advocated by the German geneticist Richard Goldschmidt (1940). The existence of such a link between evolution and the modification of regulatory genes was pursued by Mary-Claire King and Allan Wilson (King and Wilson 1975; Wilson, Carlson, and White 1977; Wilson, Maxson, and Sarich 1974), Stephen J. Gould (Gould 1977) and, later, by Rudolf Raff and Thomas Kaufman (1983). Like Goldschmidt before them, but in contrast to Jacob, Wilson and King correlated the modifications in the developmental regulatory systems to gene and chromosome rearrangements. By doing this they made the model heterodox and unacceptable by most molecular biologists (Wilson, Sarich, and Maxson

1974; Wilson et al. 1975). Further development of this work was limited by the complete absence of molecular tools allowing the isolation and characterization of these regulatory genes. This probably explains the caution of Jacob, as well as his personal reluctance to revive a ghost from the past – the macromutations – and to directly fight with the neo-Darwinists (Morange 1998, pp. 243–252).

A First Step: The Concept of Selector Gene

One of the geneticists who best understood the implications of this distinction between structural and regulatory genes was Antonio Garcia-Bellido (Garcia-Bellido 1975, 1977; Garcia-Bellido, Lawrence, and Morata 1979; Garcia-Bellido, Ripoll, and Morata 1973; Morange 1996). The existence of regulatory genes offered a simple explanation for the puzzling observations on the transdetermination of the cells of the imaginal disks discovered by Ernst Hadorn (Hadorn 1968) and studied by Garcia-Bellido during his stay in Hadorn's laboratory in Zurich. From these data, a genetic model with bistable control circuits was proposed by Stuart A. Kauffman (1973). In 1975, after a stay in California with Ed Lewis, Garcia-Bellido formulated a sophisticated and complete model of the genetics of development of *Drosophila* (Garcia-Bellido 1975, 1977). One of the main components of this model was the existence of a family of genes, the selector genes, that controlled the formation of compartments during *Drosophila* development. This model was publicized by Peter Lawrence and Francis Crick (Crick and Lawrence 1975; Morata and Lawrence 1977). The complexity of the model and the complex way it was presented, its theoretical value in the absence of any structural data on the selector genes, contributed to its having been forgotten by the time the tools of genetic engineering allowed the isolation of selector genes (and in particular of the homeotic genes) and the discovery of their extraordinary conservation during the evolution of living beings (Gehring and Ruddle 1998). This should not mask the fact that the models of Garcia-Bellido offered the first "genetic framework for *Drosophila* development" (Baker 1978) and paved the way for the future work of Walter Gehring, Christiane Nüsslein-Volhard, and Eric Wieschaus on the characterization of the early developmental genes (Keller 1996; Nüsslein-Volhard and Wieschaus 1980).

A Limited Number of Highly Conserved Developmental Genes

It has often been stated that the importance of Nüsslein-Volhard's work stems from her characterization of the maternal genes affecting early development of the embryo. Such a work would have constituted a sort of revenge of the cytoplasm on the nucleus (Keller 1995). In fact, the Nobel Prize was awarded to Nüsslein-Volhard and Wieschaus in 1995 for their work on the characterization and classification of the genes responsible for the segmentation of the *Drosophila* embryo, when they were at European Laboratories in Heidelberg. Most of these genes are expressed after fertilization in the embryo, by the zygotic genome, rather than by the oocyte. The importance of this work lay in the methodology used, saturation mutagenesis, that allowed the authors to demonstrate that the number of genes responsible for segmentation was limited to a few dozen. Early *Drosophila* development was controlled by a well-defined number of genes.

In the late 1970s and early 1980s, all the studies converged to suggest that development was due to the action of a limited number of genes controlling the activity of a battery of structural genes. Does this mean that the first molecular characterization of a master gene in 1984 was simply a verification of this well-accepted model? The previous presentation of the data may be misleading, because the main discovery responsible for the rapid formation and acceptance of the developmental gene concept was the demonstration that these genes had been highly conserved during evolution. This conservation was totally unexpected and never mentioned in previous articles published by developmental biologists: The development of the *Drosophila* fly seemed to be totally unrelated to the development of a vertebrate. What was sought instead was the conservation of a logic of development, the structure of a developmental program, and not the conservation of the genes active in this developmental process. This appears obvious when one considers the animals chosen in the 1960s as model systems to study development. *Caenorhabditis elegans,* adopted by Sydney Brenner, is very distantly related to vertebrates and mammals, and its pathway of development is totally different. In the 1960s and even later, this was not considered a handicap because only a strategy of development, not a history, was sought (Brenner 1984).

The existence of a structural element, the homeobox, conserved between different genes involved in *Drosophila* development was revealed by two different groups in 1984 (McGinnis 1994; McGinnis and Levine et al. 1984; McGinnis and Garber et al. 1984; Scott and Weiner 1984). This conservation came as no surprise, since Ed Lewis had already suggested that the different homeotic genes had been formed by successive duplication of an ancestral gene. The demonstration that this conserved motif was found in distantly related organisms – vertebrates such as *Xenopus,* mouse and humans – was more surprising (Carrasco et al. 1984; McGinnis and Garber et al. 1984; McGinnis and Hart et al. 1984). It is possible that the first experiments were performed as negative controls: The presence of this structural element initially was associated in the mind of the experimenters with the segmental nature of insects and arthropods. In the same year it was discovered that the conserved motif corresponded to a DNA-binding structure, allowing the proteins in which it was located to regulate the expression of other genes by binding to their promoters (Laughon and Scott 1984; Sheperd et al. 1984).

It was soon shown that this high level of conservation was not limited to the homeotic genes. Other motifs, related but not identical to the homeobox found in homeotic genes, were discovered in other *Drosophila* developmental genes and shown to have also been conserved during evolution (De Robertis 1994). These included the *otx* genes that play a fundamental role in the formation of the head (Acampora et al. 1995; Ang et al. 1996; Finkelstein and Boncinelli 1994; Matsuo et al. 1995), the *pax* genes (Gruss and Walther 1992), the *Lim* genes (Shawlot and Behringer 1995), and many other genes. The conservation was not limited to the DNA-binding motif, although it was often higher in this motif than in the surrounding parts of the gene. It also extended to the functions of these genes. These can be successfully exchanged between different organisms such as, for instance, between *Drosophila* and mammals (Malicki, Schughart, and McGinnis 1990; McGinnis, Kuziora, and McGinnis 1990). Other developmental genes unrelated to the homeobox-containing genes were also characterized and shown to have been conserved during evolution. These genes coded for molecules participating in cell-to-cell communication, such as secreted proteins *hedgehog* in *Drosophila* and related *hedgehog* genes in mice and humans, *wingless* in

Drosophila and the corresponding *wnt* genes in mammals, members of the transforming growth factor families (*decapentaplegic* in *Drosophila* and its homolog *BMP-4* in *Xenopus*), or cell membrane receptors (such as *Notch*), etc. Finally, from *in vitro* studies of cell differentiation and commitment, another family of developmental genes, including *MyoD,* was isolated (Gilbert, this volume). Perhaps even more important than the conservation of developmental genes between very different organisms such as *Drosophila* and mouse is the conservation of entire pathways or networks of regulatory genes between these organisms (Artavanis-Tsakonas, Matsuno, and Fortini 1995; Gaunt 1997; Goodrich et al. 1996).

THE SIGNIFICANCE OF THE DEVELOPMENTAL GENE CONCEPT

A Fundamental, Unexpected Discovery

Today, biologists widely share the conviction that they have identified at least some of the genes involved in the control of embryogenesis. Even if different names are used to designate these genes, biologists agree on their central significance, on the fact that their characterization is the best way to unravel the complexity of embryogenesis in all animals. The developmental gene is clearly different from the classical gene, as emphasized by Scott Gilbert in this volume.

As we saw previously, the characterization of the developmental genes was the result of a long process, from the initial demonstration of the involvement of genes in development to the isolation, among the thousands of genes harbored in an organism, of a small group of master genes. But, most of all, the importance attributed to the developmental genes resulted from the unexpected, unprecedented, and unexplained conservation of these genes. The discovery of developmental genes is a good example of a discovery due to a new technology, but also furnishes an historical event that is not explainable by strategies of the participants. As nicely described by Scott Gilbert, John Opitz, and Rudolf Raff (1996), the recent rediscovery of morphological fields in biology that had been marginalized by the development of genetics and the Modern Synthesis of genetics and evolution was important but was more a consequence of the interest in developmental genes than its cause. It would probably not have been sufficient to give rise to the new present-day developmental biology, if the extraordinary structural and functional conservation of developmental genes had not convinced biologists of their importance.

The Difficulties to Assign Precise Functions to Developmental Genes

Yet the conservation of these developmental genes during evolution, which is certainly the best argument in favor of their primordial role in the development of living organisms, raises a paradox: The same funcionally equivalent genes control the development of very different organisms which are built along different pathways and which have different plans of organization (Kenyon 1994).

A second difficulty arose from the study of the expression of these genes. Their importance in development was first confirmed in different experimental systems by the correlation between their expression and the inductive processes described more than 50 years ago by Hans Spemann and his coworkers (Saha 1991; Spemann 1938). However, this correlation did not remain as simple as it was originally. For instance, the first studies of Eddy M. De Robertis suggested that the homeobox-containing gene *goosecoid* played a central role in the functions of Spemann's organizer (Cho et al. 1991). Further studies revealed the intervention of many other proteins and genes and confirmed the complexity of the inductive processes and the division of Spemann's organizer into independently regulated organizing centers (De Robertis 1995).

A similar complexification of the initially illuminating data occurred each time the developmental biologists made an advance in the characterization of the different genes involved in a specific developmental process.

We will mainly focus our discussion on homeobox-containing genes of the homeotic complexes: They were discovered first, they are among the best conserved developmental genes, and their organization in the genome has been conserved as well. Today, more information has been collected on them than on any other developmental gene. The previously described difficulties explain the rapid succession of models proposed to interpret the role of these genes in mammals. They are gathered in four homeotic gene complexes, resulting from a twofold duplication of the complexes present in *Drosophila*. The first model proposed that these genes had the same functions as in *Drosophila* and were involved in the segmentation of the mammalian embryo, more precisely in the determination of the nature of its segments (Dressler and Gruss 1988). This model was supported by some Knock-Out (K.O.) experiments in mice showing that the genes' inactivation led to homeotic transformations (Le Mouellic, Lallemand, and Brûlet 1992). This first model assumed that segmentation in vertebrates is homologous to segmentation in insects and obeys the same mechanisms, a hypothesis that is still controversial (De Robertis 1997; Hogan, Holland, and Schofield 1985). When homeotic gene complexes were discovered in nonsegmented animals such as *C. elegans* (Kenyon and Wang 1991), it was proposed that homeotic genes only played a role in the antero-posterior patterning of the organisms. The observation that homeotic genes are also expressed in the limbs, and are essential for their formation, led to the elaboration of more sophisticated models in which homeotic genes were responsible for regional specification and/or axial patterning, constituting a positional code used during development (Hunt and Krumlauf 1992; Kessel and Gruss 1991; McGinnis and Krumlauf 1992; Wolpert 1995): Homeotic genes might simply be generators of spatial diversity. Recently, these models have been criticized in favor of a new vision in which homeotic genes bear the temporal information controlling cell proliferation and the consequent building of the organism: Vertebral specification, as well as digit size and number, are determined by the dose of a Hox protein rather than by a qualitative *Hox* code (Dollé et al. 1993; Duboule 1995; Zakany et al. 1997).

The same difficulties are met by developmental biologists when they try to correlate the modifications in homeotic genes with the morphological modifications that occurred during evolution, for instance during the evolution of vertebrates. Preliminary observations were very attractive: Modifications of homeotic gene expression led, in some cases, to atavistic transformations, to the reappearance of phenotypes that had disappeared during further evolution (Kessel, Balling, and Gruss 1990; Lufkin et al. 1992). Since homeotic genes and their organization have been extensively conserved during vertebrate evolution, transformations can result from a change in either the regulation of homeotic gene expression, or in the nature of the genes that are regulated by these homeotic genes (Carroll 1995; Gellon and McGinnis 1998; Shubin, Tabin, and Carroll 1997). For instance, the transition between fins and limbs is explained by an alteration of genes located upstream or downstream of homeotic genes (Sordino, van der Hoeven, and Duboule 1995). Similarly, the evolution of crustacean appendages is correlated with changes in *Hox* gene expression (Averof and Patel 1997). These interpretations attribute to the homeotic genes a badly defined role in development, a kind of empty framework (see Gilbert, this volume, for a tentative explanation of this paradox).

Regardless of which interpretation will ultimately be adopted to explain the role of the homeotic genes, two hard facts remain. The first is the central role of these genes, revealed by the (often) drastic effects resulting from their inactivation. It is rather surprising, however, that few spontaneous mutations in the homeotic genes of either humans or mice have ever been described (Mortlock, Post, and Innis 1996; Muragaki et al. 1996) whereas many mutations have been discovered in other developmental genes, such as the *pax* genes. One possibility is that the redundancy of homeotic genes is responsible for the scarcity of visible mutations. The second observation is the conservation of the structural organization of the homeotic gene complex in the genome during evolution. From *Drosophila* to man, the position of a gene inside the complex(es) has remained linked with its level and timing of expression in the embryo. The organization of the homeotic genes violates the particulate, "classical" conception of genes. The supragenetic organization of homeotic genes might play a role in the control of homeotic gene functioning. Denis Duboule has correlated this structural constraint to the embryological constraint that makes similar the development of all vertebrate embryos after the onset of the gastrulation phase (Duboule 1994). Are the structural constraints of development located in the structure of the genome, in the linear nature of the DNA molecule?

Homology and the Conservation of Developmental Genes

The conservation of developmental genes has revealed a hidden conservation of the body plans and structures between distantly related living beings. Probably Geoffroy Saint-Hilaire was right and Cuvier wrong: The bodies of insects and vertebrates are built according to similar instructions and differ only by an inversion of the dorso-ventral body axis (Arendt and Nübler-Jung 1994; De Robertis and Sasai 1996; Jones and Smith 1995). In a similar way, the development of the highly diverse arthropod limbs is based on common genetic mechanisms (Panganiban et al. 1995). The analysis of the genes involved in the construction of the eyes reveals a community (and relationship) of structures that was unexpected by biologists. It is commonly accepted that eyes have been invented many times during evolution (Salvini-Plawen and Mayr 1977). These eyes can adopt similar structures by convergence (such as in the cephalopod and vertebrate eyes). On the other hand, the compound eye of the insects is constructed according to principles different from those involved in the construction of vertebrate or cephalopod eyes. Despite these well-established facts, similar, related master genes seem to be involved in the construction of these different types of eyes. The modification of one of these genes, belonging to the *pax* gene superfamily, leads to abnormalities in the formation of eyes in *Drosophila* (eyeless), human (aniridia), or mouse (small eye) (Quiring et al. 1994). Overexpression and ectopic expression of this gene leads to the formation of ectopic eyes in *Drosophila* (Halder, Callaerts, and Gehring 1995a). In these experiments, the *Drosophila* gene can be replaced by the mouse gene, resulting in the formation of ectopic *Drosophila* eyes! A possible explanation for the conserved function of this gene is that it was initially essential for the formation of an ancestral sensitive organ, and only later in evolution was it recruited as the master gene for eye development (Burian 1997; Deutsch and Le Guyader 1995; Halder, Callaerts and Gehring 1995b), i.e., that the different vision systems in living organisms are both divergent from a primitive photosensitive system and convergent to higher forms of organization such as the eyes.

The developmental genes are the tools with which living beings have tinkered during evolution. The characterization of structural proteins in distantly related organisms had already shown that the same proteic components have been used many times during evolution, for different purposes. For instance, proteins with enzymatic functions have been recruited to form the crystallins that give the eye its transparency (Piatigorsky 1992). The same intracellular signaling pathways are used in *C. elegans* to regulate the formation of the vulva, in *Drosophila* to regulate the formation of the photoreceptors, and in mammals to control the rate of cell division (Egan et al. 1993). In other cases, the same protein has been used several times to fulfill the same purpose: The protein rhodopsin is responsible for light collection in primitive bacteria as well as in the highly organized vertebrate eye, as if, once this molecule was invented, it was used over and over, each time it was needed. Conservation of the developmental genes is perhaps no more significant than the conservation of these proteic components: Some specific characteristics of these genes were invented once and used repeatedly during evolution to face similar constraints (Dickinson 1995). In agreement with this hypothesis, it is already known that some of the developmental genes are used at many places and times in an embryo to regulate functionally similar processes. Importantly, developmental genes are rarely acting alone, but more frequently as "junta" (Desplan 1997). This redundancy of developmental genes may be seen as an efficient way to safeguard the complex process of development (Tautz 1992).

Were Developmental Genes Involved in the Cambrian Explosion?

It is tempting to correlate this conservation of developmental genes to what is known on the formation of metazoans. It was hypothesized since Darwin that the formation of metazoans probably occurred very rapidly during the early Cambrian geological age. The characterization of well-conserved faunas of this period has confirmed the reality of the Cambrian explosion and narrowed it in time (Gould 1989; Morris and Whittington 1979). The studies of faunas have shown that the diversity (disparity) resulting from the Cambrian explosion was even greater than the diversity that has been retained by further evolution.

It was hypothesized that the diversity of the living beings that formed during these crucial times resulted from the rapid acquisition by the first metazoans of the molecular tools, developmental genes, allowing a multicellular organism to build different and complex structures. The animals – but not the green plants since the developmental genes controlling the homeotic transformations in green plants are of a different nature (Weigel and Meyerowitz 1994) – would thus be defined by the existence of these highly conserved developmental genes, that would constitute what Slack calls the zootype (Slack, Holland, and Graham 1993).

More recent phylogenetic reconstructions do not completely support the previous scheme: The kit of developmental genes seems to have been present long before the Cambrian explosion (Balavoine and Adoutte 1998; Wray, Levinton and Shapiro 1996). The latter studies argue in favor of an external, environmental, cause for the Cambrian explosion.

Philosophical Meaning of the Developmental Gene Concept

The significance of developmental genes remains puzzling. If the developmental genes are really the master genes that are responsible for the formation of body plans and different organs, we can truly hope that their characterization will tell us something about the logic driving the formation of the presently known multicellular organisms. But it is also possible that the developmental genes are only the molecular components used to build the organisms: Their study will not reveal any principle of construction. Building of an organism is similar to the assembly of a nest by wasp colonies; it results only from the responses of individuals to local configurations and it is written nowhere (Theraulaz and Bonabeau 1995). It remains to be seen whether such a molecular description of development will be considered as a satisfactory explanation of the structures adopted by the organisms.

Moreover, the very notation of genes as developmental genes appears more and more problematic: If a gene codes for a proteic component involved in the regulation of cell division by growth factors, but is also essential for the formation of the vulva in C. elegans or the photoreceptors in Drosophila, can we call this gene a developmental gene? How must we consider a gene involved in the control of the most central part of metabolism, the regulation of glycogen metabolism, that also plays a major role in the definition of the primary morphogenetic axis of the embryo (He et al. 1995; Pierce and Kimelman 1995; Welsh, Wilson, and Proud 1996)? As we saw, the value of the developmental gene concept is already vanishing to the benefit of the developmental gene pathway concept. Will this new vision better support the wealth of data provided by the tools of molecular biology? Maybe the repeated use of a developmental gene pathway during evolution is simply a higher form of tinkering (Jacob 1977) and, no more than the conservation of individual genes, a mark of homology relationships (Gaunt 1997).

NOTES

1. We are indebted to Rosemary Sousa Yeh, François Jacob, André Adoutte, and Hervé Le Guyader for critical reading of the manuscript. We thank Peter Beurton, Scott Gilbert, and Stéphane Schmitt for providing bibliographic references and interesting suggestions.

REFERENCES

Acampora, D., S. Mazan, Y. Lallemand, V. Avantaggiato, M. Maury, A. Simeone, and P. Brûlet. 1995. Forebrain and midbrain regions are deleted in $Otx2^{-/-}$ mutants due to a defective anterior neuroectoderm specification during gastrulation. *Development* 121: 3279–3290.

Allen, G. 1974. Opposition to the Mendelian-chromosome theory: The physiological and developmental genetics of Richard Goldschmidt. *Journal of the History of Biology* 7: 49–92.

Allen, G. E. 1978. *Thomas Hunt Morgan: The man and his science.* Princeton: Princeton University Press.

Ang, S.-L., O. Jin, M. Rhinn, N. Daigle, L. Stevenson, and J. Rossant. 1996. A targeted mouse *Otx2* mutation leads to severe defects in gastrulation and formation of axial mesoderm and to deletion of rostral brain. *Development* 122: 243–252.

Arendt, D., and K. Nübler-Jung. 1994. Inversion of dorsoventral axis? *Nature* 371: 26.

Artavanis-Tsakonas, S., K. Matsuno, and M. E. Fortini. 1995. Notch signaling. *Science* 268: 225–232.

Averof, M., and N. H. Patel. 1997. Crustacean appendage evolution associated with changes in Hox gene expression. *Nature* 388: 682–686.

Baker, W. K. 1978. A genetic framework for Drosophila development. *Annual Review of Genetics* 12: 451–470.

Balavoine, G., and A. Adoutte. 1998. One or three Cambrian radiations? *Science* 280: 397–398.

Balkaschina, E. L. 1929. Ein Fall der Erbhomöosis (die Genovariation "Aristopedia") bei Drosophila melanogaster. *Wilhelm Roux's Archiv für Entwicklungsmechanik* 115: 448–463.

Brenner, S. 1984. Nematode research. *Trends in Biochemical Sciences* 9: 172.

Brenner, S., W. Dove, I. Herskowitz, and R. Thomas. 1990. Genes and Development: Molecular and logical themes. *Genetics* 126: 479–486.

Bridges, C. B. 1917. Deficiency. *Genetics* 2: 445–465.

Bridges, C. B., and T. Dobzhansky. 1933. The mutant "proboscipedia" in Drosophila melanogaster – a case of hereditary homoösis. *Wilhelm Roux's Archiv für Entwicklungsmechanik* 127: 575–590.

Burian, R. M. 1997. On conflicts between genetic and developmental viewpoints – and their resolution in molecular biology. In *Structure and Norms in Science*, edited by M. L. Dalla Chiara, K. Doetz, D. Mundici and J. van Bentham. Dordrecht: Kluwer, pp. 243–264.

Burian, R. M., J. Gayon, and D. T. Zallen. 1991. Boris Ephrussi and the synthesis of genetics and embryology. In *A Conceptual History of Modern Embryology*, edited by S. F. Gilbert. New York: Plenum, pp. 207–227.

Carrasco, A. E., W. McGinnis, W. J. Gehring, and E. M. De Robertis. 1984. Cloning of an *X. laevis* gene expressed during early embryogenesis coding for a peptide region homologous to Drosophila homeotic genes. *Cell* 37: 409–414.

Carroll, S. B. 1995. Homeotic genes and the evolution of arthropods and chordates. *Nature* 376: 479–485.

Chesley, P. 1935. Development of the short-tailed mutant in the house mouse. *Journal of Experimental Zoology* 70: 429–459.

Cho, K. W. Y., B. Blumberg, H. Steinbeisser, and E. M. De Robertis. 1991. Molecular nature of Spemann's organizer: The role of the Xenopus Homeobox gene *goosecoid. Cell* 67: 1111–1120.

Crick, F. H. C., and P. A. Lawrence. 1975. Compartments and polyclones in insect development. *Science* 189: 340–347.

De Robertis, E. M. 1994. The homeobox in cell differentiation and evolution. In *Guidebook to the Homeobox Genes*, edited by D. Duboule. Oxford: Oxford University Press, pp. 13–23.

De Robertis, E. M. 1995. Dismantling the organizer. *Nature* 374: 407–408.

De Robertis, E. M. 1997. The ancestry of segmentation. *Nature* 387: 25–26.

De Robertis, E. M., and Y. Sasai. 1996. A common plan for dorsoventral patterning in the Bilateria. *Nature* 380: 37–40.

Desplan, C. 1997. Eye development: Governed by a dictator or a junta? *Cell* 91: 861–864.

Deutsch, J., and H. Le Guyader. 1995. Le fond de l'oeil: l'oeil de la drosophile est-il homologue de celui de la souris? *Médecines/Sciences* 11: 1447–1452.

Dickinson, W. J. 1995. Molecules and morphology: Where's the homology? *Trends in Genetics* 11: 119–121.

Dollé, P., A. Dierich, M. LeMeur, T. Schimmang, B. Schuhbaur, P. Chambon, and D. Duboule. 1993. Disruption of the *Hoxd-13* gene induces localized heterochrony leading to mice with neotenic limbs. *Cell* 75: 431–441.

Dressler, G. R., and P. Gruss. 1988. Do multigene families regulate vertebrate development? *Trends in Genetics* 4: 214–219.

Duboule, D. 1994. Temporal colinearity and the phylotypic progression: a basis for the stability of the vertebrate Bauplan and the evolution of morphologies through heterochrony. *Development* (Supplement): 135–142.

Duboule, D. 1995. Vertebrate *Hox* genes and proliferation: An alternative pathway to homeosis? *Current Opinion in Genetics and Development* 5: 525–528.

Egan, S. E., B. W. Giddings, M. W. Brooks, L. Buday, A. M. Sizeland, and R. A. Weinberg. 1993. Association of Sos Ras exchange protein with Grb2 is implicated in tyrosine kinase signal transduction and transformation. *Nature* 363: 45–51.

Finkelstein, R., and E. Boncinelli. 1994. From fly head to mammalian forebrain: The story of *otd* and *Otx. Trends in Genetics* 10: 310–315.

Garcia-Bellido, A. 1975. Genetic control of wing disc development in *Drosophila.* In *Cell Patterning,* edited by S. Brenner. Amsterdam: Associated Scientific Publishers, pp. 161–182.

Garcia-Bellido, A. 1977. Homeotic and atavic mutations in insects. *American Zoologist* 17: 613–629.

Garcia-Bellido, A., P. A. Lawrence, and G. Morata. 1979. Compartments in animal development. *Scientific American* 241 (1): 90–98.

Garcia-Bellido, A., P. Ripoll, and G. Morata. 1973. Developmental compartmentalisation of the wing disk of *Drosophila. Nature. New Biology* 245: 251–253.

Gaunt, S. J. 1997. Chick limbs, fly wings and homology at the fringe. *Nature* 386: 324–325.

Gehring, W. J., and F. Ruddle. 1998. *Master Control Genes in Development and Evolution: The Homeobox Story.* New Haven: Yale University Press.

Gellon, G., and W. McGinnis. 1998. Shaping animal body plans in development and evolution by modulation of *Hox* expression patterns. *BioEssays* 20: 116–125.

Gilbert, S. F. 1988. Cellular politics: Ernest Everett Just, Richard B. Goldschmidt and the attempt to reconcile embryology and genetics. In *The American Development of Biology,* edited by R. Rainger, K. Benson and J. Maienschein. New Brunswick: Rutgers University Press, pp. 311–346.

Gilbert, S. F. 1991a. Induction and the origins of developmental genetics. In *A Conceptual History of Modern Embryology,* edited by S. Gilbert. Baltimore: The Johns Hopkins University Press, pp. 181–206.

Gilbert, S. F. 1991b. Commentary: Cytoplasmic action in development. *The Quarterly Review of Biology* 66: 309–316.

Gilbert, S. F., J. M. Opitz, and R. A. Raff. 1996. Resynthesizing evolutionary and developmental biology. *Developmental Biology* 173: 357–372.

Gluecksohn-Schoenheimer, S. 1938. The development of two tailless mutants in the house mouse. *Genetics* 23: 573–584.

Gluecksohn-Schoenheimer, S. 1940. The effect of an early lethal ($t°$) in the house mouse. *Genetics* 25: 391–400.

Gluecksohn-Schoenheimer, S. 1949. The effects of a lethal mutation responsible for duplications and twinning in mouse embryos. *Journal of Experimental Zoology* 110: 47–76.

Goldschmidt, R. 1938. *Physiological Genetics*. New York: McGraw-Hill.

Goldschmidt, R. 1940. *The Material Basis of Evolution*. New Haven: Yale University Press.

Goodrich, L. V., R. L. Johnson, L. Milenkovic, J. A. McMahon, and M. P. Scott. 1996. Conservation of the *hedgehog/patched* signaling pathway from flies to mice: Induction of a mouse *patched* gene by Hedgehog. *Genes and Development* 10: 301–312.

Gould, S. J. 1977. *Ontogeny and Phylogeny*. Cambridge, Mass.: Belknap Press of Harvard University Press.

Gould, S. J. 1989. *Wonderful Life: The Burgess Shale and the Nature of History*. New York: Norton.

Gruss, P., and C. Walther. 1992. Pax in development. *Cell* 69: 719–722.

Hadorn, E. 1968. Transdetermination in cells. *Scientific American* 219 (November): 110–120.

Halder, G., P. Callaerts, and W. J. Gehring. 1995a. Induction of ectopic eyes by targeted expression of the *eyeless* gene in Drosophila. *Science* 267: 1788–1792.

Halder, G., P. Callaerts, and W. J. Gehring. 1995b. New perspectives on eye evolution. *Current Opinion in Genetics and Development* 5: 602–609.

Harwood, J. 1993. *Styles of Scientific Thought: The German Genetics Community. 1900–1933*. Chicago: University of Chicago Press.

He, X., J.-P. Saint-Jeannet, J. R. Woodgett, H. E. Varmus, and I. B. Dawid. 1995. Glycogen synthase kinase-3 and dorsoventral patterning in *Xenopus* embryos. *Nature* 374: 617–622.

Hogan, B., P. Holland, and P. Schofield. 1985. How is the mouse segmented? *Trends in Biochemical Sciences* 10: 67–74.

Hunt, P., and R. Krumlauf. 1992. *Hox* codes and positional specification in vertebrate embryonic axes. *Annual Review of Cell Biology* 8: 227–256.

Jacob, F. 1977. Evolution and tinkering. *Science* 196: 1161–1166.

Jacob, F. 1981. *Le jeu des possibles*. Paris: Fayard.

Jacob, F., and J. Monod. 1959. Gènes de structure et gènes de régulation dans la biosynthèse des protéines. *Comptes Rendus de l'Académie des Sciences de Paris* 249: 1282–1284.

Jacob, F., and J. Monod. 1961. Genetic regulatory mechanisms in the synthesis of proteins. *Journal of Molecular Biology* 3: 318–356.

Jacob, F., and J. Monod. 1963. Genetic repression, allosteric inhibition and cellular differentiation. In *Cytodifferentiation and Macromolecular Synthesis*, edited by M. Locke. New York: Academic Press, pp. 30–64.

Jones, C. M., and J. C. Smith. 1995. Revolving vertebrates. *Current Biology* 5: 574–576.

Kauffman, S. A. 1973. Control circuits for determination and transdetermination. *Science* 181: 310–318.

Keller, E. F. 1995. *Refiguring Life: Metaphors of Twentieth-Century Biology*. New York: Columbia University Press.

Keller, E. F. 1996. *Drosophila* embryos as transitional objects: The work of Donald Poulson and Christiane Nüsslein-Volhard. *Historical Studies in the Physical and Biological Sciences* 26: 313–346.

Kenyon, C. 1994. If birds can fly, why can't we? Homeotic genes and evolution. *Cell* 78: 175–180.

Kenyon, C., and B. Wang. 1991. A cluster of *Antennapedia*-class homeobox genes in a nonsegmented animal. *Science* 253: 516–517.

Kessel, M., R. Balling, and P. Gruss. 1990. Variations of cervical vertebrae after expression of a *Hox-1.1* transgene in mice. *Cell* 61: 301–308.

Kessel, M., and P. Gruss. 1991. Homeotic transformations of murine vertebrae and concomitant alteration of *Hox* codes induced by retinoic acid. *Cell* 67: 89–104.

King, M.-C., and A. C. Wilson. 1975. Evolution at two levels in humans and chimpanzees. *Science* 188: 107–116.

Kohler, R. E. 1991. Systems of production: Drosophila, neurospora and biochemical genetics. *Historical Studies in the Physical and Biological Sciences* 22: 87–130.

Landauer, W. 1944. Length of survival of homozygous creeper fowl embryos. *Science* 100: 553–554.

Laughon, A., and M. P. Scott. 1984. Sequence of a *Drosophila* segmentation gene: Protein structure homology with DNA-binding proteins. *Nature* 310: 25–31.

Le Mouellic, H., Y. Lallemand, and P. Brûlet. 1992. Homeosis in the mouse induced by a null mutation in the *Hox-3.1* gene. *Cell* 69: 251–264.

Lewis, E. B. 1951. Pseudoallelism and gene evolution. *Cold Spring Harbor Symposia on Quantitative Biology* 16: 159–174.

Lewis, E. B. 1963. Genes and developmental pathways. *American Zoologist* 3: 33–56.

Lewis, E. B. 1978. A gene complex controlling segmentation in Drosophila. *Nature* 276: 565–570.

Lewis, E. B. 1992. Clusters of master control genes regulate the development of higher organisms. *Journal of American Medical Association* 267: 1524–1531.

Li, J.-C. 1927. The effect of chromosome aberrations on development in *Drosophila melanogaster. Genetics* 12: 1–58.

Lufkin, T., M. Mark, C. P. Hart, P. Dollé, M. LeMeur, and P. Chambon. 1992. Homeotic transformation of the occipital bones of the skull by ectopic expression of a homeobox gene. *Nature* 359: 835–841.

Malicki, J., K. Schughart, and W. McGinnis. 1990. Mouse *Hox-2.2* specifies thoracic segmental identity in Drosophila embryos and larvae. *Cell* 63: 961–967.

Matsuo, I., S. Kuratani, C. Kimura, N. Takeda, and S. Aizawa. 1995. Mouse *otx2* functions in the formation and patterning of rostral head. *Genes and Development* 9: 2646–2658.

McGinnis, N., M. A. Kuziora, and W. McGinnis. 1990. Human *Hox-4.2* and Drosophila *Deformed* encode similar regulatory specificities in Drosophila embryos and larvae. *Cell* 63: 969–976.

McGinnis, W. 1994. A century of homeosis, a decade of homeoboxes. *Genetics* 137: 607–611.

McGinnis, W., R. L. Garber, J. Wirz, A. Kuroiwa, and W. J. Gehring. 1984. A homologous protein-coding sequence in Drosophila homeotic genes and its conservation in other metazoans. *Cell* 37: 403–408.

McGinnis, W., C. P. Hart, W. J. Gehring, and F. H. Ruddle. 1984. Molecular cloning and chromosome mapping of a mouse DNA sequence homologous to homeotic genes of Drosophila. *Cell* 38: 675–680.

McGinnis, W., and R. Krumlauf. 1992. Homeobox genes and axial patterning. *Cell* 68: 283–302.

McGinnis, W., M. S. Levine, E. Hafen, A. Kuroiwa, and W. J. Gehring. 1984. A conserved DNA sequence in homoeotic genes of the *Drosophila* Antennapedia and bithorax complexes. *Nature* 308: 428–433.

Monod, J., and F. Jacob. 1961. Teleonomic mechanisms in cellular metabolism, growth and differentiation. *Cold Spring Harbor Symposia on Quantitative Biology* 26: 389–401.

Morange, M. 1990. Le concept de gène régulateur. In *Histoire de la génétique: pratiques, techniques, théories,* edited by J.-L. Fischer and W. H. Schneider. Paris: ARPEM et Sciences en situation, pp. 271–291.

Morange, M. 1996. Construction of the developmental gene concept. The crucial years: 1960–1980. *Biologisches Zentralblatt* 115: 132–138.

Morange, M. 1998. *A History of Molecular Biology.* Cambridge, Mass.: Harvard University Press.

Morata, G., and P. A. Lawrence. 1977. Homeotic genes, compartments and cell determination in *Drosophila. Nature* 265: 211–216.

Morgan, T. H. 1934. *Embryology and Genetics.* New York: Columbia University Press.

Morris, S. C., and H. B. Whittington. 1979. The animals of the Burgess shale. *Scientific American* 241 (July): 110–120.

Mortlock, D. P., L. C. Post, and J. W. Innis. 1996. The molecular basis of hypodactyly (Hd): A deletion in *Hoxa13* leads to arrest of digital arch formation. *Nature Genetics* 13: 284–289.

Muragaki, Y., S. Mundlos, J. Upton, and B. R. Olsen. 1996. Altered growth and branching patterns in synpolydactyly caused by mutations in HOXD13. *Science* 272: 548–551.

Nüsslein-Volhard, C., and E. Wieschaus. 1980. Mutations affecting segment number and polarity in *Drosophila*. *Nature* 287: 795–801.

Panganiban, G., A. Sebring, L. Nagy, and S. Carroll. 1995. The development of crustacean limbs and the evolution of arthropods. *Science* 270: 1363–1366.

Piatigorsky, J. 1992. Lens crystallins: Innovation associated with changes in gene regulation. *Journal of Biological Chemistry* 267: 4277–4280.

Pierce, S. B., and D. Kimelman. 1995. Regulation of Spemann organizer formation by the intracellular kinase Xgsk-3. *Development* 121: 755–765.

Poulson, D. F. 1937. Chromosomal deficiencies and the embryonic development of *Drosophila melanogaster*. *Proceedings of the National Academy of Sciences USA* 23: 133–137.

Quiring, R., U. Walldorf, U. Kloter, and W. J. Gehring. 1994. Homology of the *eyeless* gene of *Drosophila* to the *Small eye* gene in mice and *Aniridia* in humans. *Science* 265: 785–789.

Raff, R. A., and T. C. Kaufman. 1983. *Embryos, Genes, and Evolution: The Developmental-Genetic Basis of Evolutionary Change*. Bloomington: Indiana University Press.

Saha, M. 1991. Spemann seen through a lens. In *Developmental Biology*, edited by S. F. Gilbert. New York: Plenum Press.

Salvini-Plawen, L. V., and E. Mayr. 1977. On the evolution of photoreceptors and eyes. *Evolutionary Biology* 10: 207–263.

Sapp, J. 1983. The struggle for authority in the field of heredity, 1900–1932: New perspectives on the rise of genetics. *Journal of the History of Biology* 16: 311–342.

Sapp, J. 1987. *Beyond the Gene: Cytoplasmic Inheritance and the Struggle for Authority in Genetics*. New York: Oxford University Press.

Scott, M. P., and A. J. Weiner. 1984. Structural relationships among genes that control development: sequence homology between the Antennapedia, Ultrabithorax, and fushi tarazu loci of *Drosophila*. *Proceedings of the National Academy of Sciences USA* 81: 4115–4119.

Shawlot, W., and R. Behringer. 1995. Requirement for *Lim1* in head-organizer function. *Nature* 374: 425–430.

Sheperd, J. C. W., W. McGinnis, A. E. Carrasco, E. M. De Robertis, and W. J. Gehring. 1984. Fly and frog homoeo domains show homologies with yeast mating type regulatory proteins. *Nature* 310: 70–71.

Shubin, N., C. Tabin, and S. Carroll. 1997. Fossils, genes and the evolution of animal limbs. *Nature* 388: 639–648.

Slack, J. M. W., P. W. H. Holland, and C. F. Graham. 1993. The zootype and the phylotypic stage. *Nature* 361: 490–492.

Sordino, P., F. van der Hoeven, and D. Duboule. 1995. *Hox* gene expression in teleost fins and the origin of vertebrate digits. *Nature* 375: 678–681.

Spemann, H. 1938. *Embryonic Development and Induction.* New Haven: Yale University Press.

Tautz, D. 1992. Redundancies, development and the flow of information. *BioEssays* 14: 263–266.

Theraulaz, G., and E. Bonabeau. 1995. Coordination in distributed building. *Science* 269: 686–688.

Thieffry, D. L. 1996. *Escherichia coli* is a model system with which to study cell differentiation. *History and Philosophy of the Life Sciences* 18: 163–193.

Waddington, C. H. 1939. Preliminary notes on the development of the wings in normal and mutant strains of Drosophila. *Proceedings of the National Academy of Sciences USA* 25: 299–307.

Waddington, C. H. 1962. *New Patterns in Genetics and Development.* New York: Columbia University Press.

Weigel, D., and E. M. Meyerowitz. 1994. The ABCs of floral homeotic genes. *Cell* 78: 203–209.

Welsh, G. I., C. Wilson, and C. G. Proud. 1996. GSK3: A shaggy frog story. *Trends in Cell Biology* 6: 274–279.

Wilson, A. C., G. L. Bush, S. M. Case, and M.-C. King. 1975. Social structuring of mammalian populations and rate of chromosomal evolution. *Proceedings of the National Academy of Sciences USA* 72: 5061–5065.

Wilson, A. C., S. S. Carlson, and T. J. White. 1977. Biochemical evolution. *Annual Review of Biochemistry* 46: 573–639.

Wilson, A. C., L. R. Maxson, and V. M. Sarich. 1974. Two types of molecular evolution: Evidence from studies of interspecific hybridization. *Proceedings of the National Academy of Sciences USA* 71: 2843–2847.

Wilson, A. C., V. M. Sarich, and L. R. Maxson. 1974. The importance of gene rearrangement in evolution: Evidence from studies on rates of chromosomal, protein, and anatomical evolution. *Proceedings of the National Academy of Sciences USA* 71: 3028–3030.

Wolpert, L. 1995. Bet on positional information. *Nature* 373: 112.

Wray, G. A., J. S. Levinton, and L. H. Shapiro. 1996. Molecular evidence for deep pre-Cambrian divergences among metazoan phyla. *Science* 274: 568–573.

Zakany, J., C. Fromental-Ramain, X. Warot, and D. Duboule. 1997. Regulation of number and size of digits by posterior *Hox* genes: A dose-dependent mechanism with potential evolutionary implications. *Proceedings of the National Academy of Sciences USA* 94: 13695–13700.

4

Conceptual Perspectives

10

Gene Concepts

Fragments from the Perspective of Molecular Biology[1]

HANS-JÖRG RHEINBERGER

"[It is] the vague, the unknown that moves the
world."
(Bernard 1954, 26).

ABSTRACT

The paper is divided into three parts. In the first part, I argue for an epistemology of the imprecise and try to characterize the historical and disciplinary trajectory of gene representations as the trajectory of an exemplar of a boundary object. In the second part, I follow the apparently simple solution of the gene problem to which early molecular biology gave rise, and I retrace some of the steps and events through which the later development of molecular biology came to explode this simple notion. The last part derives some conjectures from this story and seizes upon the notion of *integron* developed by François Jacob in order to establish a symmetrical perspective on genomes and phenomes.

INTRODUCTION

In what follows, I intend to cast some light on the changing epistemic and experimental dispositions through which molecular biology came to deal with genes. The paper is meant neither as a systematic assessment nor as a critique of the way molecular biology has appropriated this concept. Nor will I be able to retrace the history of the gene as an object of experimentation in molecular biology in all its complexity. My concern in this overview is to point out, in a loose and associative fashion, some questions that I think will have to be addressed if we wish to understand where the second half of the twentieth century has taken us with respect to that unit that Herman

J. Muller, on the occasion of the fiftieth anniversary of the rediscovery of Gregor Mendel's pea work, described with the following words: "[T]he real core of gene theory still appears to lie in the deep unknown. That is, we have as yet no actual knowledge of the mechanism underlying that unique property which makes a gene a gene – its ability to cause the synthesis of another structure like itself . . . in which even the mutations of the original gene are copied. . . . We do not know of such things yet in chemistry" (Muller 1951, 95–96). These remarkable sentences were written in 1950. What has molecular biology taught us since then about those "unique properties which make a gene a gene"? We have come a long way since the days of Muller. Yet we will see that instead of solving the riddle of the gene and rescuing it forever from the deep unknown, molecular biology has managed to redefine its properties and its boundaries, and it has continued to change our conception of this strange entity repeatedly and almost beyond recognition.

EPISTEMOLOGY: FLUCTUATING OBJECTS AND IMPRECISE CONCEPTS

If there are concepts endowed with organizing power in a research field, they are embedded in experimental operations. The practices in which the sciences are grounded engender epistemic objects, epistemic things as I call them, as targets of research. Despite their vagueness, these entities move the world of science. As a rule, disciplines become organized around one or a few of these "boundary objects" that underlie the conceptual translations between different domains (Star and Griesemer 1988). For a long time in physics, such an object has been the atom; in chemistry, the molecule; in classical genetics, it became the gene. It is the historically changing set of epistemic practices that gives contours to these objects. According to received accounts, that I need not question here in depth, the boundary object of classical genetics has worked as a formal unit: That which, in an ever more sophisticated context of breeding experiments, accounts for the appearance or disappearance of certain characters that can be traced through subsequent generations. Accordingly, what has made classical genetics different from nineteenth-century inquiries in heredity, is that its practice allowed combining the notion of character discreteness, rooted in the Darwinian and early de Vriesian traditions, with Weismann's distinction between germ plasm and body substance. The result, read back into Mendel's experiments, was a deliberate distinction between genetic units and unit characters, taken in their entirety, between genotype and phenotype, respectively.

This is a reminder, and trivial to that extent. Let me give now an equivalent caricature of what molecular genetics has contributed to the field. At the beginning, molecular genetics, with its set of biochemical practices and genetic manipulations, was characterized by switching from higher plants and animals to bacteria and phages as model organisms. First, it transformed its boundary object, the gene, into a material, physicochemical entity. Second, it has made a unit endowed with informational qualities from the object. The first transformation provided a solution to the problem that classical genetics had with the stability of its units. The answer was: Genes consist of metastable macromolecules of such as nucleic acids. The second transformation provided a solution to the problem that classical genetics had with its units' mode of reproduction, and the connection between genotype and phenotype. The answer was: Nucleotide sequences, and DNA in particular, can be replicated specifically and faithfully by virtue of the stereochemical properties of their building blocks. In addition, DNA stretches specify traits by virtue of the ordered sequence of nucleotides they contain for being translated, with the help of a complex cytoplasmic machinery, into corresponding sequences of amino acids that yield structural proteins or enzymes catalyzing all sorts of metabolic reactions.

What I would like to stress against this excessively schematic and simple-looking outline, so suggestive in its clarity and distinctness after the event, is that the fruitfulness of boundary objects in research does not depend on whether they can be given a precise and codified meaning from the outset. Stated otherwise, it is not necessary, indeed it can be rather counterproductive, to try to sharpen the conceptual boundaries of vaguely bounded research objects while in operation. As long as the objects of research are in flux, the corresponding concepts must remain in flux, too (Elkana 1970). Boundary objects require boundary concepts. The fruitfulness of such concepts depends on their operational potential. "All definitions of the gene require operational criteria" (Portin 1993, 174). It is these criteria that make them work as definitions, if at all.

The spectacular rise of molecular biology has come about without a comprehensive, exact, and rigid definition of what a gene is. As I will trace in the historical section of this essay, this claim can be substantiated for both aspects distinguishing the gene concept of molecular biology from that of classical genetics: the aspect of representing a material entity, and that of being a carrier of information (Sarkar 1996). The meaning of both of these notions has remained fuzzy and tied to the experimental spaces that the new biology was going to explore, from the identification of DNA as the hereditary material in bacteria in 1944 to the genome sequencing projects of the late 1980s. I am even inclined to postulate that attempts at precise definitions generally have worked as epistemological obstacles (Bachelard 1957), as theoretical artifacts at best, such as the early efforts to provide a purely quantitative definition of biological information in the sense of information theory (Kay, forthcoming). Just to give one more example: On the basis of his bacteriophage mapping experiments, Seymour Benzer tried to clarify the messy field in 1955 and dissolve the gene into three units of hereditary expression or function, of recombination, and of mutation. He called these units *cistron, recon,* and *muton,* respectively (Benzer 1955; Holmes, this volume). No doubt, the distinctions were theoretically justified. They relied on the most advanced experiments in phage genetics. A cistron was analyzable into recons, a recon into mutons, presumably ending with the simplest possible unit, a DNA base pair. But despite their clarity, for the variegated community of molecular experimentalists, these distinctions proved to be too restrictive, partially redundant, and prone to conflation. They did not catch on in the long run.

I contend, in contrast to other authors in this volume, that it is not the task of the epistemologist either to criticize or try to specify vague concepts in the hope of helping scientists clarify their convoluted minds and do better science with them. There is an urgent need, however, to understand how and why fuzzy concepts work in science (Moles 1995). Instead of trying to codify precision of meaning, we need an epistemology of the vague and the exuberant. Boundary objects and boundary concepts operate on and derive their power from a peculiar epistemic tension: To be tools of research, they must reach out into the realm of what we do not yet know. I would like to characterize this tension as *contained excess*. François Jacob, in a similar context, has spoken of a "game of the possible" (Jacob 1981). If we look at contemporary textbooks written by researchers from the forefront of molecular biology, we find very sloppy definitions of the gene – if we find definitions at all. Quite obviously, molecular genetics is not held together by such a definition. Let us take this as a lesson about the dynamics of science instead of accusing the actors of being careless in defining the entities they work with. It is revealing and intriguing indeed that, for example, in the glossary appended to a recent book, *Secrets of the Gene*, written by the French molecular biologist and former director of the Pasteur Institute, François Gros, there is not even an entry for the term *gene* (Gros 1991). There is an entry, however, for the term *genome*, to which I will come back. This is by no means a new feature of recent literature, however. In the glossary to Leslie Dunn's 1965 classic *A Short History of Genetics* we find, at the end of the entry *gene*, the following caveat: "At present, discussions of properties to be explained are more useful than attempts at rigid definition" (Dunn 1991 [1965], 234).

Do molecular biologists need a unified and generalized gene concept? As mentioned, if we screen the pertinent literature, there appears to be no singular, unique, and rigidly determined usage of the term. What we find is context dependence (Fogle, this volume). Despite its unifying appeal, molecular biology is made up of many different contexts, in terms of disciplinary contributions, of experimental systems, and of genome reading conditions. Boundary objects and related concepts work because they are malleable and can be adapted according to the varying needs in these different contexts. But in order to grant a certain coherence to the field, these translations must remain reversible. We may say that two strategies are at work here. One consists of trying to give boundary objects sharper contours according to specific experimental contexts. Paradoxically, but as a rule, this leads to the elimination from discourse of the objects so defined. For example: The attempt to use ribosomal RNA as a template for bacterial protein synthesis led Marshall Nirenberg and Heinrich Matthaei to the characterization of a nonribosomal messenger RNA, and consequently, a nontemplate ribosomal RNA (Rheinberger 1997, chapter 13). In lucky cases such as this one, such shifts in the "reference potential" of scientific expressions (Kitcher 1982) go along with the emergence of unprecedented new objects that are at least as vague as those from which the search began. The other strategy consists of immunizing the conceptual framework against the fluctuations that result from the first strategy. In the case mentioned above, microsomal templates remained an option for all those whose work was based on eukaryotic cells. Maintaining such conceptual indeterminacy over a period of time, however, requires a certain level of imprecision, and thus of fuzzy boundaries.

Molecular biology is a hybrid science combining experimental systems from biophysics, biochemistry, and genetics, and it uses widely different model organisms in its search for biological function at the molecular level. Not surprisingly, it presents itself as conceptually hybrid as well, which is not to say that it has no consistency. Precisely as the result of its translational power, the discourse of molecular biology pervades contemporary biology as a whole, including evolution. The message I would like to disseminate is that we have to learn more about such hybrid consistencies: how they come about, how they work, and how they evolve.

Here are a few examples of fragmented perspectives on genes, both enabled and constrained by experimental systems that are situated in these different domains. For a biophysicist working with a crystalline DNA fiber and an X-ray apparatus, a gene might be sufficiently characterized by a particular conformation of a double helix. If asked, he or she might define a gene in terms of the atomic coordinates of the nucleic acid bases. For a biochemist working with isolated DNA fragments in the test tube, genes might be sufficiently defined as nucleotide polymers exhibiting certain stereochemical features and recurrent sequence patterns. The biochemist can reasonably try to give a macromolecular definition of the gene, based on the unique chemical features of DNA. For a molecular geneticist, genes might be defined as informational elements of chromosomes that eventually give rise to specific functional or structural products: transfer RNAs, ribosomal RNAs, enzymes, and proteins destined to serve other purposes. Molecular geneticists certainly will insist on considering these issues in terms of replication, transcription, and translation and will require examination of the products of hereditary units when speaking of genes. For those interested in evolution, genes might be the products of mutated, reshuffled, duplicated, transposed, and rearranged bits of DNA within a complex chromosomal environment that has evolved through differential reproduction, selection, or other evolutionary mechanisms (Beurton, this volume). Therefore, evolutionists will rely on concepts such as transmission, lineage, and historical contingency. For developmental geneticists, genes might be sufficiently described, on the one hand, as hierarchically ordered switches that, when turned on or off, induce differentiation, and on the other hand, as patches of instructions that are realized in synchrony through the action of these switches (Gilbert and Morange, this volume). Thus, developmental biologists are likely to refer to the regulatory aspects of genetic circuitry when defining a gene or a larger transcriptional unit such as an operon. We could continue and add more items to the list.

Is it necessary, or even desirable, to have a unified concept of the gene in order to tie all disciplinary specializations together and to develop them in a coordinated fashion? From a historical perspective, this has not been the case in the half century since molecular biology came into existence. Quite frankly, I do not think that it would have helped the development of the field in appreciable ways; further, I contend that an attempt to do so today would produce nothing more than a rhetorical exercise. The coherence of molecular biology is not tied into an axiomatic structure or an algorithm; it is embedded in a complex set of experimental systems, each with its generic epistemic practices, that have evolved over time and that have constrained earlier interpretations as well as allowed new ambiguities to arise. Genes as we now know them are boundary objects par excellence that are crafted, more than by any theory, by the practices and instruments that helped to create the new biology.

HISTORY: EARLY RIPPLES AND RECENT WAVES[2]

As mentioned at the beginning, the development of gene concepts looks rather straightforward from a historical vantage point from which surprising convergences can be observed *post hoc*. It can, however, be questioned whether we should reconstruct the history of genetics as a history of gene concepts. My contention is that this would amount to a fallacious epistemological artifact, if such reconstruction were not tied to a history of the usages of these concepts. The following staccato account cannot do full justice to this claim. It inevitably cuts short laborious and errant experimental explorations that proceeded from widely different starting points, and sometimes, were not even motivated by genetic questions at the outset (for a more detailed account see Morange 1998).

At the beginning of molecular genetics as distinguished from classical genetics stands the search for a materially grounded, molecular constitution of the units of heredity. In the late 1920s, H. J. Muller as well as Lewis Stadler showed that X-rays can be used to mutate genes (Muller 1927; Stadler 1928). But for the majority of the classical geneticists involved in extended crossing and breeding experiments, at least in the first three decades of the past century, the material constitution of the gene was a problem that needed no answer within the realm of their own experimental regimes. For physicist Max Delbrück, who turned to biology in the 1930s, however, this became the central target to be addressed. As far as his reasoning went, genes could be assumed to be autocatalytic proteins that had the capacity to become permanently altered through physical intervention. After some early attempts at determining the gene as an elementary physical unit (Timoféeff-Ressovsky, Zimmer, and Delbrück 1935), he set out to investigate phages that he imagined as being the smallest equivalents of genes independently occurring in nature (Delbrück 1942). It is one of the many ironies of the history of molecular biology that phage as an experimental system became as formal as the systems of classical genetics. What physics contributed was a set of rigorous measurement techniques. Only after a long detour, and in conjunction with other developments, did phage research finally come to contribute to the physical elucidation of the gene.

A milestone in the transition from classical genetics to molecular genetics was reached with George Beadle and Edward Tatum's one-gene – one-enzyme hypothesis. It emerged from the hybrid discipline of biochemical genetics based on the mold *Neurospora crassa* as a model organism (Beadle 1945). The one-gene – one-enzyme hypothesis is quite different from the one-gene – one-character relation of early classical genetics (Schwartz, this volume). As a rule, enzymes intervene in the metabolic production of characters, or phenes. One particular character can depend on a whole cascade of enzymatic reactions, something envisaged by Alfred Kühn as "gene action chains": genes acting on "substrate chains" via ferment systems (Kühn 1941). One mutable entity, in turn, can affect many characters if the respective enzyme (or gene product) acts at the basis of metabolic bifurcations, thus explaining the long-observed phenomenon of pleiotropy. Still, neither Beadle and Tatum nor Kühn, in approaching genes by tracing their biochemical products, came anywhere near to illuminating the physical agency of the genes they were investigating.

The new genetics had to be confronted with Oswald Avery's transformation experiments with different types and strains of *Pneumococcus* (Avery, MacLeod, and McCarty 1944) as well as with chemist Erwin Chargaff's analysis of the specific base composition of deoxyribonucleic acid toward the end of the 1940s before genes started to be addressed as made of DNA instead of representing autocatalytic proteins (Vischer, Zamenhoff, and Chargaff 1949). The one-gene – one-enzyme hypothesis now took the form of one segment of chromosomal DNA – one proteinic enzyme.

At this point, determining the molecular structure of DNA became a desirable, if not urgent task. I have to cut short here the long biochemical and physical story of a molecule known to be a constituent of the nucleus since the last decades of the nineteenth century (Portugal and Cohen 1977). Until the end of the 1940s, however, this story had been hardly, if at all, connected to the discourse of genetics. It belonged to structural chemistry instead and was part of textile fiber research. When James Watson and Francis Crick and Maurice Wilkins and Rosalind Franklin presented their evidence for a double helix model of DNA in 1953, a straightforward mode of replication for a specified class of biological macromolecules could be envisaged. With that, the riddle of *autocatalysis,* that is, the replication of genes, could be conceived in terms of base pairing via hydrogen bonds, a stereochemical feature that became known as base complementarity (Olby 1994).

Heterocatalysis on the other hand, that is, the mechanism by which genes give rise to their products, would occupy molecular biologists, geneticists, and especially biochemists for the next twenty years. Another strand of inquiry until then completely disconnected from genetics, the biochemical analysis of how proteins are fabricated in the cell, began to be woven into the texture of molecular biology. Despite many efforts to solve the "problem of coding" theoretically (Sarkar 1996), it is largely in the context of biochemical experimental systems that, in the course of the 1950s, the notion of genetic information and genetic information transfer took on a generic biological meaning, resulting in what Francis Crick eventually codified as the central dogma of molecular biology: DNA is transcribed into RNA, RNA is translated into protein. This "flow of information" was envisaged as being irreversible.

Around 1960, the principle of colinearity between a particular sequence of DNA and a corresponding sequence of amino acids had gained plausibility and became crucial for defining what a gene is on the molecular level: a finite, linear sequence of nucleotides that carries the instruction for a corresponding, finite, and linear amino acid sequence of a polypeptide. The contiguous array of nucleotides was envisaged to be translated according to a nucleotide letter code. The demonstration of a strict colinearity between gene (polynucleotide) and gene product (polypeptide) (Yanofsky et al. 1964) was indeed essential for establishing the experimental regime that finally, around 1966, led to the completion of the genetic code. Without the postulate of colinearity, *code* would have remained an empty concept. Colinearity became a notion of operational power – temporarily, until the unexpected complexity of the eukaryotic genome came to be recognized.

With Marshall Nirenberg and Heinrich Matthaei's finding in 1961 that a uniform stretch of nucleic acid consisting of uridine residues can be translated into a uniform stretch of a protein consisting of phenylalanine residues in a test tube protein synthesis system, the new molecular genetics had reached the heyday of its precocious simplicity (Nirenberg and Matthaei 1961). The experimental strategies of molecular biology appeared to have come to fruition. The gene had first become a physicochemical *molecule,* and subsequently a carrier of sequence *information.* During that same year, however, matters had already started to appear more intricate and more entangled. François Jacob and Jacques Monod presented their operon model for the regulation of the production of a couple of *Escherichia coli* sugar metabolizing enzymes (Jacob and Monod 1961). The news was that genes come in two classes, one structural, the other regulatory, and that operator regions had to be envisaged on parts of the chromosomal DNA that did not code for polypeptides, but nevertheless were essential for the regulation of gene expression.

Since then, these noncoding, but specific, regulatory DNA elements have proliferated almost beyond enumeration. There are promoter and terminator sequences; upstream and downstream activating elements in transcribed or nontranscribed, translated or untranslated regions; there are leader sequences; externally and internally transcribed spacers before, between, and after structural genes; there are interspersed repetitive elements and tandemly repeated sequences such as satellites, LINEs (long interspersed sequences), and SINEs (short interspersed sequences) of various classes and sizes whose function is still far from being understood (Fischer 1995). Are these batteries of elements to be counted as part and parcel of genes, or not?

On the level of posttranscriptional modification, the picture has become equally diffuse, messy, and complicated. Soon it was realized that DNA transcripts such as transfer RNA and ribosomal RNA had to be trimmed and matured in a complex enzymatic manner to become transformed into functional molecules, and that messenger RNAs of eukaryotes underwent extensive posttranscriptional modification both at their 5'-ends (capping) and their 3'-ends (polyadenylation) to be ready to go into the translation machinery. In the late 1970s, to the surprise of everybody, molecular biologists had to acquaint themselves with the idea that eukaryotic genes were composed of modules, and that, after transcription, introns were cut out and exons spliced together to yield a functional message. This was one of the first major scientific offshoots of recombinant DNA technology. With that, the colinearity postulate that had been so crucial for the early experimental history of the code receded into the background. What was a gene? Was it the contiguous DNA stretch from which the primary transcript was derived, or was it the spliced messenger that sometimes comprised a fraction as little as ten percent or less of the primary transcript? Since the late 1970s, we have become familiar with various kinds of RNA splicing: autocatalytic splicing;

alternative splicing of one single transcript to yield different messages; and even transsplicing of different primary transcripts to yield one hybrid message. The egg laying hormone of *Aplysia*, for example, is telling in this respect: One and the same stretch of DNA finally gives rise to eleven protein products involved in the reproductive behavior of this snail (Gros 1991, 492–499).

Let me mention here one last mechanism that has recently been found to operate on the level of RNA transcripts. It is called *messenger RNA editing* (for review, see Adler and Hajduk 1994). In this case, which meanwhile has turned out not just to be an exotic curiosity of some trypanosomes, the original transcript is not merely cut and pasted, but its nucleotide sequence is systematically altered after transcription. The nucleotide replacement happens before translation starts, and is mediated by various guide RNAs and enzymes that excise old and insert new nucleotides in order to yield a product that is no longer complementary to the DNA stretch from which it was originally derived. What is a gene?

The trouble with the gene in molecular biology continues on the level of translation. There are findings such as translational starts at different start codons on one and the same messenger RNA. There are instances of "obligatory" frameshifting within a given message without which a nonfunctional polypeptide would result. There is posttranslational protein modification such as removing amino acids from the amino terminus of the translated polypeptide. There is an observation that has been made within the last few years and which amounts to what is now being called *protein splicing:* Portions of the original translation product have to be catalytically or autocatalytically cleaved and joined together in a new order before yielding a functional protein (Cooper and Stevens 1995). The latest news from the translational field is that a ribosome can manage to translate one single polypeptide by accommodating two different messenger RNAs: A case of *transtranslation,* to use Raymond Gesteland's term (Atkins and Gesteland 1996).

So, how shall we define the gene? Its autocatalytic property has been relegated to the DNA at large. Which are the properties that define the heterocatalytic entities? Which sequence elements are to be included, which ones excluded? On the middle ground of transcription, ambiguities multiply. On the other side of the big divide between putative genes and correlated phenes, that is, the proteins, the ambiguities do not come to an end, as we have seen. There seems to be an end, however, in that the final protein products function after all. François Gros, after a long life in molecular biology, has come to the rather paradoxical and heretical sounding conclusion that the genes are specified, if at all, by "the products that result from [their] activity," that is, the functional RNA molecules and proteins to which they give rise (Gros 1991, 297). But is such a retrodefinition satisfactory? Fogle (this volume) argues to the contrary. Furthermore, what are we going to do with all the nonfunctional products caused by DNA mutation, by transcriptional and by translational errors?

Let me come to one last point. With eukaryotic model organisms moving center stage, the genome as a whole has assumed a more and more flexible and dynamic configuration. Not only have the mobile genetic elements, that Barbara McClintock characterized some fifty years ago in maize, gained currency in the form of transposons that regularly and irregularly become excised and inserted all over the genome. There are also other forms of gene shuffling at the DNA level. A gigantic amount of somatic gene tinkering and DNA splicing is involved in organizing the immune response, that is, in giving rise to the production of potentially millions of different antibodies. No genome would be large enough to cope with this task if parceling of genes and permutation of their parts had not been invented. Gene families contain silenced genes (sometimes called pseudogenes); we find jumping genes; and polymorphism on the DNA level, that is, multiple genes and isoforms. In short, there is a whole battery of mechanisms and entities constituting what could be called hereditary respiration, or breathing.

Molecular biologists have scarcely started to understand this flexible genetic apparatus in terms of evolution and embryonic development. In order to arrive at an adult stage or to give rise to viable offspring, these gene-phene or phene-gene complexes have to be reproduced in their proper context, their genomic as well as their cellular and intercellular environment. The development of molecular biology itself, that enterprise so often described as an ultimately reductionistic conquest, has made it impossible to think of the gene as a continuous piece of DNA matter colinear with a piece of protein matter and defined by a ready-made instruction laid down in its nucleotide sequence and contained between a precise start and end point. Today, it has become more reasonable, and it even might turn out to be sufficient, to speak of genomes, at least of "genetic material" (Kitcher 1982, 357) instead of genes, in the developmental as well as in the evolutionary dimension. It has become evident that the genome is a dynamic body of ancestrally tinkered pieces and forms of iteration. Genome sequencing combined with intelligent comparison may bring out more of this structure.

If there is a chance to understand evolution beyond the classical synthesis, it is from the perspective of this dynamic configuration. The purported elementary building block of this complex machinery, the simple point mutation, is not the prime element of the genetic process, but solely one component in a huge arsenal of DNA tinkering. Likewise, the comparatively simple arrangement of genes in the bacterial chromosome may not reflect a primitive simplicity, but may rather be the byproduct of some billion years of streamlining. We have come a long way with molecular biology from genes to genomes. There is still a way to go from genomes to organisms that will need the efforts of a new generation of molecular developmental biologists, and the path from there to populations and communities, and vice versa, will not be shorter and left for still another generation.

The foundering of "The Gene" seems to be well under way (Burian 1985; Carlson 1991; Falk 1984, 1986; Fischer 1995). Recently, Jürgen Brosius and Stephen Jay Gould have come up with a new terminology. They propose to abandon the gene concept altogether and to call any segment of DNA that has a recognizable structure and/or function (such as a coding segment, a repetitive element, a regulatory element) *nuon*. By duplication, amplification, recombination, retroposition, and the like, nuons can give rise to *potonuons*, that is, entities potentially recruitable as new nuons. These in turn have either the chance of being transformed into *naptonuons*, that is, dissipate their nonadaptive former information without acquiring new ones, or else into *xaptonuons*, that is, elements exapted to a new function (Brosius and Gould 1992, 1993). As tempting as this evolutionary genome terminology may be, its operationality remains to be probed. One of its drawbacks may be that it remains completely restricted to the level of DNA adaptation.

INTEGRONS AND SOME CONJECTURES

We are farther than ever from the gene as a simple and single bit of DNA carrying the information for a simple and single, colinear polypeptide chain. There is one thing, however, that appears to have survived all the recent turmoil around the foundering of the gene – *"le gène éclaté"* (Gros 1991, chapter 7) – and it conveys a certain robustness to the unbroken gene talk of genetic engineering with its naturalistic and often deterministic overtones. This is the so-called central dogma of molecular biology as defined by Francis Crick some forty years ago. The postulate claims that the flow of genetic information is strictly unidirectional and irreversible; it goes from DNA to protein and not the other way round. Whatever may happen on that way, on the DNA level, on the RNA level, or on the protein level, "once 'information' has passed into protein it cannot get out again" (Crick 1958, 153). Although there is a complicated machinery of enzymes and DNA binding proteins capable of replicating, recombining, and manipulating DNA, and despite considerable evidence for directed mutation, there is no hard evidence for an environmentally or somatically guided, "intelligent" alteration of and retroaction on the genetic material. Despite epigenetic inheritance mechanisms that can carry on patterns of gene activation and repression such as methylation patterns into the next generation(s) (Jablonka and Lamb 1995; Russo, Martienssen, and Riggs 1996), a functionally improved protein that could bring about a corresponding sequence change in its DNA region has so far not been found. Although this Lamarckian dream has never ceased to be dreamed, until today nobody has had an inkling how such a mechanism could work in a biochemically specifiable manner. There appears to have evolved no protein code for producing sequence-specific strings of DNA.

But this hard core of Weismann's legacy notwithstanding, time has come to treat the genome and the phenome as reciprocally important partners from the perspective of an integrated ensemble. With Evelyn Fox Keller, we can state that "owing in large part to the reemergence of developmental biology, molecular biology may well be said to have 'discovered the organism'" again. But we should also take seriously her reminder that "the subject of the new biology, however whole and embodied, is only a distant relative to the organism that had occupied an earlier generation of embryologists" (Keller 1995, 117). It has become a body pervaded by a whole bunch of linguistic metaphors, above all: information, code, signaling, and communication.

I propose to come back to a suggestion that François Jacob made some twenty-five years ago to speak of *integrons* instead of genes and gene products (Jacob 1970). Think of a piece of DNA that gives rise to a DNA-polymerizing enzyme, and forget for a moment that this process already involves a complex transcription and translation machinery; the polymerase in turn gives rise to more DNA of the same kind. Any mutation in this DNA that raises the efficiency of the product (polymerase making more DNA) will lead to the selection of this basic integron among its competitors under conditions in which space and resources are limited. Given such a "hypercycle" (Eigen and Schuster 1979), the information said to be contained in the gene takes on meaning. In its simplest form, this meaning is confined to the capacity of proliferation. There is more involved, even in this simple case, than just information in the quantitative sense that information theory conveys to this notion. Whatever escape we may try, genetic information cannot be dissociated from biological meaning. It has to make sense in terms of a biological function. Even when rejecting the strictly hierarchical aspects of Michael Polanyi's theory of boundary conditions, we can retain his claim that there is a "semantic relation" between the genetic level and the organismic level in that the latter conveys "meaning" to the former (Polanyi 1969, 236). In assuming such a relation, however, we are in the realm of the symbolic. After all, organisms may be viewed as symbolical machines, even if their only basic signifying activity is to further signify, meaning, to reproduce. Just as there is no text without context, "a molecule does not become a message because of any particular shape or structure. [A] molecule becomes a message only in the context of a larger system of physical constraints which I have called a 'language'" (Pattee 1969, 8). But let us remain cautious. Biologists have always widely borrowed from other fields in shaping their vocabulary. Ultimately, however, the objects of their study have also always forced them to transcend analogies. Neither natural languages nor technical information processing systems are isomorphic with the generation and transmission of information in living systems. Finally it is these differences that count. Let us assume that meaning is something transiently generated at the intersection between different chains of signifiers. If we look for meaning in the organism, we must look, not at its genes, but at the multiple interfaces between the genome and the body (Griesemer, in press).

In order to avoid the ambivalence of talking about information, some prefer to speak about instruction – "Bauanleitung" in the words of Manfred Eigen (Eigen 1987). But in using the performative idiom, one has to be cautious not to equate instruction with the notions of "blueprint" or genetic "program." For, as Jacob reminds us, in order to be read and executed, the "program" of a living being needs always also the products of its execution in order to be read and executed (Jacob 1970, 318). Henri Atlan, therefore, is inclined to reverse the metaphor and see the whole metabolic machinery of the organism as constituting a "program" and the genes as "data" fed into the machinery (Atlan and Koppel 1990). Where such a suggestion might lead is open for debate. More is open for debate. The biosphere as a whole might be conceived as an ensemble of bio-semiotic regimes on many different levels, from biochemical signaling to human and machinic communication (Eder and Rembold 1992). In this perspective, the hereditary process finally might be viewed as just one – although a very basic – form, to turn around the title of J. L. Austin's classic, of *How to Do Words with Things*.

You may expect me to come up with a nice solution to this meandering story of the gene at the end. As far as the scientific story goes, there is none. As to an epistemological take-home lesson, I have one. Alas, it is a very disappointing message for nondeconstructivists. I might say in paraphrasing virologist André Lwoff, the early mentor of François Jacob at the Pasteur Institute in Paris, and teacher of Gunther Stent: A gene is a gene is a gene.[3] This is a strong claim (see also Kitcher 1992). Taken seriously, it means that in science every presumed referent is turned into a future signifier.

The gene has been a powerful epistemic entity in the history of heredity, in all the vagueness that is characteristic for such entities. It is tempting to generalize this statement and assume that fruitful scientific concepts are bound to be polysemic. I will resist this temptation and assure my critics that I do by no means deny the value of precision in science. But precision itself has historically changing boundaries. Assessing what it means to be fuzzy, instead of eliminating vagueness altogether and implementing precision, has become a major concern in fields such as AI-research. Lofti Zadeh claims that "there is a rapidly growing interest in inexact reasoning and processing of knowledge that is imprecise, incomplete, or not totally reliable. And it is in this connection that it will become more and more widely recognized that classical logical systems are inadequate for dealing with uncertainty and that something like fuzzy logic is needed for that purpose" (Zadeh 1987, 27). On a methodological level and in contrast to the technical solution that the syntax of fuzzy logic has to offer, the question arises whether we need, in order to understand conceptual tinkering in research, more rigid metaconcepts than those first-order concepts that we, as epistemologists, analyze. I am inclined to deny this. Why should historians and epistemologists be less imprecise, less operational, and less opportunistic after all, than scientists? But do not let our narratives be less coherent than our protagonists' stories either! Let us only be aware that "there is an incompatibility between precision and complexity. As the complexity of a system increases, our ability to make precise and yet nontrivial assertions about its behavior diminishes" (Zadeh 1987, 23). The history of the gene could contribute to the exploration of this epistemological principle of uncertainty and show how to manage complexity, not by striving for a "Theory of Everything," but by looking for patterns of "moderate compressibility" (Coveney and Highfield 1995, 39).

NOTES

1. I thank Raphael Falk, Michel Morange, and Sahotra Sarkar for their close reading of earlier drafts of this paper and for valuable suggestions.
2. This heading has been inspired by the title of a review by Paul Zamecnik on "Protein synthesis – early waves and recent ripples" (Zamecnik 1976).
3. André Lwoff concluded his Marjorie Stephenson Memorial Lecture in April 1957 with, as he put it, the "prosy, coarse and vulgar" statement: "Viruses should be considered as viruses because viruses are viruses" (Lwoff 1957, p. 252).

REFERENCES

Adler, B. K., and S. L. Hajduk. 1994. Mechanism and origins of RNA editing. *Current Opinion in Genetics and Development* 4: 316–322.

Atkins, J. F., and R. F. Gesteland. 1996. A case for *trans* translation. *Nature* 379: 769–771.

Atlan, H., and M. Koppel. 1990. The cellular computer DNA: Program or data. *Bulletin of Mathematical Biology* 52: 335–348.

Avery, O. T., C. M. MacLeod, and M. McCarty. 1944. Studies on the chemical transformation of pneumococcal types. *Journal of Experimental Medicine* 79: 137–158.

Bachelard, G. 1957. *La formation de l'esprit scientifique.* Paris: Vrin.

Beadle, G. W. 1945. The genetic control of biochemical reactions. *The Harvey Lectures* 40: 179–194.

Benzer, S. 1955. Fine structure of a genetic region in bacteriophage. *Proceedings of the National Academy of Sciences USA* 41: 344–354.

Bernard, C. 1954. *Philosophie. Manuscrit inédit.* Edited by J. Chevalier. Paris: Editions Hatier-Boivin.

Brosius, J., and S. J. Gould. 1992. On "genomenclature": A comprehensive (and respectful) taxonomy for pseudogenes and other "junk DNA." *Proceedings of the National Academy of Sciences USA* 89: 10706–10710.

Brosius, J., and S. J. Gould. 1993. Molecular constructivity. *Nature* 365: 102.

Burian, R. M. 1985. On conceptual change in biology: The case of the gene. In *Evolution at a Crossroads: The New Biology and the New Philosophy of Science*, edited by D. J. Depew and B. H. Weber. Cambridge, Mass.: MIT Press, pp. 21–42.

Carlson, E. A. 1991. Defining the gene: An evolving concept. *American Journal of Human Genetics* 49: 475–487.

Cooper, A. A., and T. H. Stevens. 1995. Protein splicing: Self-splicing of genetically mobile elements at the protein level. *Trends in Biochemical Science* 20: 351–356.

Coveney, P., and R. Highfield. 1995. *Frontiers of Complexity: The Search for Order in a Chaotic World.* London: Faber and Faber.

Crick, F. H. C. 1958. On protein synthesis. *Symposia of the Society for Experimental Biology* 12: 138–163.

Delbrück, M. 1942. Bacterial viruses (bacteriophages). *Advances in Enzymology* 2: 1–32.

Dunn, L. C. 1991 [1965]. *A Short History of Genetics.* Ames, Iowa: Iowa State University Press.

Eder, J., and H. Rembold. 1992. Biosemiotics – a paradigm of biology. *Die Naturwissenschaften* 79: 60–67.

Eigen, M. 1987. *Stufen zum Leben.* München: Piper.

Eigen, M., and P. Schuster. 1979. *The Hypercycle – A Principle of Natural Self-organization.* Heidelberg: Springer.

Elkana, Y. 1970. Helmholtz' 'Kraft': An illustration of concepts in flux. *Historical Studies in the Physical Sciences* 2: 263–298.

Falk, R. 1984. The gene in search of an identity. *Human Genetics* 68: 195–204.

Falk, R. 1986. What is a gene? *Studies in the History and Philosophy of Science* 17: 133–173.

Fischer, E. P. 1995. How many genes has a human being? The analytical limits of a complex concept. In *The Human Genome*, edited by E. P. Fischer and S. Klose. München: Piper, pp. 223–256.

Griesemer, J. R. in press. The informational gene and the substantial body: On the generalization of evolutionary theory by abstraction. In *Varieties of Idealization*, edited by N. Cartwright and M. Jones. Amsterdam: Editions Rodopi.

Gros, F. 1991. *Les secrets du gène.* Paris: Editions Odile Jacob.

Jablonka, E., and M. J. Lamb. 1995. *Epigenetic Inheritance and Evolution: The Lamarckian Dimension.* Oxford: Oxford University Press.

Jacob, F. 1970. *La logique du vivant.* Paris: Gallimard.

Jacob, F. 1981. *Le jeu des possibles.* Paris: Fayard.

Jacob, F., and J. Monod. 1961. Genetic regulatory mechanisms in the synthesis of proteins. *Journal of Molecular Biology* 3: 318–356.

Kay, L. E. Forthcoming. *Who Wrote the Book of Life? A History of the Genetic Code.* Stanford: Stanford University Press.

Keller, E. F. 1995. *Refiguring Life: Metaphors of Twentieth-Century Biology.* New York: Columbia University Press.

Kitcher, P. 1982. Genes. *The British Journal for the Philosophy of Science* 33: 337–359.

Kitcher, P. 1992. Gene: Current usages. In *Keywords in Evolutionary Biology,* edited by E. F. Keller and E. A. Lloyd. Cambridge, Mass.: Harvard University Press, pp. 128–131.

Kühn, A. 1941. Über eine Genwirkkette der Pigmentbildung bei Insekten. *Nachrichten der Akadmie der Wissenschaften in Göttingen,* Mathematisch-Physikalische Klasse: 231–261.

Lwoff, A. 1957. The concept of virus. *Journal of General Microbiology* 17: 239–253.

Moles, A. A. 1995. *Les sciences de l'imprécis.* Paris: Seuil.

Morange, M. 1998. *A History of Molecular Biology.* Cambridge, Mass.: Harvard University Press.

Muller, H. J. 1927. Artificial transmutation of the gene. *Science* 66: 84–87.

Muller, H. J. 1951. The development of the gene theory. In *Genetics in the 20th Century,* edited by L. C. Dunn. New York: Macmillan, pp. 77–99.

Nirenberg, M. W., and J. H. Matthaei. 1961. The dependence of cell-free protein synthesis in E. coli upon naturally occurring or synthetic polyribonucleotides. *Proceedings of the National Academy of Sciences USA* 47: 1588–1602.

Olby, R. 1994. *The Path to the Double Helix.* 2nd ed. New York: Dover.

Pattee, H. 1969. How does a molecule become a message? *Developmental Biology Supplement* 3: 1–16.

Polanyi, M. 1969. Life's irreducible structure. In *Knowing and Being: Essays by Michael Polanyi*, edited by M. Grene. Chicago: University of Chicago Press, pp. 225–239.

Portin, P. 1993. The concept of the gene: Short history and present status. *The Quarterly Review of Biology* 68: 173–223.

Portugal, F. H., and J. S. Cohen. 1977. *A Century of DNA*. Cambridge, Mass.: MIT Press.

Rheinberger, H.-J. 1997. *Toward a History of Epistemic Things: Synthesizing Proteins in the Test Tube*. Stanford: Stanford University Press.

Russo, V. E. A., R. A. Martienssen, and A. D. Riggs. 1996. *Epigenetic Mechanisms of Gene Regulation*. New York: Cold Spring Harbor Laboratory Press.

Sarkar, S. 1996. Biological information: A skeptical look at some central dogmas of molecular biology. In *The Philosophy and History of Molecular Biology: New Perspectives*, edited by S. Sarkar. Dordrecht: Kluwer Academic Publishers, pp. 187–231.

Stadler, L. J. 1928. Genetic effects of x-rays in maize. *Proceedings of the National Academy of Sciences USA* 14: 69–75.

Star, S. L., and J. R. Griesemer. 1988. Institutional ecology, 'translations' and boundary objects: Amateurs and professionals in Berkeley's Museum of Vertebrate Zoology 1907–39. *Social Studies of Science* 19: 387–420.

Timoféeff-Ressovsky, N., K. Zimmer, and M. Delbrück. 1935. Über die Natur der Genmutation und der Genstruktur. *Nachrichten von der Gesellschaft der Wissenschaften zu Göttingen, Mathematisch-physikalische Klasse, Fachgruppe VI, Nachrichten aus der Biologie*, N.F. 1: 190–245.

Vischer, E., S. Zamenhoff, and E. Chargaff. 1949. Microbial nucleic acids: The desoxypentose nucleic acids of avian tubercle bacilli and yeast. *Journal of Biological Chemistry* 177: 429–438.

Yanofsky, C., B. C. Carlton, J. R. Guest, D. R. Helinski, and U. Henning. 1964. On the colinearity of gene structure and protein structure. *Proceedings of the National Academy of Sciences USA* 51: 266–272.

Zadeh, L. 1987. Coping with the imprecision of the real world. In *Fuzzy Sets and Applications: Selected Papers by Lofti A. Zadeh*, edited by R. R. Yager, S. Ovchinnikov, R. M. Tong, and H. T. Nguyen. New York: John Wiley, pp. 9–28.

Zamecnik, P. C. 1976. Protein synthesis – early waves and recent ripples. In *Reflections in Biochemistry*, edited by A. Kornberg, B. L. Horecker, L. Cornudella, and J. Oro. New York: Pergamon Press, pp. 303–308.

11

Reproduction and the Reduction of Genetics

JAMES R. GRIESEMER[1]

ABSTRACT

In this essay I develop a new, unified perspective on genetics, development, and reproduction. I suggest a heuristic use of theory reduction to address an issue of contemporary theoretical importance: the framing of a theory of developmental units. I claim that the gene concept, properly understood, *is* a concept of developmental unit and suggest that the historiography of genetics should reflect this fact. There is no denying that genetics has been a successful science. If its relation to development could be adequately expressed, genetic theory might also provide clues to a theory of development. Conventionally, the mechanisms of development are expected to be explained in terms of mechanisms of genetics. Development is treated as an epigenetic process and, if theory reduction is possible, a theory of development is expected to reduce to a general theory of genetics. This expected direction of reduction from development to genetics depends on the conventional understanding of genetics and its relations. I argue for a reversal of this expectation by reconceptualizing genetics and development as fields describing aspects of the process of reproduction. The relation of these aspects is not one of simple parts to a whole. They are deeply entwined, as my analysis of reproduction will show. Once certain features of scientific reduction are identified, the new perspective can be used to pursue reductionism heuristically, to use what we know about the theoretical units of genetics to speculate about units of a general theory of development. Thus, reductionism may be scientifically useful even though the conditions for formal reduction are not met.

THE PROCESS OF REPRODUCTION

I urge a change of perspective on genetics and gene concepts. The fundamental entities of biology are processes rather than structures or functions. Genetics is about genetic processes. Development is about developmental processes. Whether, when, and how entities of a particular structural or functional kind serve such processes are empirical questions. Whether there are genes besides the nucleic acid molecular kind (structure) or besides the replicator kind (function) are empirical questions requiring an ontology of biological processes for answers. Various kinds of objects might serve in genetic units of heredity or development. The functions that are carried by such structures when they serve a process of reproduction are determined by the relation between heredity and development in the process: Functions need not be invariant to changes of structure and structures need not be invariant to changes of function in order to serve reproduction. Thus, necessary and sufficient conditions on structure or function are bound to fail as definitions for units of process.

In a process view of evolution, relations among genetic and developmental *processes* may be distinguished from relations among *objects* serving these processes. In the traditionally interpreted structural hierarchy "biological individuals" serving developmental processes such as cells and organisms are thought to be at a higher level of organization than objects serving genetic processes such as genes. In the process perspective I endorse, developmental processes are more fundamental: Genetic processes contain developmental processes, so development is not *epi*genetic and hierarchical structural order is not equated with processual order.

The prospects for reductionism and mechanistic explanation are altered in the process perspective. We need not reject theory reduction as an alternative philosophy to mechanism (see Brandon 1996, chapter 11) if theories are about processes while mechanisms are function-bearing, structured things that cause behaviors (see Glennan 1996) that we assign to processes.[2] We can use what we know about structures and functions of mechanisms in the special case of genetic processes as heuristic constraints to construct theories of the general case of reproduction processes.

Genetic processes can be treated as special cases of reproduction processes that contain developmental processes. Thus, genetic processes contain development. The direction of reduction is changed from the conventional expectation because the special theory of genetic processes should reduce to a general theory of reproduction that explicitly incorporates development. Reduction of special case

241

theories to more general theories is a central feature of the traditional view of theory reduction (Nagel 1961; Sarkar 1998). My use of theory reduction is anything but traditional, however. I distinguish between the *image* of a reduction model, meaning the expected direction of reduction between scientific realms and thus which realm is to count as more fundamental, and the *mapping relations* that connect items across realms.[3] Reversal of the biological image of reduction from development to genetics can drive a reductionistic research program to establish new mapping relations by working at the level of a general theory of reproduction without already having in hand an articulate, formal theory of development. This reductionism is a scientific research strategy, not an *a posteriori* philosophical analysis (Wimsatt 1976). To pursue this line of thought, I will first give a general characterization of biological reproduction and then suggest more precisely how genetic processes are special cases of reproduction processes. Once done, the heuristic program and its historiographic implications for the gene concept can be sketched.

What Is Replication?

I do not think that the most general philosophical treatments of evolutionary theory to date have interpreted genetic processes adequately. Dawkins (1976, 1982) and Hull (1980, 1981, 1988), for example, invoke "copying" and "replication" in their definitions of the "replicator," but fail to give satisfactory accounts of these processes. Others invoke properties such as heritability in their accounts of units of selection, but are not clear about the ontological status of such properties and their relationship to the processes that carry them. For example, heritability has been claimed to be a necessary property of a unit of selection (e.g. Lewontin 1970). Heritability is not a sufficient expression for an inheritance *process* because heritability expresses a capacity, not a process.[4]

My purpose in analyzing reproduction as a process is to formulate an alternative to the notion of replicator. The replicator concept appeals to certain examples of replication as definitive. This procedure threatens circularity: A replicator is "an entity that passes on its structure largely intact in successive replications" (Hull 1988, 408). But what is replication? It is tempting to say either that replication is the process by which replicators pass on their structure or simply refer the reader to a molecular genetics textbook, implying that replication is anything "sufficiently like" nucleic acid replication. The failure to analyze process is widespread in the units of selection literature and reflects the fact that analysts of the structuralist perspective have often been more interested in units that map neatly onto hierarchies of structural organization than in tracing the ontological implications for evolution as a process. Those following the functionalist perspective of G. C. Williams (1966) as well as Dawkins and Hull have returned the emphasis to process, but in a way that has tended to exclude developmental processes as central to the integration of replication and interaction by selection. The exclusion is symptomatic of strategies of theory generalization by abstraction. One cannot appeal to exemplars of replication (e.g., DNA) that have been identified in terms of the structural hierarchy and be confident that their properties are necessary and sufficient for replicators in general without a general analysis of the replication process.

Progeneration: Multiplication with Material Overlap

Reproduction is a process with two aspects: *progeneration* and *development*.[5] Progeneration is the multiplication of entities with *material overlap* of old (parent) and new (offspring) entities. Material overlap means that some of the parts of the offspring were once parts of the parents. If you cut a loaf of bread in half, you progenerate it, making two new things out of one old one; the parts of the new half-loaves were formerly parts of the old loaf. If a cell reproduces by fission, it divides into two new cells, all of whose parts used to be parts of the parent. Reproduction by fission involves progeneration. Progeneration is distinguished from *copying* and *multiplication* by its requirement of material overlap.[6]

The term *progeneration* evokes an impression of self-motion or self-reproduction. Progenerating organisms are agents of reproduction, not merely material or efficient causes. They may be called *reproducers*. Bread loaves must be cut in pieces by an agent with a knife while cells divide *themselves* in reproduction. The contrast between agent and patient should not be taken seriously without a precise notion of biological agency, but whatever that notion turns out to be, my concern is with reproduction by biological agents.[7]

Progeneration is universal in biological reproduction. We can, of course, imagine that DNA replication might have been fully conservative rather than semiconservative. Furthermore, it is not typical for modern DNA strands to be transmitted to new units of reproduction or to play a role in gene expression without the material overlap of more inclusive entities, the cells. So even if it turned out that DNA replication per se did not satisfy the material overlap condition for biological reproduction, DNA replication would be part of, rather than a kind of, reproduction. What about single stranded RNA viruses? They do not materially overlap their daughter strands except by accidental nucleotide incorporation. Does viral "replication" count as reproduction? No. Viruses replicate only by virtue of the development of their hosts. Viral nucleic acid replication is best understood as a kind of process in terms of virus development, which is part of host development, reproduction and genetics.

The distinction between replication and reproduction and between their associated units comes down to the role of development in the respective processes: The replicator is a concept abstracted from development while the reproducer is a developmental concept. If there is something we would count as a case of biological reproduction that does not involve material overlap, I suggest that is because we have abstracted the concept of reproduction from its developmental aspects and thus mistakenly identified replication with copying rather than with reproduction.

If DNA replication might not have been reproduction, if viral "replication" is not reproduction, then is the material overlap distinction between progeneration and copying mere semantic hairsplitting? If the analysis of reproduction only added precision to the use of words like *replication, copying,* and *reproduction,* perhaps the philological effort would not be worth the scientific gain. However, if the evolutionary dynamics of the flow of matter differ from the dynamics of the flow of information or structural pattern, then there may well be a scientific significance to the distinction.[8] Indeed, why *is* DNA replication semiconservative while viruses can get away without it? If there is an evolutionary cost of material overlap, what evolutionary benefit could counterbalance that cost? One possibility concerns transmission fidelity. Chemically bonded or bounded matter is a way to "transmit structure largely intact" without having to depend too much on the contingent stability of the environment. In general, information flow is reliable only to the extent that the channel conditions supply stability to messages. If fidelity is low, replication rate must be high if selection is to operate effectively (Wimsatt 1980, 245). Material overlap may buffer information in fidelity-limiting environments.

Besides distinguishing copying from progeneration in terms of material overlap, it is critical to be clear about what types of progenerating entities count as reproducing entities. That clarification is a problem for a general theory of development. Roughly, when we assert that some thing has reproduced, we mean that that thing has produced another thing of a certain type or kind. However, cutting a loaf of bread in half is not reproduction, no matter how much the half loaves resemble the whole "parent" loaf, so the relevant kind is a very special property. To count as biological reproduction, the half loaves would have to be capable of indefinite progeneration. The half loaves can certainly be cut again, but there comes a point when it is impossible to divide the tiny cut pieces into things that are still bread, let alone loaves. Even if matter turns out to be infinitely divisible, a point will undoubtedly be reached when no physical process divides the matter further. The capacity for indefinite possibility of progeneration (i.e., that a lineage of progenerants could be indefinitely long) sounds similar to Dawkins' account of replicators as things with sufficient longevity, fecundity, and fidelity to be the units of selection and the beneficiaries of evolution (Dawkins 1976). However, in process perspective the replicator can be understood to be a very special case of something much more general, the reproducer.[9]

Development: Acquisition of the Capacity to Reproduce

In general, the type of thing that is reproduced by progeneration is the reproducing type. It is not the case that resemblance in just any respect (i.e., having any property in common) is sufficient to turn cases of progeneration into cases of reproduction. Progeneration must result in more things that can reproduce for that progeneration process to count as an aspect of a reproduction process. In other words, there is exactly one property in which parents and offspring *must* be correlated for their progeneration to count as reproduction: the capacity to reproduce. The production of offspring that do not have the capacity to reproduce is not reproduction, but only progeneration.

Typically, the capacity to reproduce must be *acquired* or built-up; things are not born with it. Even cleavage cells in an early embryo that do not take up nutrients or synthesize new RNA must go through some internal rearrangement in order to successfully divide again. At the very least, they must move chromosomes to the metaphase plate and pull them apart in order to divide in a way that yields a reproductively capable offspring cell. Things can be born with the capacity to progenerate, but if they are to be reproductive, they must also be born with the capacity to develop, that is, with the capacity to acquire the capacity to reproduce.

Evolutionary theory requires, minimally, that reproductive capacity be "transmitted" from generation to generation. The means for this have typically been the transfer, in material overlap, of particular *mechanisms* that confer the capacity to develop the capacity to reproduce. It is not essential that an offspring develop or reproduce in the same way as its parents in order that the essential property – the capacity to reproduce – be instantiated in offspring. Some insect species alternate between generations of winged and nonwinged forms, so offspring are rather unlike parents in those structures and functions. Haploid male hemiptera do not resemble their female parent in a number of respects. Moss and fern gametophytes do not resemble their sporophyte parents in many respects that are essential to sporophyte development. Successful academics do not always produce academic offspring (if they have offspring at all), nor do nonacademic parents always succeed in keeping their children out of the ivory tower. These examples dress in processualist clothes the familiar point that heritability of complex, adaptive traits or parts is rarely perfect. The more fundamental point is that heritability is a developmental concept because reproduction must transfer "pieces of development" from parent to offspring, i.e., parts conferring the capacity to develop.[10]

Critiques of heritability as a concept of genetics (e.g., Lewontin 1974a; Sarkar 1998) often point out that heritability is a phenotypic concept. That critique relies on the partition of structures into phenotype and genotype. From the process perspective, heritability is a developmental concept. However, even from this perspective there is no full partitioning of development and progeneration except by imposing an explicit model of structural organization in which reproduction can be interpreted as operating at a single level of structural organization. The reason for this difference between structuralism and processualism has to do with the recursive relation of progeneration and development in reproduction.

To count as reproduction, a progeneration process must result in entities with the capacity to reproduce. The acquisition of the capacity to reproduce is the process of development.[11] Since development is the acquisition of the capacity to reproduce, we can say that reproduction is progeneration of entities with the capacity to develop the capacity to reproduce. This analysis of reproduction is therefore recursive: Entities with the capacity to develop the capacity to reproduce are entities with the capacity to develop the capacity to develop the capacity to reproduce, and so on. The recursion stops or "bottoms out" when something is progenerated which has the capacity to reproduce and does not need to acquire it through development. This automatic capacity to reproduce may be called *null development.*

Genetics in a Process Perspective

The analysis shows that development is a critical component of reproduction, equal in significance to the progeneration of new individuals, and indicates what is essential to parent-offspring resemblance. Reproduction is neither mere multiplication of individuals nor mere copying. It requires that multiplication involve material overlap and it bounds the specification of development. The analysis can be put to use to reverse the direction of the image expressing the expectation that a theory of development reduces (in principle) to a theory of genetics and to show how, in process perspective, appeal to mechanisms of genetics can explain developmental phenomena. To do so, we first need to reinterpret genetics in a process perspective.[12]

The analysis of reproduction subsumes two special cases: inheritance processes and replication processes. *Inheritance processes* are the subject of classical genetics, in which no special knowledge of the mechanistic basis for particular developmental mechanisms causing reproductive capacity is required. Mendelian genetics is a special case theory of inheritance in which certain developmental mechanics are controlled by experimental design. Mendel expressed his concern for such cases when he characterized his theory as one of the "development of hybrids" (Mendel 1866 [1965], 21). *Replication processes* are the subject of classical molecular genetics, which assumes special knowledge of the Watson-Crick coding mechanisms. Genetics is the field centrally concerned with both of these types of process, so I will call inheritance and replication types of *genetic processes*.

Replication processes can be interpreted as special cases of inheritance processes, which in turn are special cases of reproduction processes. Since reproduction involves both progeneration and development, genetic processes also involve these. Because genetic processes are special kinds of reproduction process, and if theory reduction is possible, theories of genetic processes should reduce to the theory of reproduction processes in the sense that special case theories should be derivable from general theories.

It is important to recognize that, from the conventional standpoint of Weismannism, progeneration and development are logically separable component processes that can be studied in isolation from one another. Progeneration is the physical process described in terms of hereditary character "transmission" in genetic models. Although transmission sounds like a process, its treatment in genetics and evolutionary theory is in terms of an abstraction: the genotype-to-genotype or phenotype-to-phenotype mapping among generations (Lewontin 1974b, 1992). Development is the process represented by the genotype-to-phenotype map within generations. Thus, the traditional Weismannist separation of germ and soma, that provides a cytological basis for Mendel's factor-character distinction and Johannsen's genotype-phenotype distinction, seems untouched conceptually by the analysis of reproduction. But this is not so because progeneration and development are recursively entwined. There is no progeneration without development and no development without progeneration, except in the special case of null-development. Weismannism demands idealizations and abstractions in *both* the scientific domains of genetics and developmental biology, in which developmental aspects of heredity are ignored by geneticists and in which hereditary aspects of development are ignored by developmental biologists. These idealizations and abstractions must be recognized precisely for what they are – heuristics of scientific research – in order to go beyond them to a general, unifying theory of heredity/development that does not limit the science of genetics to classical Mendelian or molecular genetics.[13]

Genetics is distinguished from other fields concerned with reproduction by its particular interest in problems of heredity. Heredity is a *relation* of similarity between parents and offspring that is caused by reproduction processes. Geneticists measure the *degree* of heredity relations in terms of trait correlations among relatives. A trait is any measurable property of parents and offspring. Trait correlation is a measure of the association between trait "values" of the relatives being compared. Traits are classes of trait values. We speak of height as a trait. Being two meters tall is a value of the trait. Heritability, as noted above, is the quantitatively expressed *capacity* to cause relations of heredity that is carried by reproduction processes. The *measures* that are called narrow-sense and broad-sense heritability in the technical literature of quantitative genetics are statistics on populations of individuals that carry the heritability capacity. Because there must be a correlation in the capacity to develop for a process to be a reproduction process, heritability, as noted above, is a developmental concept.

Reproduction processes are the cause of genealogical relations. *Genetic processes*, specifically inheritance and replication processes, are specially structured reproduction processes. The field of genetics as it has developed historically is concerned with reproduction processes in which heredity relations are caused in particular ways. Some genetic problems do not require a detailed knowledge or representation of the mechanisms that cause heredity, but only with measuring the degree of the relation. Darwinian evolution before Mendel could do with a simple inequality: Offspring resemble their parents more (on average) than they resemble other members of the population into which they are born. Classical biometry was concerned to measure empirical correlations among relatives, although the biometricians took varying positions on what could be learned of the "underlying" mechanisms from these empirical values. More recent genetics addresses the chromosomal and molecular mechanisms of heredity and the statistical consequences of their operation among populations of hereditary units.

Despite the variety of historically changing problems, interests, and theoretical needs of genetics, the nature of genetic processes can be characterized simply in terms of reproduction processes. Reproduction processes cause heredity relations as a result of the entwined effects of progeneration and development. *Inheritance processes* are reproduction processes in which there are *evolved* causal mechanisms "for" producing heredity relations in development in particular respects and degrees. Development is the acquisition of the capacity to reproduce. Acquiring a capacity for something usually requires constructing a mechanism – a complex system of component parts – to generate it, so the process of development can be understood as constructing mechanisms for particular hereditary relations in reproductive capacity. *Replication processes* are inheritance processes with a special kind of evolved causal mechanism, namely *coding* mechanisms.[14] Coding mechanisms, like other evolved mechanisms of development, produce heredity in particular respects and degrees; they may also be for the *control* of the degree of heredity relations.

Many molecular geneticists study replication processes in the special sense defined here, although they would characterize their work in structural or functional terms. Other geneticists study nonreplicative inheritance processes such as the so-called epigenetic inheritance systems or EISs (Jablonka and Lamb 1995). Some evolutionary biologists study group reproduction (Wade and Griesemer 1998), although it is not clear to what extent interdemic reproduction processes are genetic (see Griesemer and Wade, 2000). EISs involve inheritance mechanisms that are probably not coding mechanisms (or at least not precise coding mechanisms compared to the nucleic acid system). Interdemic reproduction may or may not include evolved mechanisms of group development, but this way of thinking about group reproduction gives new meaning to Van Valen's aphorism that evolution is the control of development by ecology.

The cytosine methylation system is perhaps the EIS closest to a genetic coding system known (Jablonka and Lamb 1995, chapter 4). It is dependent on nucleic acid coding in the sense that cytosine methylation codes gene activation states in terms of the degree of methylation of cytosine residues in DNA sequences: More methylation means lower gene activity. The relation is quantitative, but appears not to be exact. Methylation states have been shown to be transmissible in mitosis and in some cases in meiosis for up to a few tens of generations, providing a possible mechanism for *Dauermodifikationen.*

Other EISs include structural inheritance systems such as cortical inheritance in paramecia in which surgical rearrangements of cilia patterns have been shown to be transmitted from parent to offspring through many generations. Heredity of these patterns can be precise and quantitative, but there is no *coded* information in the sense that: (1) the information is not about something other than the cortical pattern itself and, (2) is not specified by a combinatorial set of elements that is small relative to the size of the sequences coded for.[15]

Still less coded (or "digital" or "unlimited" [see Szathmáry 1994]), are metabolic steady state systems of inheritance. When cells reproduce by fission, they divide cytoplasm into two not necessarily equal parts. Cytoplasm is the substance of much of the metabolic machinery of the cell. The division of cytoplasm can preserve structure and hence cause correlations between parent and offspring in that structure. Because these EISs typically depend on chemical diffusion and molecular transport processes to equalize metabolic patterning throughout the cell, they may be of lower fidelity than the nucleic acid coding system. But just because a mechanism is not a coding mechanism does not mean that it is utterly incapable of causing heredity relations. Heredity, like correlation, is a relation that ranges in degree from −1 to +1.

The reproduction of demes by the progeneration and development of propagules (composed of one to many organisms) that colonize new habitat, may, in some ecological contexts, be another sort of steady state inheritance system, albeit at a much higher level of spatial organization than metabolic steady state systems composed of autocatalytic chemical cycles within cells. Group or deme reproduction involves sampling organisms from a parent deme, i.e., deme progeneration, and constructing new offspring demes from it, i.e., deme development, resulting in parent deme – offspring deme correlations and measurable group heritabilities, g^2 (McCauley and Wade 1980; Slatkin 1981; Wade and Griesemer 1998; Wade and McCauley 1980).

This kind of sampling process, whether in nature or in the laboratory, satisfies the analysis of reproduction given above. Parent demes progenerate by material overlap of organisms as parts. Offspring flour beetle demes studied in the laboratory must acquire the capacity to reproduce by means of an appropriate balance of organismal processes of development, reproduction, cannibalism, medium-poisoning, and so forth resulting in persistence on an environment (i.e., experimenter) imposed demic generation time.[16] The recursive nature of the analysis of reproduction expresses the fact that group reproduction depends on organism reproduction as part of group development, not merely because groups can only reproduce because organisms reproduce. Group reproduction is more than group subdivision, just as organismal reproduction is more than organismal fission. From the process perspective, controversy over interdeme selection can be understood to concern whether certain systems, such as the flour beetle laboratory system, model inheritance systems with evolved mechanisms of development for particular heredity relations at the group level, or are only artificially constructed group progeneration systems with little prospect for group level evolution of mechanisms of group development.

Material overlap per se is the material basis for all mechanistic theories of genetics that explain how and in what respects and degrees offspring resemble their parents. The issue is no different in its general aspects whether the structural level is alleles transferred by mitosis to offspring cells or organisms transferred in group reproduction to offspring demes. The mechanisms as well as the population and evolutionary consequences may be quite different, but they are instances of the same kind of process.

Geneticists may complain that this characterization of heredity is too broad to be of any scientific use. Geneticists typically are not interested in low-longevity, unstable, low-fidelity inheritance systems. But lo-fi inheritance systems were probably important before the evolution of more precise mechanisms of inheritance (Dyson

1985; Gánti 1971). They may be involved in major evolutionary transitions to new levels of organization (Maynard Smith and Szathmáry
1995). They may also play a role in evolutionary dynamics now
through interaction between epigenetic and replication processes
(Jablonka and Lamb 1995). Heredity among reproducers emerging at
new levels of organization created in major evolutionary transitions
need only be of sufficient fidelity to drive evolution of mechanisms
for development that stabilize them to disintegration from below,
due to the competing genetic "interests" of their constituents. Thus,
evolutionary transition may be characterized as the origin and evolution of new levels of reproduction rather than replication (see also
Szathmáry and Maynard Smith 1997).

In general, a reproduction process in which there is a material
overlap of mechanisms for development is what is minimally required for an inheritance process. Reproduction transfers from parents to offspring the mechanisms that cause heredity relations to
arise in development. A theory of replication processes must interpret these as well as the structure and function of the genetic code
abstracted from its material embodiment in physical processes.
Differently put, a theory of genetic processes cannot merely be a
theory of genetic structure or relations. The transmission of the genetic code itself depends on transfer of a set of aminoacyl-tRNAs and
the rest of the translation machinery in cell division, not just the
transfer of genes. General theories of genetic transmission cannot
take this for granted, or else the origin and evolution of genetic
systems cannot be explained.

A HEURISTIC USE OF PHILOSOPHY

On the Scientific Utility of History and Philosophy of Science

The perspective on reproduction formulated in the previous section
suggests a scientific research program for constructing an integrated
theory of heredity, development, and evolution that accounts for the
way heredity and development are entwined in reproduction processes. Populations of genealogically connected, varying reproduction processes are the material basis of evolutionary processes. Exist-

ing accounts of heredity and development each tend to "black box" the other: Development is ignored in transmission genetics and transmission is ignored in the development of the phenotype. The entwinement of these processes, made clear by the analysis of reproduction in the process perspective, might serve as a conceptual scaffold for constructing an integrated theory without these black boxes. However, there is a problem: There are well-developed theories of genetics, but there are no comparably developed theories of development. So how can the integrated perspective help us if there are no corresponding theories to integrate? We must seek a theory of development comparable in scope, detail, and power to existing theories of genetics.[17] The theory we seek should not, however, be constructed in isolation from genetic theories if it is to contribute to a theory of reproduction that can integrate phenomena of heredity, development, and evolution.

Developmental genetics, of course, aims to explain developmental phenomena in terms of the behavior of genetic mechanisms. However, my theoretical project is not simply to theorize developmental genetics. I propose an expanded view of genetics from a process perspective that transgresses the boundaries of function and structure delimited by the tradition of genetic research in the twentieth century. Nevertheless, it would be pointless to ignore tradition and march off in an entirely new direction because the resulting theory would not bear significantly on the large body of empirical work and thus would integrate nothing. We seek a theory of development in the process perspective that is "backward compatible" with the scientific research programs, if not the theories, of today and yesterday.

Historians and philosophers of science have spent the last fifty years studying how scientists construct, test, and replace theories and how scientific methods and products are best classified and interpreted. History may thus provide conceptual and empirical resources for constructing new scientific theories. Philosophy may suggest formal strategies for interpreting these new constructs in terms of familiar ones. In this section, I suggest such a strategy for constructing a theory of development by arguing for a heuristic, scientific use of philosophical analyses of scientific reductionism. In the next section, I suggest that some clues to the units of development may be found in the history of genetics.

Constructing a Theory of Development by Heuristic Application of Reductionism

Reductionism is a complex subject. Its analyses and varieties are too numerous to survey here (Sarkar 1998). Some scientists characterize their research as reductionistic, though it is not always clear what they mean (Brandon 1996, chapter 11; Wimsatt 1976). Philosophers of science have long interpreted relationships among theories, laws, models, fields, properties, and phenomena in terms of formal and substantive accounts of reduction. There is an important contrast between scientists' descriptions of scientific research strategies or heuristics and philosophers' descriptions of formal and informal criteria for what is to count as a reduction. Scientists have sometimes drawn on philosophical work to characterize their activity, for example, when they endorse Popper's view of scientific method as hypothesis-testing and falsification rather than induction and verification as the best description of their work. Similarly, philosophical work on reductionism can be put to use scientifically. I am not merely noting, with philosophers since Nagel (1961), that while formal analyses of reduction do not fully describe scientific norms and practices some broader account of reduction would. I am not aiming to describe scientific practice at all. Rather, I propose a way to do science that makes use of philosophical ideas about the nature of reductionism. In other words, I propose to put philosophical ideas about reduction into scientific practice, as heuristic guides to theory construction.

The basic idea flows from the results of the previous section. I rejected the structural hierarchy as the starting point for interpreting the relation between heredity and development. As has already been noted, in the process perspective, replication processes are a special case of inheritance processes, which are a special case of reproduction processes. Reproduction requires the progeneration (multiplication with material overlap) of the evolutionary minimum mechanisms of development, namely mechanisms conferring the capacity to reproduce. Mechanisms conferring a complete set of species-typical traits provide an evolutionary maximum specification of development. (Anything more would make the origin of new species impossible.) Therefore, general theories of development that can be integrated with evolutionary theory are bounded by this spectrum of developmental mechanisms. Inheritance requires progeneration of evolved component mechanisms of development. These are mechanisms of development that increase transmission fidelity in particular respects and degrees over what can be achieved at the bare evolutionary minimum. Replication requires progeneration of evolved coding mechanisms of development. So, in process perspective, theories of genetic processes are special cases of theories of reproduction processes. Genetic theories should therefore be reducible to a theory of reproduction in Nagel's formal sense: Special case theories – with terms connectable to terms of a more general theory as well as laws and principles deducible from the general theory – are reduced by the more general theory.

Many philosophers have criticized Nagel's formal criteria for reduction on a variety of grounds. For present purposes, the most important of these criticisms is that his formal criteria of connectability and deducibility only supply analytical tools for interpreting the success or failure of a reductionistic research program *after the fact*. They are philosophers' tools, not scientists' tools. One explanation of this limitation is that Nagel's "syntactic" view of theory structure, in which scientific theories are axiomatizable in first order predicate logic, is not a realistic model of scientific theories in action. In retrospect, it may be possible to express theories this way, but in the heat of scientific theory-building at the cutting edge of empirical work this kind of presentation is rarely possible. Theories are more typically presented in models, not axioms (Lloyd 1988; van Fraassen 1980). Philosophers who have described reductionism as a kind of scientific activity and theorizing as a kind of model-building provide tools that can be put to heuristic, scientific use.[18]

However, to have a general theory of reproduction, we need theories of both "component" processes of progeneration and development.[19] Progeneration has a variety of structural models at different levels of the organizational hierarchy (autocatalysis for molecules, cell division for cells, sexual and asexual modes for organisms, inter- and intrademic models of group reproduction), but these are not complete models of progeneration because we lack a general theory of development, the process that supplies the units that get progenerated in reproduction processes. In contrast, there are explicit structural models of the units of progeneration that cause development.

Whether these models, for example, of gene-based developmental genetics or of epigenetic inheritance systems, are sufficient for a general theory of development, is unknown. Differently put, the entwinement of the processes of progeneration and development means that a theory of development should supply us with the units of progeneration and a theory of progeneration should supply us with the units of development. Thus, models of either process that ignore the other component of reproduction must be incomplete.

The heuristic application of reductionism aims to take advantage of the asymmetry of theoretical resources in genetics and development to bootstrap one class of general models of development. We do have theories of genetic processes and their units, albeit with the familiar array of idealizations, abstractions, heuristic assumptions, and simplifications in the models that black box development and much else as well. These idealizations blocked traditional interpretations of the relationship between genetics and development as one of reduction because the formal criteria for reduction, term-connectability, and law/principle-derivability could not be strictly met (reviewed in Sarkar 1998, chapter 2). Used heuristically, however, as a means to theory construction rather than philosophically (as an argument that reduction of Mendelian to molecular genetics or of development to genetics failed), the idealizations of genetics at least suggest classes of models for development. A serious reductionistic research program in theoretical developmental biology must explore other classes of models as well if the robustness of the theory to idealizing assumptions of its genetic models[20] is to be tested.

Since genetic theories are special cases of reproduction theory, they must at least implicitly involve development. In other words, genetic units of reproduction are also developmental units, but models in genetics tend to treat the developmental aspects in such a degree of idealization and abstraction that developmental properties are unspecified. One idea of how development has been implicitly assumed in the history of genetics will be suggested in the next section, that genetic units – genes – are modeled as "developmental invariants" – things that play a role in development but are not changed by the process of development.

Indeed, I propose a heuristic use of theory reduction to take advantage of this conception of developmental invariance implicit in

257

genetic theories. The heuristic is this: We have theories of genetics. We seek a theory of development. Treat genetic theory as a theory of development as well as of heredity. Use the implied units of development in an explicit, albeit limited model for the general case of a reproduction process. Then examine the limitations of the model to account for developmental phenomena outside the range for which it was originally constructed in order to bootstrap richer models of developmental units. In the process, the relationship between the original genetic model of developmental units and the bootstrapped models should be clear enough to relate traditional genetic theory to the general theory of reproduction. In other words, treat genetics as a theory of development in order to discover the limitations of that kind of theory of development in explaining developmental phenomena. This is reductionism in Wimsatt's engineering sense: You perform the reductive analysis in order to explore how the artificially constructed system or model fails to explain the phenomena. Then you systematically study the failure in order to build a better system (Wimsatt 1976; Wimsatt 1987; Wimsatt in press).

The heuristic approach aims to construct models of development using genetic theory we already have, subject to the constraint that the resulting theory of reproduction must reduce the theory of genetics we use to construct it. The reduction constraint is what makes genetic theory, for all its idealizations and abstractions, a useful guide to formulating models of development. Scientific interest is not likely to center on the success of the reduction, which is only a heuristic means to a theoretical end, but rather in the discovery of what is outside the explanatory scope of a theory of development constructed this way. Many biologists working on the evolution of development already have a good sense of what is left out of account (see Buss 1987; Jablonka and Lamb 1995). The theory of reproduction suggests one line of work toward a theory of development: filling the development gap between genetics and reproduction.

Substantive Criteria for the Reduction of Genetics

Following Sarkar (1998, chapter 3), I distinguish between formal and substantive criteria for reductive explanation. Substantive criteria concern the assumptions made in such an explanation, while formal criteria such as Nagel's concern the logical form of explanation. Sarkar (1998, 43–44) lists three main substantive criteria: fundamentalism, abstract hierarchy, and spatial hierarchy.

In terms of the views expressed in the first section, Sarkar's fundamentalism criterion concerns the image and mapping relations of a reduction. The expected direction of a reduction relation between realms is determined by what counts as more fundamental. Abstract hierarchy applies to the properties bracketed by the evolutionary minimum and maximum specification of development as well as the condition of material overlap in progeneration. Insofar as a given reproductive process involves properties of component processes of progeneration and development, the latter will be at lower levels. The recursive nature of reproduction generates a hierarchy of processes with null development at the lowest level. Spatial hierarchy does not apply to the general theory of reproduction except in the weak sense that some spatial model is called for by any application of the theory because of the universality of material overlap. Material parts are spatial parts. However, the general theory does not specify a particular spatial model or class of models. In that sense, reduction of genetics to reproduction would be a case of "abstract hierarchical reduction" (Sarkar 1998, 44) because the case satisfies the criteria of fundamentalism and abstract hierarchy but not spatial hierarchy. Particular applications of the general theory of reproduction would likely involve some kind of spatial model, for example, of molecular genes, in which case it would be "strong reduction" (Sarkar 1998, 45). Cases in which only fundamentalism is satisfied, for example, the explanation of phenotypes in terms of heritability properties, are cases of "weak reduction" (Sarkar 1998, 44).

A Genetic Model of Developmental Units: Genes as Developmental Invariants

In the process perspective, the reduction image that pictures development as epigenetic is reversed. The arrow of reduction does not point from development as reduced theory toward genetics as reducing theory, but rather the other way, toward development as an aspect of reproduction processes in general, from the special case theories of genetic processes.[21]

The reversed image of reduction seems wrong-headed from the structure perspective because genetics and development are treated there as theories of processes operating on or between different levels of the spatial hierarchy. In the process perspective, genetics and reproduction are at different levels of generality, but since genetic processes always involve developmental processes, they operate at the same structural levels together.

In the functionalist replicator-interactor perspective, as with genotype-phenotype distinctions more broadly, autonomy of genetic processes from development is the basis for the individuation of levels: There are only the two levels of replicators and interactors. Thus, the functionalist perspective inherits the Weismannist constraint from the structuralist perspective. Dawkins-Hull replicators are things that pass on their structure intact, invariant in the developmental processes that construct new vehicles each generation that are exposed to the buffeting of selection. In the functionalist perspective, interactors are characterized in terms of selective interaction, though development is clearly the process that gives rise to interactors. However, development does not figure in the analysis of either replicators or interactors.

Indeed, in structure perspective, development is explained by treating genetic units as *developmental invariants*, things not changed by development but that pass through and control it. As constants in the cycling developmental system, genes are desirable theoretical entities because they have stable properties that can be followed through life cycles and are thus always there to hang explanations on. Genes are not only the only things to which evolutionary benefit can be traced (according to Dawkins' account of the units of selection); they are also the only things that go through development without developing. They epitomize self-reproduction: They are null developers born with the capacity of replication.

In the next section, I suggest that historical challenges to claims about the stability of genes provoke controversy about the nature of the gene because they raise doubt about whether genes can explain characters arising in development. This puts controversy about the concept of a gene on a par with "the species problem" as a perennial conceptual problem in science. Just as with species, scientists debating gene concepts argue strongly for particular conceptions and at the same time wonder whether the concept is even necessary to the science. In any case, the concept of the gene as a developmental invariant suggests a new historiography of genetics may be in order. If developmental invariance is central to the gene concept's history, then it may not be appropriate to interpret the history of relations between genetics and embryology as a divorce. Perhaps genetics never really left embryology after the origination of the gene theory, (cf. Gilbert 1978). Classical geneticists, even Muller in his search for the physical atoms of heredity, really did care about development (Falk 1997).

We can use this property of developmental units of genetics in the process perspective to speculate about theories of development. Genetic theories are theories of development because genetic processes are special cases of reproduction processes, but they are theories expressed in terms of developmental invariants. Are there other developmental invariants besides genes? Can the stability of developmental units be attributed to features of environment rather than internal, structural features? Can the apparatus of theoretical genetics (even if not Mendel's laws) be used to describe other developmental invariants? Are there ways to formulate a theory of development in terms of variable rather than invariant components or quantities?

Asking these questions in a scientifically productive way relies on theory reduction as a heuristic constraint. We seek a theory of development. We know that development is an aspect of reproduction, and we know that genetic processes are a special case of reproduction. Genetic theories describe genetic processes, so we look for the ways genetic theories handle the developmental aspects of "genetic" processes, for example, by treating the units of the theory as developmental invariants. We generalize those ways to a theory of reproduction processes and assess the limitations and failings of developmental theories reached by such generalizations, subject to the heuristic constraint that the formal theory of reproduction must reduce the special case theories of genetics. If we could succeed in constructing such a theory, it might help guide us to mechanisms causing developmental invariance beyond the special cases (genes) that we already know about. Or it may help show us what mechanisms of development are left out by a theory of developmental invariants.

A PROCESS HISTORIOGRAPHY OF THE GENE CONCEPT?

Genetics tradition treats reduction as a problem ordered by the structural hierarchy, a *strong reduction* in the terminology of the previous section. Reductionist theories and explanations concern entities lower in the hierarchy than those at the level to be explained, while holist, emergentist, and teleological theories and explanations are couched in terms of entities higher in the hierarchy. Geneticists early in the twentieth century generally favored "nuclear monopoly," explaining heredity and development in terms of the controlling role of nuclear substance (Sapp 1987), while embryologists often debated whether the cytoplasm, the cell, or the organism as a whole was the seat of developmental control (Gilbert 1991). Thus, disciplines as well as phenomena were sorted by the structural hierarchy.

As we have seen, the commonly expected direction of reduction linking development and genetics also depends on the structural hierarchy. Although genetics can be treated in a purely phenomenological way, concerned only with parent-offspring resemblance, most theoreticians of heredity beginning with Darwin (gemmules), Mendel (factors), Weismann (biophores, determinants, ids, idants), Galton (stirp), Nägeli (idioplasm), and de Vries (intracellular pangenes) interpreted hereditary phenomena in terms of material, subcellular bodies (Robinson 1979). These entities are constituent parts of cells, just as cells are constituent parts of (multicellular) organisms. Development (at least its classical aspect in embryogenesis) has often been taken to apply only to the determination, differentiation, and growth of cells of multicellular bodies from such constituents. As a cellular or supercellular phenomenon of generation out of components, development necessarily takes place at a higher compositional level than genetics because genes are parts of cells.

The rigidity of the structural hierarchy guarantees a direction of reduction. Images of reduction based on structure cannot easily be reversed because parts and wholes are not likely to trade assigned places in the hierarchy. It is extremely unlikely that genes will ever turn out to be wholes of which cells are parts or that organisms are really parts of cells. Yet that is the kind of reversal that would be required to keep the structural hierarchy as the proper framework for reduction and also try to reduce genetics to reproduction. There are, of course, genes that at times are not parts of cells such as genes inside viral phage particles, genes spilled out of lysed cells, and artificial genes manufactured in vitro. If Dawkins (1982) is right about the extended phenotype, there can be genes that control phenotypes of bodies in which they do not reside, placing such genes "outside" the cells and cell complexes bearing "their" phenotypic expressions. But these cases do not come close to what would be required to recognize that cells are parts of genes. The radical change of perspective required to make *that* reversal would involve a dramatic change in much more than the gene concept.[22]

Weismannism

An important source of the image that development reduces to genetics is the view that genes are the causes of development as well as the causes of heredity. This is a version of Weismannism.[23] Weismann (1892) espoused the view that germ-plasm is continuous while somato-plasm is discontinuous. The nuclear plasma of cells destined to become germ cells passes untouched through cell lines to form a line of hereditary continuity from zygote to germ cell in development. The parts of the plasma of other cell line nuclei, destined to become the differentiated cells of the body, were thought to be parceled out to the cells of different types such that the qualitative division explains observed cellular differentiation in embryogeny. Related views of heredity, though not of development, are implicit in Mendel and Johannsen (Lewontin 1992). This similarity was noted by Johannsen's contemporaries early in the history of genetics. For example, E. G. Conklin, in his 1914 Harris Lectures, wrote: "This contrast between the germ and the body, between the undeveloped and the developed organism, is fundamental in all modern studies of heredity. It was especially emphasized by Weismann in his germ-plasm theory and recently it has been made prominent by Johannsen under the terms 'genotype' and 'phenotype'" (Conklin 1919, 126–127).

Although T. H. Morgan rejected Weismann's theory of development (understood by him as the Roux-Weismann mosaic theory of qualitative division), Morgan endorsed the view of heredity common to Weismannism and his own factorial theory of heredity, which "rests on the assumption that the germ plasm contains a host of elements, that are independent of each other in the sense that one allelomorph may be substituted for another one *without alteration of either*, and that these allelomorphs will now perpetuate themselves *unchanged* although in company of different factors" (Morgan et al. 1915, 277; *emphasis added*).

Some prominent claims attributed to Weismann should more accurately be attributed to those of his interpreters who inspired the formation of genetics as a distinct discipline, notably E. B. Wilson (Griesemer and Wimsatt 1989). Wilson interpreted Weismann's view as one of "nuclear monopoly" on biological causation (Sapp 1987, 1991). The germ-plasm is the cause both of new germ-plasm and of the soma in development (Wilson 1896, figure 4). In modern guise, genes are double causal agents while characters are epiphenomenal, acausal entities. The noncausal role of the body in modern molecular terms is best shown in Maynard Smith's "rosetta stone" diagram comparing Wilson's interpretation of Weismann to Crick's central dogma of molecular genetics. DNA gives rise to new DNA and to protein, just as the germ gives rise to new germ and to the soma (Maynard Smith 1975, figure 8).

An alternative to strong reduction to lower levels of the spatial hierarchy is abstract hierarchical reduction of special to general cases of reproduction processes. Generalization, like spatial composition, is a hierarchy-forming relation. This is a common sense of classical theory reduction in the physical sciences, as in the case of the reduction of Kepler's laws of planetary motion to Newton's laws of mechanics. If general cases are described by fundamental laws of nature and special cases by phenomenological laws, rules, regularities, or singular observations or events, then reduction may be interpreted as the theoretical enterprise of explanation by fundamental laws.

The purpose of this section is to urge a new historiography of the gene concept, in process perspective, in which the character of the gene as a developmental unit is as central to the gene concept as is the gene's role as a unit of hereditary transmission among generations. As suggested above, from the point of view of genetics, the gene is a unit of reproduction that is invariant in development, stable in heredity in the short term and variable in evolution over the long run (through mutation and other processes of genetic change). The main goal in urging a new historiography is scientific: to gain heuristic purchase on a general theory of development by viewing theories of genetics as theories of development based on invariant units. For that reason, the historical observations below are offered only to illustrate my conviction that historical defenses of the gene as a developmental invariant, as an entity stable in development that explains hereditary transmission and somehow explains development, will be found wherever there is controversy about the nature of the gene.

The proposal that genes are developmental invariants implies a more radical view of the history of genetics than Scott Gilbert's argument that genetics originated in embryology and then superseded it, at least in "theological" rhetoric (Gilbert 1978). Instead, I suggest that genetics never left its ancestral home in embryology. Geneticists of all periods from Mendel onward have expressed direct interest and concern with development as well as hereditary transmission, with the gene-character relation as well as gene-gene and character-character correlation among generations. What has happened historically is that interpretation of heredity has shifted from a developmental aspect of parent-offspring relations in reproduction to genes as autonomous master molecules controlling development (Keller 1995; Sapp 1987). Geneticists have always been intensely interested in development. The historical change of rhetoric masks the developmental nature of the gene.

Mendelian Factors

In a sense, classical transmission genetics beginning with Mendel is the first modern theory of development because it identified the first developmental invariant, the Mendelian factor. Invariant entities are useful theoretical units because they are units one can calculate with and that one can use as tools of experimental intervention. They are conceptually and empirically "solid," "tame," and "domesticated." Entities that change while one is using them are difficult to work with, hard to manage and keep track of, hard to defend. Mendel described factor segregation in terms of passage unchanged through the hybrid organisms he created and controlled in his experimental garden.

Mendel's paper on hybridization in peas illustrates the notion of a gene (factor) as a developmental invariant (Mendel 1866). The law of segregation expresses an invariance principle. Mendelian factors in hybridization experiments segregate. They can be brought together in hybrids and separated again without any effect on their capacity to act subsequently as factors: "among the progeny of the hybrids constant forms appear, and that this occurs, too, in respect of all combinations of the associated characters" (Mendel 1866, 20). Dominant and recessive factors are both recovered in subsequent generations, meaning in developmental contexts similar to those of the parental generation. Segregation shows that putting a factor into the developmental environment of a hybrid, that differs from that of the true-breeding lineage from which it was extracted, does not alter factors. Segregating factors show themselves to be stable over multigenerational inheritance involving processes of syngametic combination in zygotes and assortment in reproduction that cause developmental contexts to fluctuate. Of course, Mendel characterized these processes and contexts crudely; he considered whether pollen and egg cells were of "like" or "unlike" character and described their combinatorial possibilities without attempting to explain how the factors get from parent cells to offspring cells or how they play a role in the development of offspring traits.

While recessive traits are not expressed in the hybrid, recessive factors carry the capacity for recessive trait expression even in developmental contexts where the recessive trait is not actually expressed. This is revealed when traits equivalent to those of the purebred parentals appear in the second generation bred from the hybrids (grandoffspring). Although systems may be such that a gene "is expressed" in one developmental context and not in another, this does not mean that the gene fails to carry a capacity into different developmental contexts and is therefore not developmentally invariant. However, some theories of dominance and recessiveness, such as Bateson's presence-absence theory, challenge the stability of the recessive capacity. That theory interprets recessive traits as due to the absence of a gene and dominant traits to the presence of a gene. If true, this theory would imply that the capacity for expression of recessive traits is carried, not by a particular material factor, but by the genome, nucleus, cell, or organism lacking dominant factors. Morgan, for one, attacked Bateson's theory on precisely these grounds (Morgan 1926).

Factors are not altered by passage through different kinds of hybrids. Trait expression in Mendel's diploids depends on factor pairs (or more complex collections), not individual factors, so changes in expression with change of developmental context do not constitute changes in the gene. The gene is identified and characterized solely in terms of its invariant qualities. If it also has developmentally variable qualities, these will be judged accidental with respect to its status as a gene. That judgment is made explicit by the use of phrases like "non-Mendelian" to label "genes" that, because of their variability, are by definition aberrant. The label does not signal a belief that non-Mendelian factors must be relatively rare to be counted aberrant. Instead it signals the belief that what it means to be a Mendelian gene includes the invariance specified by Mendel's laws.

It is clear from Mendel's text that his theory was developmental as well as hereditary in nature. In the section "The Reproductive Cells of the Hybrids," Mendel offers a hypothesis to explain his results, that "So far as experience goes, we find it in every case confirmed that constant progeny can only be formed when the egg cells and the fertilizing pollen are of like character, so that both are provided with the material for creating quite similar individuals, as in the case with the normal fertilization of pure species. We must therefore regard it as certain that exactly similar factors must be at work also in the production of the constant forms in the hybrid plants" (Mendel 1866, 20). He goes on to offer an assumption that clearly expresses the joint developmental-hereditary nature of the theory: "In point of fact it is possible to demonstrate theoretically that this hypothesis would fully suffice to account for the *development of the hybrids in the separate generations*, if we might at the same time assume that the various kinds of egg and pollen cells were formed in the hybrids on the average in equal numbers" (Mendel 1866, 21, *emphasis added*).

Subsequent interpreters of Mendel's achievement characterized this stability as the hallmark of his theory. In *Heredity and Environment*, for example, Conklin characterizes stability in terms of the "purity" of the germ cells of hybrid organisms despite their origin from hybrid parent cells (Conklin 1919, 88). The principle of segregation, which describes the maintenance of purity through crosses and the development of hybrid offspring, is "the most important part of Mendel's law" (Conklin 1919). Offspring ratios not predicted by simple Mendelian assumptions threaten the concept of the factor as stable in development, unless they can be interpreted in an expanded Mendelian sense, for example, by multiple factors, partial dominance, or maternal effects (Conklin 1919, 107–112).

Castle's argument in the 1910s that natural selection might alter Mendelian factors and that he had experimental evidence in support of this interpretation was dealt with severely by the Morgan school, particularly by Muller. The case is discussed by Provine (1971, chapter 4; 1986, chapter 2) and need not be repeated here. The lesson, though, is the same: If selection alters the gene, then the gene is not invariable in the developmental process described by Mendelism. Such challenges must be met if genetics is to be built on Mendelian units.

In *The Theory of the Gene*, Morgan wrote that "Mendel's theory of heredity postulates that the gene is stable. It assumes that the gene that each parent contributes to the hybrid remains intact in its new environment in the hybrid" (Morgan 1926, 292). Morgan offered a series of examples to show the variety of evidence that the central Mendelian assumption is true, despite the apparent challenge to the stability of the gene implicit in unexpected ratios.

Despite much diversity of behavior in development, the stability of genes is never allowed to remain at issue for long. Even mutation gets interpreted as a stable change in gene state. Whenever the stability of an alleged genetic unit becomes doubtful, its status as a gene does also. Genes are the epitome of Salmon's notion of a causal process: Genes can transmit "marks" as the result of causal interactions (Salmon 1984). Salmon (1994) argues that invariance is central to the general concept of a causal process (see also van Fraassen, 1989; Woodward 1993). Here, invariance with respect to a developmental frame of reference is central: It is what allows us to interpret a reproduction process as genetic, as having the capacity to transmit particular kinds of developmental marks, meaning mutations, along replication processes.

An additional basis for thinking invariance is central to the gene concept lies in the use of phrases like *a gene for X*. This expression is applied carefully in classical genetics, in situations where hybridization experiments have been done to reveal in what respect the gene is a differential, in other words a cause that distinguishes between organisms with and without the gene. This differential quality of the gene concept has been emphasized by many geneticists throughout the history of the field (see Schwartz, this volume). Without the possibility of hybridization and its requirement of distinguishable phenotypes, genes that are developmentally invariant as specified by Mendelism cannot be detected. This is because, as the classical geneticists understood, characters are determined by many or all the genes and each gene may have many effects. The experimental geneticist relies on the developmental invariance of the gene to discover its presence through hybridization, so invariance is embedded in the instrumental notion of Mendelian factors as well as in its conceptual analysis. For Mendel, the relevant contexts were the zygotic environments from which the factor is extractable by further crossing. Extractability is the experimental sign of invariance.

A process historiography of the gene concept would show that challenges to the stability or invariance of the gene cut deeply because they challenge not only concepts, but the epistemological basis of experimental utility of the gene concept. A careful history of controversies over gene concepts should show that each time apparent variability in the gene itself was discovered, geneticists faced a familiar array of options. They could defend the gene concept, showing how challenging cases, for example, of "non-Mendelian" ratios, could be given Mendelian explanations, thereby vindicating purity of the germ and stability of the factor. Or, they could narrow their gene concept to that part or property of the gene as previously conceptualized that remains developmentally invariant in the new phenomena (essentially refining general descriptions of genetic phenomena). If pseudoallelism or fine structure mapping reveals that genes can no longer be thought of as units of structure, function, mutation, *and* recombination, then perhaps genes can be interpreted more narrowly as units of structure that must team up to constitute units of function. If the operon theory forces a divorce between units of structure and function, then perhaps genes can be kept on as units of structure whose functions are explained in terms of cell-level regulation. In other words, challenges to gene concepts throughout history might be understood in terms of such a list of options if we interpret the gene as something that is supposed to be developmentally invariant, in other words, something defined with respect to the developmental process. Finally, geneticists could alter the meaning of *development* in such a way that their original gene concept is of an invariant in the modified sense of development.

The history of repeated defense against challenges of instability may lend, ironically, a sense of progress. If narrowing the gene concept to ever smaller sequences preserves the developmental invariance of the concept, it is a march in step with the progress of mechanism toward lower and lower levels of organization. But what stops progress from going too far, to the level of single nucleotides? Geneticists' sense that single nucleotides are chemical rather than biological objects that create a counter-pressure to make the gene concept more inclusive. Otherwise, explanatory power would be handed over, along with the gene concept, to chemists. On the other hand, expanding the gene concept leads to holism – the genome as unit of heredity – which geneticists have typically, though not invariably (remember Goldschmidt), regarded as unproductive because of their heuristic bias in favor of strong reduction and mechanism leading down the biological hierarchy.

The important point is that successful defense of the Mendelian gene concept does not thereby *explain* development. The Mendelian gene is an entity that is invariant in the frames of reference defined by Mendelian experimental construction of those developmental processes. Mendelian methods supply no account of how genes are involved in the development of the organism's traits. Mendel's concept does not address the "paradox of development."

The Paradox of Development

The fundamental problem of development is to explain how heterogeneity comes from homogeneity. It is a hard problem because the sources, nature, and mechanisms of change must all be understood and articulated. It poses a dilemma for biologists who have committed to a particular approach or point of view if they do not see how their chosen field, theory, and methods can solve the problem directly. Either the failing is internal to their point of view, requiring internal changes in theory or methods, or the failing is external in the sense that a solution requires theories or methods from outside the chosen approach or point of view. Either way, the committed scientist who fails must give up something: either explanatory power by sticking to the point of view but failing to solve the problem, or explanatory control by admitting fundamental units from other fields or points of view.[24] Few scientists are willing to accept either horn of the dilemma and instead try to solve the problem head on by articulating, modifying, or otherwise advancing their paradigm – activities Kuhn called "normal science."

Genetics faces the fundamental problem of development because, as I have argued, genetics is about development, even if practitioners have tried to insulate their field from embryology and developmental biology. The paradox of development poses a dilemma for genetics (Gilbert 1994; Sapp 1991). How can cell differentiation, the basis of form in multicellular organisms and hence of variation in evolution, be explained by genes, since the genes of all somatic cells are identical? The dilemma was familiar to Morgan when he wrote in *Embryology and Genetics* that "At first sight it may seem paradoxical that a guinea pig that can develop areas of black hair should have white areas of hair if, as is the case, the cells of both areas carry all the genes" (Morgan 1934, 134). The paradox is the geneticist's way of stating the fundamental problem of development. The genetical version became urgent after the defeat of the Roux-Weismann mosaic conception of development, that did explain cellular differentiation out of genetical homogeneity.

The geneticists' paradox is a consequence of assumptions about the gene as an invariant. Genes must be stable in heredity in order to explain the basic phenomenon of heredity by Mendelian means: Offspring tend to resemble their parents. Deviations from this tendency are explained in terms of changing collections of genes rather than alterations of genes themselves (except for rare mutations). The basic tendency to heredity is explained as a result of the hereditary stability of the genetic units. But the experiments which originally showed this hereditary stability, Mendelian hybridization experiments, also required genes to be developmental invariants: Genes could not change as a function of developmental context and be detected by Mendelian means. However, if somatic cells are genetically equivalent and also cause development as dictated by Weismannism then they must change in order to cause differentiation. Commitment to Mendelian methods *and also* to genetic explanations of development generates the paradox: Genes must be developmentally invariant because otherwise they could not be shown to be hereditarily stable, but they must be developmentally variable if they are to be the cause of differentiation in development. The paradox of development is a paradox of genetics.[25]

Homogeneity of genes among cells of a developing organism is an aspect of developmental invariance: The genes don't change in quantity or quality as they are passed from cell to cell in the dividing organism. But what must geneticists give up in order to address the paradox? Should they stick to their classical genes and Mendelian methods and give up the power to explain development? Should they modify their gene concept in the hope that some other notion of what makes a gene a stable unit in hereditary transmission will also explain the gene's role in expression and differentiation? Or should they refine the description of the developmental invariance property of the gene in such a way as to cede explanatory control but not give up the classical gene or its explanatory power?

Morgan explored all three of these strategies for explaining development by genes. He outlined them in the introduction to *Embryology and Genetics* (9). First, all genes might act all the time in the same way. That would leave differentiation completely unexplained. Still, the dilemma would be solved by putting one's head in the sand, treating differentiation as a black box to be described phenomenologically for the purpose of character analysis. Second, it may be that "different batteries of genes come into action as development proceeds" (Morgan 1934, 10). This solution, that Morgan interprets along the lines of the Roux-Weismann hypothesis, is contradicted, he claims, by results of experiments compressing embryos to alter the sequence of cleavage planes. Third, Morgan speculates that differences of protoplasm might influence the "growth of the chromatin" and affect "the substances manufactured by the genes,"

272

so that "The initial differences in the protoplasmic regions may be supposed to affect the activity of the genes. The genes will then in turn affect the protoplasm, which will start a new series of reciprocal reactions" (Morgan 1934). Since the initial protoplasmic differences will trace to the action of maternal genes, we have a genetic solution to the problem of development. This last explanatory strategy is most interesting. In effect, it narrows the nature of the developmental invariance of the gene to its structural properties, allowing its functioning to depend in complex and changing ways on its interactions with protoplasm. As a result, control of the explanation of development is partially ceded to the students of protoplasm – cytologists, embryologists, and colloidal chemists – but the nature of the gene as an invariant in development, though narrowed, is preserved. Genes remain the locus of explanatory power because it is their role in interaction that is explanatory.

Operons

Weismann's speculative hypotheses of genomic spatial hierarchy and mosaic development were being rejected at the same time that Weismannism's causal structure was spreading as part of the central theoretical perspective of the emerging field of genetics. The genetic equivalence of all the cells of the body, including germ-cells, only exacerbated the problem of the paradox of development for geneticists who accepted the developmental invariance of the gene resting on Weismannism's theory and Mendel's methods. Morgan's speculations in the 1930s, trying to hold on to the classical Mendelian gene concept *and* the goal of a genetic explanation of development, far outran both theory and experimental data.

Support for Morgan's third explanatory strategy came only in the 1960s after genetics began to go molecular, when it was recognized that cellular difference could be interpreted in terms of differences in states of molecular gene activity rather than differences of gene presence/absence, structure, or arrangement (position effect, inversion). The operon theory of Jacob, Monod, and others provided an interpretation of molecular feedback mechanisms by which differences in cellular and external environments could cause different states of gene activation. As Sapp put it, "The operon scheme solved

the problem of cellular differentiation without invoking environmentally directed gene mutations and without invoking a distinction
between somatic heredity or development (cytoplasm) and sexual
heredity (nucleus)" (Sapp 1991, 246). The operon model avoids ceding explanatory power to other fields, theories, or methods in the
sense that the gene remains the central explanatory unit. However,
while the operon theory provides a genetical solution to the paradox
of development, it sustains the essential tension of the paradox insofar as genes remain the cause of development. In the operon model,
explanatory control is ceded to developmental biology or (horrors!)
nutritional biochemistry by making gene activation states the central
concern rather than gene structure. This opens the door to other cell
constituents as possible units of explanation for development. Moreover, activation states are not properties of classical genes, so the
status of genes as developmental invariants, and thus the nature of
the gene concept, is muddied.

The Jacob-Monod operon theory of bacterial gene regulation is on
the opposite side of the molecular biology revolution from Mendel.
Part of the triumph of molecular genetics is due to narrowing of the
neoclassical gene concept, identifying genes with nucleotide sequences. This move eliminated much uncertainty about the nature of
genes due to their complex role in development. The Watson-Crick
molecular gene narrows the sense in which a gene is a developmental invariant to that of a structural invariant – the nucleotide sequence does not change in development even if its activation state
does. However, narrowing the gene concept in this way does not
avoid ceding explanatory control. Even though genes code for proteins, proteins could not be invoked to explain differentiation and
thus, by transitivity of causes, used as an interpretation of how genes
cause development. The reason is that sequence invariance does not
explain how different cells come to have different constellations of
proteins that cause differentiation.

The operon concept of the gene illustrates options for resolving
challenges to the notion of the gene as a developmental invariant. As
discussed above, neoclassical replies to challenges to the classical
Mendelian gene showed that non-Mendelian genes – genes that exhibited apparent developmental variability – have Mendelian explanations. In contrast, molecular solutions to challenges to the power

274

of classical genes to explain development involve either modification of the genetic properties that are developmentally invariant or a changed interpretation of developmental invariance that grounds the gene concept. The option of modifying the gene concept was not open to geneticists of the Morgan era for social-institutional reasons. Genetics would have withered with that decision just as embryology appears to have done as it pursued one unit concept after another (organizers, morphogenetic fields, and so forth). Histories of the period between the chromosome theory and the operon should supply further evidence of genes as invariants (e.g., Dietrich, this volume).

The operon theory interprets genes as collections of nucleotide sequences having different functions.[26] Some sequences – promoters and operators – are regulatory, providing binding sites for molecules that regulate gene expression. Other sequences – structural genes – code for amino acid sequences of polypeptides. RNA polymerases bind promoters to begin transcription of downstream structural DNA sequences into RNA. After transcription, RNA molecules can be processed, transported out of the nucleus, and translated on ribosomes into amino acid sequences. Transcription can be blocked if a regulator molecule binds the operator, preventing RNA polymerase from moving down the line to the structural gene. Transcription can be (re)started if effector molecules bind the regulators to prevent them from blocking transcription. The ordered collection of regulatory and structural sequences, regulatory molecules, and effectors (that can be molecules imported from the extracellular environment), form an environmental response system altering, not the nucleotide sequences of the genes, but the state of activation of the collective. Jacob and Monod called this collective a "macromolecular society" (Sapp 1991, 245).

The operon concept can address the paradox of development genetically by expanding the gene concept to include activation states as well as sequence as part of the gene concept in order to explain development and resolve the paradox of development. But precisely by ascribing the feedback that explains developmental diversification to the state of activation and not to the sequence, the operon theory of gene regulation distinguishes the sequence as a developmental invariant from activation state as a developmental variable.

The question is, what kind of a solution to the paradox is the decision to include activation state as part of the gene concept and what sort of scientific commitments does that decision require? If the operon concept helps solve the paradox of development by expanding the gene concept to include activation state, it comes at the cost of rejecting the idea that genes are developmental invariants, or else it is on the road to genetic holism: The gene is not a molecule but rather the whole macromolecular society, that is the new locus of developmental invariance. As a result, operon-based genetics ceases to be genetics in the classical sense, a serious break with tradition.

Alternatively, if the operon concept of a gene, like the classical Watson-Crick molecular gene, equates gene with nucleotide sequence, then the problem of development and its explanation by genetics is abandoned. Instead, a new interfield theory (Darden 1991; Darden and Maull 1977) of "developmental genetics" is called for that is neither development nor genetics in the classical sense and whose domain is all that lies "between the base pairs of the genotype and the limb or eye of the phenotype" (Gilbert 1991, 136).

This story illustrates how the image of reduction is tied to the core concept of a gene as a developmental invariant by the paradox of development. Genetics cannot explain development without being developmental, but in being so, it ceases to look very "genetic." As genetics incorporated ever more sophisticated interpretations of the macromolecular regulatory environment of the genes, holding out hope for a genetic solution to the paradox of development, the fundamental idea that the genes are stable in development crumbled. Genes are the locus of developmental explanation because they are fixed points in the developmental flux. The attempt to resolve the paradox of development genetically, through the operon model of gene regulation, either narrowed the concept of the gene to that of nucleotide sequence or threatened to expand it so much that the new view ceases to be genetics from a structuralist perspective.

Developmental genetics counts as distinct from genetics if genetics adopts a narrow gene concept that is developmentally invariant. Developmental genetics counts as part of genetics if the gene concept is expanded to include the activation states that provide an account of genetic control with environmental feedback. But then the gene is no longer a developmental invariant, hence developmental genetics is not genetics in the classical sense. Thus, the paradox of development continues to be a dilemma for genetics in the molecular age. This is to be expected if what is essential to a gene concept is the idea of developmental invariance.

Developmental Genetics

The fundamental assumption of developmental genetics and its application to evolution is succinctly stated by Raff and Kaufman: "Our premise is that developmental processes are under genetic control and that evolution should be envisaged as resulting from changes in the genes regulating ontogeny" (Raff and Kaufman 1983, 2). This premise supports the traditional image of reduction. One expects development to reduce to genetics because genetics is the field whose theories describe the causal mechanisms by which genes regulate ontogeny and support causal explanations of evolutionary changes in form when coupled to the laws of evolutionary genetics.

Gilbert (1991, 1994, 38–39) traces the origins of developmental genetics to work by Gluecksohn-Schoenheimer and Waddington in the late 1930s, quoting the former's reflection of 1938 that embryology and genetics are each incomplete without the other (Gilbert 1994, 38). But for Gilbert, the significance of developmental genetics is that it puts to rest the criticism of genetics by embryologists, that mutations and Mendelizing traits in general affect only superficial adult characters and not fundamental Bauplans constructed in development. Developmental genetics claims to study the control of major Bauplans in development by genes (or gene complexes). These are significant advances into the study of the developmental role of gene action, to be sure, but they do not serve to alter the expectation that *development* reduces to genetics. Instead, developmental genetics provides richer detail than the austere formal mapping of genotype to phenotype in neoclassical genetics, which skirts the whole problem of development. Developmental genetics provides a new theoretical level in its interfield theories tracing gene action located and localized spatially and temporally in the developing organism.

From the process point of view, developmental genetics does not address the important conceptual issues because it inherits the structuralist tradition of reductionism in which development is expected to reduce to genetics. In the process perspective, developmental genetics is not the genetics of "developmental" traits, but rather is an idealized model of development, one in which the progenerative role of environmental variation is simplified for the sake of emphasizing the role of the developmental resources called molecular genes. It is dubious that an adequate theory of development can be formulated that takes only these resources as its central units, since molecular biology is daily challenging the developmental invariance of nucleotide sequences right down to the level of single nucleotides (see e.g., Fogle this volume; Moss 1992; Neumann-Held 1998). A more adequate theory of development will probably not be possible without a more radical change of perspective, such as the one urged at the outset of this paper.

NOTES

1. Part of this work was pursued while the author was a fellow of the Wissenschaftskolleg zu Berlin (1992–1993) and of the Collegium Budapest (1994–1995) and was facilitated by USDA Cooperative Agreement PNW 95–0768. I thank the Rektors and Fellows of both institutes, and the respective Biology Group Conveners, Peter Hammerstein and Eörs Szathmáry. This work would not have been done without the interest and support of Leo Buss, Eva Jablonka, and Eörs Szathmáry. I owe many thanks to the editors, especially Rafi Falk for detailed criticism and advice, and to an anonymous reviewer for valuable suggestions on how to organize this material. I am grateful to the other workshop participants, to audiences at Chicago, Duke, Northwestern, and San Diego, and especially to my former colleague Michael Dietrich for helpful discussion of the manuscript and clarification of the argument. I thank Connie, Ellen, and Kate for their support.
2. This suggests a pragmatic view of theories that adds to the semantic view the notion that our assignments of objects to processes must pick out from the network of causation those pathways of interest and concern to us. Perspectives play the role of guides to commitment to scientific action in the pragmatic view of theories parallel to the role of models as guides to acceptance and belief in the semantic view.
3. I adopt Sarkar's (1998) terminology of factors, rules and realms rather than the traditional causes, laws, and domains to avoid confusion of my use of terms like *image* and *mapping* with terms from logic and mathematics having different meanings.
4. In forthcoming work I criticize the philosophical literature on replica-

tors. Work in preparation will show how to interpret heritabilities and selection differentials as quantities for causal capacities.

5. Two authors who express similar philosophical views are Brandon (1990, chapter 3) and Burian (1992, 11). Burian equates intergenerational transmission with "replication plus development." Brandon considers the role of development in relation to that of replicators in evolution. The relation between these views and mine are too complex to enter into here. Thanks to Michael Wedin for suggesting the term *progeneration*.

6. Compare progeneration to Maynard Smith's multiplication condition for something to be a unit of evolution (Maynard Smith 1988). He includes the notion of "same kind" in his definition of multiplication, but provides little elaboration and does not acknowledge the general requirement of material overlap. Copies need not materially overlap their originals to count as copies, but progenerants must materially overlap their parents if they are to count as progenerated. A photocopy of a painting does not materially overlap the painting that was laid on the copier glass, but it is nevertheless a copy. A copy of a painting resembles the original, though it may not be made of paint on canvas, let alone be made out of the very paint and canvas of the original. But a progenerant of a painting might be a torn half of the original painting. In virtue of material overlap, it will resemble the painting in some respects, but certainly not in all the same respects that copies typically do. Thus, some cases of so-called "cultural" evolution will turn out to be "biological" in the sense that they satisfy the material overlap condition for biological reproduction, which in turn is a necessary condition for evolution. Other cases will only satisfy weaker copying criteria. This differentiation of cases of cultural evolution disagrees with Dawkins' (1976) memetic theory as well as with some developmental systems theories (Griffiths and Gray 1994), that treat all cases of cultural evolution on a par due to their lack of distinction of replication and reproduction (Griesemer in press). Genetic reductionist views also tend to treat all cases of cultural evolution as reducible to biology in the same way. In my view, some cases of cultural evolution may be genuine cases of biological evolution because they are genuine biological reproduction processes, whereas others may only be analogous to reproduction processes and therefore not reducible in a theory of biological reproduction.

7. Things get interesting as we push back to the origin of life. Do autocatalytic chemical cycles count as biological agents? Must we invoke an idea of chemical agency to explain biological agency? Is a loaf of bread an agent if it "invites" a hungry knife-wielder to cut it, as Dawkins suggests when he characterizes a sheet of paper as an "active" replicator (meme) if the words written on it cause someone to put it into the photocopier and make more copies? Actors and agents must be dis-

tinguished if we are to understand the biological autonomy of reproducers (Eilhu Gerson, personal communication).

8. This is an argument I first made in Griesemer (in press).

9. The reproducer concept generalizes the replicator in parallel to the theory of reproduction processes generalizing genetic processes. In
Griesemer (in press), I argued that the reproducer concept redresses
grievances from the developmental systems perspective against reductionism (e.g., Gray 1992; Griffiths and Gray 1994; Oyama 1985). It also
argues against the divorce of genetics and development reinforced by
Dawkins (see Sterelny, Smith, and Dickison 1996).

10. Thanks to Walter Fontana for the wonderful phrase *pieces of development.*

11. In work in preparation, I argue that this is the minimal specification of
development needed for a general theory of evolution, in which the only
relevant essential properties of offspring are whether they can reproduce
and the extent to which they vary in that capacity. That is enough for so-
called "replicator dynamics" to apply to reproducers. The maximal specification of development for evolutionary theory is the acquisition of
species-typical traits. All traits with evolutionary biofunction can be
thought of as ways and means to the minimal developmental end.

12. I say *a* process perspective because mine is not the only one. Developmental systems theorists also offer a process perspective. See Griesemer
(in press).

13. In short, Weismann was no Weismannian (Griesemer 1994; Griesemer
and Wimsatt 1989).

14. Following Glennan (1996, 52), mechanisms are things – physical objects
or processes – which have parts that interact in such a way as to cause a
specifiable behavior: "A mechanism underlying a behavior is a complex
system which produces that behavior by the interaction of a number of
parts according to direct causal laws." One might also interpret mechanisms for heredity in terms of Cartwright's notion of "nomological machines" (Cartwright 1997). Saying that a mechanism is "for" heredity
means that it is efficient at producing heredity (with respect to some
trait) compared to processes that do not have the mechanism. Most such
mechanisms will be efficient in virtue of being evolutionary adaptations.
In forthcoming work, I analyze the concept of coding in terms of the
process perspective suggested here. Discussion of coding is beyond the
scope of this paper.

15. Coding, in other words, may consist in excess combinatorial information
(Gánti 1971 [1986]). The number of nucleotide (amino acid) kinds is
much smaller than the number of sequences of nucleotides (polypeptides) coded for. There are four basic kinds of nuclotides and a vast
number of possible sequences of chromosome length. Heredity is "limited" when the number of possible states specifiable by the system is
comparable to the number of actual units realizing states of such a sys-

tem, whereas heredity is "unlimited" when the number of possible states vastly exceeds the number of actual states realized by any physical system (Szathmáry and Maynard Smith 1993).

16. See Wade and Griesemer (1998) for an empirical study of propagule pool reproduction as a kind of group-level genetics and Griesemer and Wade (2000) for an interpretation of group genetics with Punnett Squares.

17. Different views of theory structure may lead to different scientific research programs than the one proposed in this section, which assumes that theories are collections of models and their robust consequences.

18. Wimsatt 1976, cf. Sarkar 1998. For a catalogue of reductionistic research heuristics, see Wimsatt (1980). The papers in Wimsatt (in press) articulate his philosophy of theorizing as model-building.

19. 'Component' is in scare quotes because I have not specified an ontology of processes. Whether and how progeneration and development can be genuinely interpreted as parts of reproduction processes must be worked out. This is an important conceptual problem for the heuristic application of reductionism proposed here because it bears on whether substantive reduction criteria are met by the theory of reproduction presented in the first section.

20. That is, models of development framed from the point of view of traditional genetics. A certain amount of violence to traditional language is unavoidable. In the process perspective, genetics is much broader than its traditional meaning because it applies to all the developmental aspects of the special classes of reproduction processes here called inheritance and replication.

21. It should be clear that the reversal is not simply one of turning the arrow around. There is also a change of basis in the switch from the structure or spatial hierarchy perspective, in which genetic units are lower in level than phenotypic units, to the abstract hierarchy of the process perspective in which genetic processes are special case but higher *level* processes of reproduction as compared to the general case but lower level processes of progeneration and development.

22. Faced with the choice of holism or part-whole reversal, developmental systems theorists (Gray 1992; Griffiths and Gray 1994; Oyama 1985) chose a form of holism by identifying whole developmental systems as units of replication (cf. Sterelny, Smith, and Dickison 1996). I argue instead that genetic systems (replicators) reduce to developmental systems, i.e. to reproducers.

23. For recent literature describing and analyzing Weismannism, see Churchill (1985, 1987); Gilbert (1991); Griesemer (1994) ; Griesemer and Wimsatt (1989); Maienschein (1987); Maynard Smith (1989); Mayr (1985); Sapp (1991).

24. Obviously, this is not an exhaustive list of options. Scientists can change

their basic commitments, lines of work or even careers in order to pursue a problem. The point is that the problem compels choice.
25. I do not have space to characterize how this paradox is generated and averted in embryology, but see Sapp (1991) for further discussion.
26. See the excellent review of the operon theory in the context of a discussion of the embryological origins of the gene theory by Gilbert (1994, chapter 2).

REFERENCES

Brandon, R. 1990. *Adaptation and Environment.* Princeton: Princeton University Press.
Brandon, R. 1996. *Concepts and Methods in Evolutionary Biology.* Cambridge: Cambridge University Press.
Burian, R. M. 1992. Adaptation: historical perspectives. In *Keywords in Evolutionary Biology,* edited by E. Fox Keller and E. Lloyd. Cambridge, Mass.: Harvard University Press, pp. 7–12.
Buss, L. 1987. *The Evolution of Individuality.* Princeton: Princeton University Press.
Cartwright, N. 1997. Where do laws of nature come from? *Dialectica* 51: 65–78.
Churchill, F. 1985. Weismann's continuity of the germ-plasm in historical perspective. *Freiburger Universitätsblätter* 24: 107–124.
Churchill, F. 1987. From heredity theory to Vererbung, the transmission problem, 1850–1915. *Isis* 78: 337–364.
Conklin, E. 1919. *Heredity and Environment in the Heredity of Men.* Revised 3rd ed. Princeton: Princeton University Press.
Darden, L. 1991. *Theory Change in Science, Strategies from Mendelian Genetics.* New York: Oxford University Press.
Darden, L., and N. Maull. 1977. Interfield theories. *Philosophy of Science* 44: 43–64.
Dawkins, R. 1976. *The Selfish Gene.* Oxford: Oxford: W.H. Freeman.
Dawkins, R. 1982. *The Extended Phenotype.* Oxford University Press.
Dyson, F. 1985. *Origins of Life.* Cambridge: Cambridge University Press.
Falk, R. 1997. Muller on development. *Theory in Biosciences* 116: 349–366.
Gánti, T. 1971. *The Principle of Life* (in Hungarian). Budapest: Gondolat. (English translation 1986. Budapest: OMIKK.)
Gilbert, S. F. 1978. The embryological origins of the gene theory. *Journal of the History of Biology* 11: 307–351.
Gilbert, S. F. 1991. Induction and the origins of developmental genetics. In *A Conceptual History of Modern Embryology,* edited by S. Gilbert. Baltimore: The Johns Hopkins University Press, pp. 181–200.
Gilbert, S. F. 1994. *Developmental Biology.* 4th ed. Sunderland: Sinauer Associates.

Glennan, S. 1996. Mechanisms and the nature of causation. *Erkenntnis* 44: 49–71.

Gray, R. 1992. Death of the gene: Developmental systems strike back. In *Trees of Life*, edited by P. Griffiths. Dordrecht: Kluwer Academic Publishers.

Griesemer, J. R. 1994. Tools for talking: Human nature, Weismannism, and the interpretation of genetic information. In *Are Genes Us? The Social Consequences of the New Genetics*, edited by C. Cranor. New Brunswick: Rutgers University Press.

Griesemer, J. R. in press. Development, culture and the units of inheritance. *Philosophy of Science* (Proceedings).

Griesemer, J. R., and M. J. Wade. 2000. Populational heritability: extending Punnett square concepts to evolution at the metapopulation level. *Biology and Philosophy* 15: 1–17.

Griesemer, J. R., and W. Wimsatt. 1989. Picturing Weismannism: A case study of conceptual evolution. In *What the Philosophy of Biology Is, Essays for David Hull*, edited by M. Ruse. Dordrecht: Kluwer Academic Publishers, pp. 75–137.

Griffiths, P., and R. Gray. 1994. Developmental systems and evolutionary explanation. *Journal of Philosophy* 91: 277–304.

Hull, D. 1980. Individuality and selection. *Annual Reviews of Ecology and Systematics* 11: 311–332.

Hull, D. 1981. The units of evolution: A metaphysical essay. In *The Philosophy of Evolution*, edited by U. Jensen and R. Harré. Brighton: The Harvester Press, pp. 23–44.

Hull, D. 1988. *Science as a Process*. Chicago: University of Chicago Press.

Jablonka, E., and M. J. Lamb. 1995. *Epigenetic Inheritance and Evolution: The Lamarckian Dimension*. Oxford: Oxford University Press.

Keller, E. F. 1995. *Refiguring Life: Metaphors of Twentieth-Century Biology*. New York: Columbia University Press.

Lewontin, R. C. 1970. The units of selection. *Annual Review of Ecology and Systematics* 1: 1–17.

Lewontin, R. C. 1974a. The analysis of variance and the analysis of causes. *American Journal of Human Genetics* 26: 400–411.

Lewontin, R. C. 1974b. *The Genetic Basis of Evolutionary Change*. New York: Columbia University Press, pp. 137–144.

Lewontin, R. C. 1992. Genotype and phenotype. In *Keywords in Evolutionary Biology*, edited by E. Fox Keller and E. Lloyd. Cambridge, Mass.: Harvard University Press.

Lloyd, E. A. 1988. *The Structure and Confirmation of Evolutionary Theory*. New York: Greenwood Press.

Maienschein, J. 1987. Heredity/development in the United States, circa 1900. *History and Philosophy of the Life Sciences* 9: 79–93.

Maynard Smith, J. 1975. *The Theory of Evolution*. 3rd ed. Middlesex: Penguin.

Maynard Smith, J. 1988. Evolutionary progress and the levels of selection. In

Evolutionary Progress, edited by M. Nitecki. Chicago: University of Chicago Press, pp. 219–230.

Maynard Smith, J. 1989. Weismann and modern biology. *Oxford Surveys in Evolutionary Biology* 6: 1–12.

Maynard Smith, J., and E. Szathmáry. 1995. *The Major Transitions in Evolution.* Oxford: W. H. Freeman.

Mayr, E. 1985. Weismann and evolution. *Journal of the History of Biology* 18: 295–329.

McCauley, D., and M. Wade. 1980. Group selection: The genetic and demographic basis for the phenotypic differentiation of small populations of *Tribolium castaneum. Evolution* 34: 813–821.

Mendel, G. 1866 [1965]. *Experiments in Plant Hybridization* [reprinted by P. Mangelsdorf]. Cambridge: Harvard University Press.

Morgan, T. H. 1926. *The Theory of the Gene.* New Haven: Yale University Press.

Morgan, T. H. 1934. *Embryology and Genetics.* New York: Columbia University Press.

Morgan, T. H., A. H. Sturtevant, H. J. Muller, and C. B. Bridges. 1915. *The Mechanism of Mendelian Heredity.* New York: Henry Holt.

Moss, L. 1992. A kernel of truth? On the reality of the genetic program. *Proceedings of the Biennial Meeting of the Philosophy of Science Association.* 1: 335–348.

Nagel, E. 1961. *The Structure of Science.* Indianapolis: Hackett Publishing Co.

Neumann-Held, E. 1998. The gene is dead – long live the gene! Conceptualizing genes the constructivist way. In *Sociobiology and Bioeconomics: The Theory of Evolution in Biological and Economic Theory,* edited by P. Koslowsky. New York: Springer Verlag, pp. 105–137.

Oyama, S. 1985. *The Ontogeny of Information.* New York: Cambridge University Press.

Provine, W. B. 1971. *The Origins of Theoretical Population Genetics.* Chicago: University of Chicago Press.

Provine, W. B. 1986. *Sewall Wright and Evolutionary Biology.* Chicago: University of Chicago Press.

Raff, R. A., and T. C. Kaufman. 1983. *Embryos, Genes, and Evolution: The Developmental-Genetic Basis of Evolutionary Change.* Bloomington: Indiana University Press.

Robinson, G. 1979. *A Prelude to Genetics. Theories of a Material Substance of Heredity: Darwin to Weismann.* Lawrence, KS: Coronado Press.

Salmon, W. 1984. *Scientific Explanation and the Causal Structure of the World.* Princeton: Princeton University Press.

Salmon, W. 1994. Causality without counterfactuals. *Philosophy of Science* 61: 297–312.

Sapp, J. 1987. *Beyond the Gene: Cytoplasmic Inheritance and the Struggle for Authority in Genetics.* New York: Oxford University Press.

Sapp, J. 1991. Concepts of organization: The leverage of ciliate protozoa. In *A*

Conceptual History of Modern Embryology, edited by S. Gilbert. Baltimore: The Johns Hopkins University Press, pp. 229–258.

Sarkar, S. 1998. *Genetics and Reductionism.* Cambridge: Cambridge University Press.

Slatkin, M. 1981. Populational heritability. *Evolution* 35: 859–871.

Sterelny, K., K. Smith, and M. Dickison. 1996. The extended replicator. *Biology and Philosophy* 11: 377–403.

Szathmáry, E. 1994. Toy models for simple forms of multicellularity, soma and germ. *Journal of Theoretical Biology* 169: 125–132.

Szathmáry, E., and J. Maynard Smith. 1993. The origin of genetic systems. *Abstracta Botanica* 17: 197–206.

Szathmáry, E., and J. Maynard Smith. 1997. From replicators to reproducers: The first major transitions leading to life. *Journal of Theoretical Biology* 187: 555–572.

van Fraassen, B. 1980. *The Scientific Image.* Oxford: Clarendon Press.

van Fraassen, B. 1989. *Laws and Symmetry.* Oxford: Clarendon Press.

Wade, M., and J. R. Griesemer. 1998. Populational heritability: Empirical studies of evolution in metapopulations. *American Naturalist* 151: 135–147.

Wade, M., and D. McCauley. 1980. Group selection: The phenotypic and genotypic differentiation of small populations. *Evolution* 34: 799–812.

Weismann, A. 1892. *Das Keimplasma: Eine Theorie der Vererbung.* Jena: Gustav Fischer.

Williams, G. C. 1966. *Adaptation and Natural Selection.* Princeton: Princeton University Press.

Wilson, E. B. 1896. *The Cell in Development and Inheritance.* London: Macmillan.

Wimsatt, W. C. 1976. Reductive explanation: a functional account. *Boston Studies in the Philosophy of Science,* 32: 671–710.

Wimsatt, W. C. 1980. Reductionistic research strategies and their biases in the units of selection controversy. In *Scientific Discovery. II: Historical and Scientific Case Studies,* edited by T. Nickles. Dordrecht: D. Reidel, pp. 213–259.

Wimsatt, W. C. 1987. False models as means to truer theories. In *Neutral Models in Biology,* edited by M. Nitecki and A. Hoffman. London: Oxford University Press, pp. 23–55.

Wimsatt, W. C. in press. *Re-Engineering Philosophy for Limited Beings: Piecewise Approximations to Reality.* Cambridge, Mass.: Harvard University Press.

Woodward, J. 1993. Capacities and invariance. In *Philosophical Problems of the Internal and External Worlds,* edited by J. Earman, A. Janis, G. Massey, and N. Rescher. Pittsburgh: University of Pittsburgh Press, pp. 283–328.

12

A Unified View of the Gene, or How to Overcome Reductionism[1]

PETER J. BEURTON

ABSTRACT

The conceptual history of the gene, beginning as an inference from the observation of Mendelian regularities in inheritance at the turn of the last century and reaching an apex with the unraveling of its biochemical structure by Watson and Crick in 1953, figured for a long time as a standard example of successful reduction in the life sciences. This has variously led to highly reductionistic interpretations of Darwinian evolution. But ever since its internal makeup was known, the gene has also become an increasingly problematic entity. Indeed, a conceptual crisis has arisen during the last twenty years as a result of the discovery of overlapping genes, alternative splicing, and so on. What once looked like a particulate gene now turns out to be scattered across parts of the genome with no hard-and-fast boundaries. Genes seem to depend on the genome's regulatory activities which, in turn, may depend on how the molecular biologist wishes to manipulate the genome in experiment. This has led to the widespread opinion that the gene is devoid of any special reality, or, is just a word. I contest this view and continue to argue for a unified gene concept. I do so by defining the gene as the genetic underpinning of the smallest possible difference in adaptation that may be detected by natural selection. Differences in adaptation among individuals, by directing natural selection toward the genetic underpinning of such differences, may be instrumental in the formation of genes. Within this scheme, the empirical evidence pointing toward the disintegration of the gene may well turn out to be evidence in favor of an evolutionary unity of the gene. Genes themselves are products of evolutionary forces at work on the population level. In such a perspective, the issue of reductionism is emptied of all content.

THE RISE OF REDUCTIONISM

Most people consider matter as something fundamentally particulate. This is an ancient notion and has its origins with Democritus.

Matter is believed to consist of "last particles" the unveiling of which will lead to disclosure of the inner most secrets of this world. Today such a body of thought is often criticized as reductionistic, but over long periods of time it proved a successful strategy in the sciences. When people became inspired by the prospects of empirical verification of such entities in the organic world, they began to coin terms like *gemmules, anlagen, pangenes, determinants,* and the like. This occurred during the second half of the last century. The recovery of Mendel's laws at the beginning of this century furnished a strong argument for the existence of such particles. They were commonly inferred from the observation of independent assortments of characters, or particulate inheritance. Finally, Johannsen (1909) proposed the word *gene* as a new name, and from then on it seemed possible to rest content with the knowledge that the Mendelian recurrence of characters was sufficient proof that there were some anlagen – genes – that must be responsible for those characters. Though Johannsen and a large community of embryologists conceived of genes in terms of a holistically designed hereditary potential that was somehow secreted by the whole organism, the idea that all things were in a deep and important sense particulate was so pervasive and overriding that it gradually took hold of large parts of the scientific community.

Between 1910 and 1920 these developments received overwhelming support from Morgan and his group who unraveled the mechanics of heredity. By 1922, Muller, who was thought to be the most gifted among Morgan's pupils, came forth with two desiderata that genes would have to satisfy to function as "ultramicroscopic particles" in the evolutionary context. First, genes had to possess an internal structure rendering them autocatalytic (or capable of producing copies of themselves). Second, genes had to be autocatalytic in the more refined sense that they also copied their own mutations (Muller 1922). A third characteristic later mentioned by Muller was the gene's heterocatalytic potential to give rise to all those products other than itself that furnish the developing organism. Thus genes were envisaged by Muller as compact little entities that, by virtue of this threefold capacity, would provide the essential key toward understanding the nature of life.

Among the most important biochemical work throughout the ensuing decades that facilitated an understanding of the fine structure of those hypothesized particles were the one-gene – one-enzyme hypothesis in the early 1940s (Beadle and Tatum 1941), the discovery of DNA as the hereditary substance (Avery et al. 1944), and finally the disclosure of the double-helical structure of DNA by Watson and Crick (1953a). This structure immediately suggested a mechanism for Muller's autocatalysis: The two strands separated from one another and then each strand functioned as a template for the assembly of a new strand through complementary base pairing. Hence, mutations (irregular substitutions, insertions, or losses of nucleotides) would continue to be copied also, ensuring Muller's second desideratum. By the early 1960s the colinear relationships between such a DNA segment and the encoded polypeptide as well as the mRNA that mediated translation were essentially elucidated, thus providing insight into Muller's third desideratum or the heterocatalytic functioning of DNA (for a short account, see Rheinberger, this volume).

Muller was enthusiastic about the vindication of his speculations:

Here at last was the down-to-earth, detailed chemical structure of the gene material . . . The chemical identity of the gene material is no longer in doubt . . . it is evident what main features of this conformation give the material its virtually unique and truly fateful three faculties – those of "reproducing itself" *and* its mutants, and of influencing other materials – the three faculties which, when in combination, underlie the possibility of all biological evolution. (Muller 1966, 504, 505)

With this reduction of the major aspects of life to one chemical substance, DNA, molecular genetics had reached its "heyday of precocious simplicity" (Rheinberger, this volume). Often there was hardly any explicit reference to genes by those immediately involved in uncovering the molecular mechanics of replication and heredity. Indeed, this concept was hardly needed for this purpose, but the implications seemed obvious. In one of a total of two places in which the word *gene* occurs in the two seminal papers by Watson and Crick of 1953, they refer to DNA as a source of "the genetic specificity of the chromosomes and thus of the gene itself" (Watson and Crick 1953b, 964). This was the implication picked up by a broader audience.

The case of the gene, then, became a standard example of a successful reduction in the life sciences. Yet this situation furnishes no unequivocal answer to the issue of *reductionism*. Reductionism refers rather to a particular way of viewing the relations between underlying particles (genes) and the objects (organisms) which are in some sense composed of, or controlled by, these particles. Strong versions of reductionistic theories hold that organisms are *totally* determined by their genes, and evidence for such a view seemed to come from early neo-Darwinian population genetics. Mathematical calculations, especially by R. A. Fisher (1930), made it likely that genes, even when only of a very slight reproductive superiority, could spread through a species population within a relatively short period of time. This gave rise to a view that all evolution could be reduced to the selective accumulation of self-contained genes in populations, each of which added a minute and independent evolutionary increment to the species. Eventually, evolution became redefined as a change in gene frequencies in populations, and phenotypes were deprived of any causal role in the whole process.

An alternative view was championed by Sewall Wright, who laid the seeds of his now-famous shifting-balance theory in the early 1930s (Wright 1931, 1932), and later also by naturalists like Mayr (1954) and Dobzhansky (1955). These evolutionists no longer thought of genes as possessing rigid selective values like inborn properties but considered these properties to depend on how each gene interacted with many other genes of many other individuals. Genes which had a positive effect against the genetic background of one individual may be harmful even in the genetic context of another individual. This kind of thinking paved the way toward understanding how different kinds of harmonious genetic interactive systems may be built up through gene interaction when natural selection acts on whole phenotypes rather than on single genes, and this is how phenotypes were rescued from evolutionary unimportance.

But the early portrayal of populations as a medium for the accumulation of self-contained genes was permeated by a self-reinforcing argument that made its proponents feel quite immune against such reasoning. For wholesale species change to take place, there always must have been a sufficient number of genes spreading species-wide. Of course it was known that such genes encountered many different kinds of interactions on the way and that due to these and other local circumstances it was impossible to ascertain a priori which ones would ultimately prevail. Nevertheless, of those that did prevail in the end it must have been true that they enjoyed an at least slightly above-average selective value right from the outset, for otherwise they would have disappeared early on. This, however, was much the same as saying that those genes which spread species-wide did so by the mutual cancellation of their interactive effects, or more precisely, by virtue of their internal goodness in the first place, or independent of interaction. This is why Fisher – though not ignoring the bearing of interaction on a gene's selective value in any one local situation – felt in a strong position when totally discounting interaction and relying solely on the genes' additive effects when offering a general explanation of the evolutionary process. Having discounted interaction from the outset as mere evolutionary noise, it became, in turn, possible for Fisher (1930, 77, 78) to speak of genes as sweeping through populations with utmost regularity and precision. This was all evolution was about. I will refer to this argument as *Fisher's cogency argument*.

More recently, Fisher's philosophy has been carried on, for example, by G. C. Williams. Though admitting that "it is [obviously] unrealistic to believe that a gene actually exists in its own world with no complications other than abstract selection coefficients and mutation rates," he sees ground to continue:

No matter how functionally dependent a gene may be, and no matter how complicated its interactions with other genes and environmental factors, it must always be true that a given gene substitution will have an arithmetic mean effect on fitness in any population. One allele can always be regarded as having a certain selection coefficient relative to another at the same locus at any given point in time. Such coefficients are numbers that can be treated algebraically, and conclusions inferred for one locus can be iterated over all loci. Adaptation can thus be attributed to the effect of selection acting independently at each locus. (Williams 1966, 56–57)

This is the perennial problem of reductionism in population genetics or Fisher's cogency argument. A decade later, Richard Dawkins emerged as the most radical spokesperson of this school of thought. His catchphrase for the self-contained gene is *the selfish gene* (Dawkins 1976). By the use of this term Dawkins wants to make it uncompromisingly clear that genes never reproduce "in the interest" of organisms and their features, but solely "in the interest" of their own. Evolution, according to Dawkins, is a process in which alternative genes (alleles) battle with each other for maximizing their own reproductive success, *and nothing else*. Genes are seen as replicators whose phenotypic effects render them successful at propagating themselves, while phenotypes are no more than survival machines for genes; "A monkey is a machine which preserves genes up trees, a fish is a machine which preserves genes in the water. . . ." (Dawkins 1976, 22). Dawkins' ruthlessly reductionistic position has met with severe criticism ever since he published his first book on the subject in 1976. At best, he is supposed to have produced a caricature of Darwinism (Gould 1997, 34). Nevertheless, I think it is fair to say that nobody yet has contradicted Dawkins on his own grounds. Though Fisher would not have used Dawkins' terminology, much of Dawkins' reasoning is implicit in Fisher's philosophy of additive genes (but not all; see the final chapter of this paper).

In view of Fisher's cogency argument population genetics per se may seem a reductionistic enterprise. Phenotypes are there for genes to compete, but in terms of evolution, lead no life of their own. As Samuel Butler said, the hen is an egg's way of making more eggs. Something must be wrong, but what? Time and again, interactionist models have been developed in population genetics to promote more balanced views, especially in the wake of Sewall Wright's work, over the last fifteen years. Notwithstanding the merits of these undertakings, my suggestion is that we not only look at the interactions in populations between ready-made genes but also take a new look at the gene itself. In fact, it has been known for a number of decades that the gene is no longer what it once was thought to be. Placed in a proper perspective, this may provide a point of departure toward a not-quite conventional solution to the problem of reductionism in population genetics.

THE DISINTEGRATION OF THE GENE

Interestingly, although Muller's original three desiderata all were put forth with reference to the unit of the gene, the very moment he was able to celebrate the vindication of his desiderata in 1966 he referred to them as properties of the "gene material" rather than of genes (see the previous quote). This hardly looks like a retraction.

Yet, with the irony of hindsight, it was in the very process of uncovering the gene structure in that "down-to-earth" manner hailed by Muller, that molecular biologists also began to lose sight of the gene. At first nearly imperceptible, these developments finally led to a situation that may be described as the far-reaching disintegration of the gene. In 1955 Benzer felt prompted to propose the terms *muton*, *recon*, and *cistron* to distinguish between the molecular biological units of mutation, recombination, and function (Benzer 1955). He did so because his successful analysis of the fine structure of a single gene in the T4 bacteriophage of *E. coli* had revealed that these units, when used to delineate a gene, referred to DNA stretches of different lengths (see Holmes, this volume).

Something of a watershed came with Jacob and Monod's (1961) model of gene regulation. Because the classical gene was inferred from the Mendelian trait it gave rise to, there always seemed to be an element of direct correspondence between genes and traits. But according to Jacob and Monod the general condition was one of "regulator genes" that controlled the expression of "structural genes." They did so in one large class of cases by giving rise to a protein that bound to a structural gene's controlling site (later designated as the promoter) thereby regulating this gene's rate of transcription. Is one then well-advised to call such a regulator sequence an independent gene? Such a terminology was justified as long as the protein could stand as a placeholder for one of those traits identified by Mendel or Johannsen. But when a regulator gene's "trait" was simply the effect it exerted on the expression of another gene's trait, it would have been more appropriate to call it part of this other gene in the first place. However, the promoter region of a structural gene often controlled the transcription, not of one, but of several adjacent structural genes. This more inclusive unit was called the "operon" by Jacob and Monod. Hence, in terms of transcription, the whole operon should be viewed as an equivalent of the gene. Unfortunately, this will do neither because in terms of translation these structural segments continue to behave as individual genes by each giving rise to a different protein. What, then, is the gene?

It became common practice among molecular biologists to substitute for the term *gene* phrases like "regulatory sequences that control expression," or "genetic elements that control coding sequences" and the like (see the list of expressions collected by Falk 1986, 165–166). In the beginning this was probably not in the interest of avoiding some nagging conceptual issues about the gene, but simply because such phrases were, at least on first sight, more informative to the molecular biologist. Yet there were many other discoveries that contributed to the blurring of the gene of which only a few will be listed here in passing. One event was the discovery of repetitive DNA by Waring and Britten (1966) and Britten and Kohne (1967) (which may have consisted of tens, hundreds, or even many thousands of repeats). For obvious reasons, highly repetitive sequences were expected to be non-coding and thus hardly fulfilled the criteria necessary for calling them genes. They may have had a well-defined location but no function and they certainly didn't give rise to Mendelian traits. Repetitive DNA looked like junk produced by an unchecked surplus in copying activity in some parts of the hereditary machine (Doolittle and Sapienza 1980). Pseudogenes (Jacq, Miller, and Brownlee 1977) were another case. These elements resembled active genes that were rendered nonfunctional, for instance by mutations that affected transcription. They could have arisen, for instance, from unsuccessful gene duplication, either as a result of unequal crossing-over or by retroposition, that is, reverse transcription of RNA intermediates into DNA. Both repetitive DNA and pseudogenes showed that the genome may be inhabited by vast amounts of "dead" genes or junk DNA of unclear evolutionary potential.

Another discovery of the 1970s of immediate bearing on the present conceptual issue came with overlapping genes (Barell, Air, and Hutchinson 1976). Two reading frames may partly overlap so that a particular DNA sequence may be double read and give rise to parts of two proteins. Such a sequence, then, stands in the services of two different genes. This foreshadowed a view in which a gene began to look like a dynamic function temporarily conferred to DNA stretches by the genome's reading apparatus for the purposes of producing a protein. At any rate, no longer did a particular DNA sequence seem uniquely responsible for a particular trait. In other cases, pieces of DNA lacked a constant location. There were genetic elements (transposons) that moved around actively in the genome modifying the expression of adjacent genes. This discovery, made by Barbara McClintock in maize plants in the 1940s (McClintock 1951) was so far ahead of its time that it met with considerable skepticism and was treated as a curiosity for a long period of time. But during the 1960s and 1970s it was found that chromosomes were often littered with mobile elements and that they were widespread among both prokaryotes and eukaryotes.

A new disclosure, made simultaneously by numerous research teams during the second half of the 1970s, was "genes-in-pieces." Rather than forming a continuous stretch of coding DNA, a gene may consist of a mosaic of coding and noncoding sequences called *exons* and *introns*, respectively (Gilbert 1978). The RNA transcript of such a gene goes through a process of splicing in which the introns are excised yielding a "mature" mRNA composed only of exons. Genes-in-pieces seems to be the rule rather than the exception in eukaryotes. The intronic regions are frequently considerably longer than the exons so that a view of the nonlocal gene emerges in which the exonic regions are truly scattered along a piece of DNA many times longer than themselves. All this is true as long as introns are no longer considered part of the gene. Carlson (1991), for instance, proposed the term *informational gene* to do justice to the fact that genes (unlike Benzer's cistron) no longer require physical integrity of the sequences in question so that introns cease to pose a problem: They are not part of the informational gene. But there is equal justification to assume that introns "even if functionless, must be included, since a mutation upsetting splicing will destroy gene function" (Gale 1990, 8). Also Carlson added in a second thought: "They [the introns] can, of course, be decisive in providing mutations that can affect the way the exons are assembled" (Carlson 1991, 478).

The most dramatic discovery at the end of the 1970s in the context of the disintegrated gene was probably *alternative splicing* (for an early review, see Lewin [1980] and for a more comprehensive one Smith, Patton, and Nadal-Ginard [1989]). During development alternative samples of exons may be pieced together from a single gene's primary transcript. A single gene, then, may yield varying assemblages of exons or different protein isoforms to meet the demands of the developmental stage in question. This condition can be compared in kind only with that of overlapping genes. With most other difficulties it was possible to hold, with some confidence, that a gene's coding sequence, however fuzzy in appearance, was either something underlying a transmissible trait or may not be called a gene proper. But overlapping genes and alternative splicing flatly contradict any one-to-one correspondence between a gene and its presumed product.

The fuzziness caused by genes-in-pieces merges with the somewhat less spectacular but even more general ambiguity about regulatory elements that are vital for a proper reading of coding sequences (a special case of which was the operon): Are they part and parcel of the gene, or should we make a difference? Carlson opts for putting all those sequences (operators, promoters, upstream and downstream regulators) into a separate bag calling them " 'accessory sequences' for gene processing." He does so "because they are universal features of all (or many) genes and are not unique features of each gene" (Carlson 1991, 478). Singer and Berg, in contrast, call for an inclusion of the regulatory elements because they wish to define the gene "as a combination of DNA elements that together comprise an expressible unit" (Singer and Berg 1991, p. 461). But there are also elements that modify a gene's expression from a distance, like the enhancers (not to mention transposons), and so they arrive at the following dilemma:

These regions are so varied in their structure, position, and function as to defy a simple inclusive name. Among them are enhancers and silencers, sequences that influence transcription initiation from a distance irrespective of their orientation relative to the transcription start site. (Singer and Berg, 1991, 462)

Fogle concludes for similar reasons: ". . . it is as if the influence of the distal elements fades into the genetic horizon as one searches farther and farther upstream of the gene. . . . a completely inclusive model is vacuous" (Fogle 1990, 360; see also Fogle, this volume). Given what is known about regulatory sequences in the broad sense, it seems as though what is called a single gene may not only be interrupted in various places by noncoding sequences but also be scattered across more inclusive parts of the genome. The question that needs to be answered is: What are the standards by which such accessory sequences count are involved in gene function?

To summarize, the general situation is somewhat reminiscent of the theory of atoms after the turn of the century: The more molecular biologists learn about the structure and functioning of the gene, the less they know what a gene really is. On the one hand, one should not be too narrow minded: "It is important that geneticists recognize the many levels at which genes can be perceived, but it is not helpful to select one of these levels and arbitrarily designate that as the universal definition of a gene" (Carlson 1991, 478). On the other hand, Carlson's counsel might be taken as an advocacy of a pluralistic approach. Pluralism, however, often dissolves perennial problems without solving them. True, if there is assemblage of gene transcripts to new functional units prior to translation and if, moreover, assemblage may differ according to developmental stage, and also, if regulator sequences determine which segments are to be processed as a single gene in the first place, then the term *gene* becomes meaningless independent of the organizing activities carried out by the genome's regulatory apparatus. Such a gene, scattered across some part of the genome, is then "gathered together" by what may be called *the clever genome*. A gene seems a genome's way of making a protein, or the clever genome temporarily brings genes into being to lever itself into the next generation. So we may also ask: How did the gene get hidden away in the genome's background?

These more recent findings of developmental genetics have also prompted the view that the gene is just a word. Genes are really anonymous stretches of DNA that the experimenter, depending on how he or she chooses to manipulate the genome, makes use of in various ways and calls genes, that is, according to his or her own ends. There seems to be an element of primacy, or rather generality, about DNA strings when compared with genes. DNA strings seem more real while genes are seen as somewhat more instrumental or operational and thus ranging more closely to the epistemological end of our inquiry into nature. Now if genes prove unreal the whole problem of reductionism so pressing in population genetic contexts collapses – for how could an unreal entity turn selfish and govern phenotypes? But this would be a premature deconstruction of the gene as the following will show.

A PROPOSAL FOR A NEW UNITARY CONCEPT OF THE GENE

In spite of the evidence to the contrary and in contrast to the present mainstream thinking I will argue in this section for a unitary concept of the gene. I will try to outline the story of genes in a setting that harmonizes with the more recent findings of molecular biology without calling the reality of genes into question. Indeed, I will do my best to put the recent molecular biological findings that seem to argue in favor of the disintegrity of genes, to the service of an argument in favor of the unity of genes. However, in view of widespread sentiment among molecular biologists that the gene is just a word, and search for the gene may even impede molecular biological investigation (Fogle, this volume), I will adopt the procedure not to talk positively of genes before having shown how nature in and of herself turns DNA strings into discrete and well-established entities which deserve such a name, that is, irrespective of any experimentalist's more immediate needs.

In pursuing this strategy, I am free to talk of DNA strings, though not yet of genes. Also, I am free to talk of differential reproduction of individuals due to underlying genetic differences. What then are these differences, if not of genes? They are caused, for instance, by mutations (minute irregular upsettings in nucleotide sequences) or by the reshuffling of chromosomes (e.g., by unequal crossing-over and other rearrangements). Crossing-over, for instance, does not respect the boundaries of genes – or rather, imagine a situation in which crossing-over finds no genes to disrespect. Both mutations and the shuffling of chromosomes may already be part of my discourse.

We have then an array of genetic variations inside the genome – but not of genes. How does this variation affect the individual during development? Naturally, the attention of scholars focused for a long time on mutations with clearcut somatic effects. The study of such mutations established the universality of Mendelian heredity, and in the evolutionary context these were the most suitable for the study of the efficacy of natural selection. It is, however, unclear to what extent this is the usual condition. According to the neutral theory of evolution (Kimura 1968, 1983), mutations without sufficient somatic expression to be detected by selection might be the rule rather than the exception. It should be noted that Kimura's emphasis was not on neutrality per se, but rather on an equivalence of functions among alleles under certain contingencies while a change in conditions might well induce selection. According to Kimura (1983,

271, 307), mutations have frequently a *latent* potential for coming under the regime of selection. They may drift in and out of selective contexts.

Because neutral or near-neutral genetic variation might play an important role in evolution, the notion of a mutation having by its very nature a straightforward effect that can be detected by selection (like causing one of those alternative characters observed by Mendel or Johannsen) might be an unwarranted assumption. The following scenario is probably more to the point: Being neutral one by one, such variations will tend to accumulate across generations until genomes are littered with different kinds of neutral mutations. Though individually subliminal in their effects, it can be assumed safely that after sufficient accumulation and through their joint action, these variations will materialize into *some* net difference in development and in the performance among individuals across their whole life-cycle sufficient for selection to detect. This is a slight difference in adaptation in some respect. Because it results from genetic variations spread across the whole genome, it is reasonable to expect that in the beginning this difference is very diffuse. It will amount to some slight difference in overall performance among individuals across their life-cycle rather than becoming manifest in an alternative expression of some well-demarcated trait. The emphasis is then at present on *differences among* individuals, not on *traits of* individuals (we will see soon how these two things hang together). However, as a net difference it is sufficiently distinct for selection to detect and to qualify as a difference in adaptation; this is the important point. Note that we are talking here already in terms of adaptation and a genetic basis of adaptation, though not yet of genes.

Selection, by discriminating among differences in adaptedness, causes differential reproduction. Differential reproduction of what? This has always been a somewhat tricky question (e.g., Keller 1987). We often talk loosely of differential reproduction of individuals. But in sexually reproducing organisms the genome is reshuffled in every generation. Individuals do not reproduce themselves literally; rather, two individuals of the opposite sex always give rise to a new unique individual. It is therefore more to the point to talk of a differential reproductive *contribution* of an individual to the gene pool (or "pool of DNA strings") of the generations to come. This, however, leaves unanswered our quest for the *unit* undergoing differential reproduction, or the unit of selection. While whole genomes are broken up in sexual reproduction, those rearrangements among a few locally contiguous nucleotides known as mutations will hardly ever be ruptured in this process. But mutations don't qualify as such units if we take for granted that they have too small an effect to be detected by natural selection. Is it not true that nucleotide upsettings have at least in some cases a sufficiently distinct phenotypic expression to serve that end? Yes, when they qualify as an alteration of a gene's properties. Genes are the smallest units of selection, not mutations. But of genes, owing to my self-imposed constraint, I must not talk yet.

I am free to say, however, that differential reproduction is a statistical bias among such *arrays* of alternative DNA variations that through their joint action produce a difference among individuals large enough for selection to detect. This is, I claim, the unit of selection we are seeking. Surely also these variations, being spread across the whole genome, are subject to constant shuffling in the sexual process. Might this not obliterate such a selective bias? Not really; all we need to assume for the formation of some constant selective bias across the generations is that this bias works in favor of all those genetic variations which happen to support through very many different kinds of combinations some such adaptation more often than other variations do. This being so, it follows that such a minute adaptive advantage will direct selection in favor of its own genetic underpinnings and cause discrimination across the generations among, say, tens of thousands of alternative variations scattered across entireties of DNA strings.

I suggest that such an array of nonlocalized DNA variations, whose reproduction comes to be controlled by some such an adaptive difference large enough for natural selection to detect, begins to qualify as a gene. Hence, a gene need not be located in any one place. All those DNA variations that, due to this common guidance (because they all share in the production of *one* adaptive difference), spread at the same rate in a population qualify as a single gene irrespective of location. It is this *sameness of reproductive rate* by which these DNA variations begin to meet the standards of being one single gene.[2] Such a gene lacks physical discreteness but nevertheless acquires some distinctness as the smallest unit of selection. This is, I suggest, the unifying element in genes, the near-invisibility of which has been constantly bedeviling us. This element possesses the potential to bring together a variety of divergent gene concepts and to synthesize them into one common view of the gene.

Now if there is some constant selection for such an array of DNA variations spread across larger genomic areas and if, moreover, there are continual rearrangements and thus redistribution of DNA segments in the genome, for reasons of economy these variations *might* gradually come to occupy a single location and acquire physical integrity. Being in one place provides, for instance, an additional safeguard against a gene's dissolution by divergent selection pressures impinging on different parts of the genome. Physical integrity would be, then, an additional outcome of selection. I propose the following terminological distinction: The gene as a *particle* presupposes the existence of the gene as an *entity*, or unit of selection, that may or may not possess physical integrity. Physical integrity of the gene is then the result of the locally contiguous relocation of all those bits of DNA throughout the genome that happen to support some *one* minute adaptive difference among individuals. As the process continues, further sophistications might be added, like initiators, terminators, and so on, although from the genetic point of view this need not be so. While in the developmental context an initiator and terminator may define a gene, in the context of evolution in populations selection, when "gathering together" a gene, may also bring such additional sophistications into being.

How do these propositions harmonize with the empirical evidence? Unsuccessful gene duplications might provide the right kind of variation to fuel such processes: "They . . . keep the genome in flux . . . they can be considered a shotgun approach of nature wherein the majority of these genetic elements are inactive and left to rot in the genomic soil. Nevertheless, some seeds will integrate near a fertile genomic environment giving rise . . . to new genes or gene domains . . ." (Brosius 1991). And genes-in-pieces might provide one set of clues about possible subunits that have merged into genes. Soon after the discovery of genes-in-pieces it was hypothesized that individual exons encode protein domains, i.e., compactly folding globular units of characteristic shape and function (Blake 1978). In the meantime, evidence seems to be accruing that it takes more than one exon to encode a single domain (for a discussion, see Doolittle 1995). If it is true that only whole domains define function, single exons would be below the selective threshold. But there never was a rigorous definition of what is a domain or a structure of characteristic shape and function in the first place. For instance, it has been claimed that what ought to be such an element was largely identified with hindsight and on "aesthetic" grounds (Maynard Smith and Szathmáry 1995, 136). Be this as it may, I refer to these arguments for general importance rather than detail: No longer is there any doubt among molecular biologists that genes may form to some extent through shuffling of subgenic elements. A "universe of exons," as envisaged by Dorit, Schoenbach, and Gilbert (1990), could then provide part of the raw material from which genes are built by natural selection. Or more generally, I propose a setting in which subfunctional particles are kept in suspense, as it were, on the interface between useless genomic rot and potentially meaningful genetic structure as the most appropriate source for the origin of ever-new units of selection, or genes.

The phrase "unit of selection" acquires a special meaning in the present context. This is not a unit *encountered* by natural selection. Rather, it is an emergent unit or one *generated* in the process of natural selection from a background of never-ceasing variation contained in the genome. Behind this stands some specific difference in adaptive performance among individuals that triggers differential reproduction and keeps it going. Difficulties in identifying genes or their locations have led frequently to the assertion that a gene is simply "what makes a difference" between any two individuals. I am turning this around and saying, a physical difference among organisms, when perpetuated through populations, is what makes a gene. Schrödinger (1944) was stunned by the genes' stability because he saw them as instrumental to the stability of organisms. It is, however, the other way around: Some constant differential reproduction of organisms due to some particular overall difference in adaptive performance is the most important source for the emergence of stable units that deserve to be called genes. Such adaptive differences, then, impose in a process of *downward causation* those distinctions within genomes we call genes. Hence, the phrase *downward causation*, originally introduced by Campbell (1974, 180) to stand for contexts in which "the laws of the higher-level selective system determine in part the distribution of lower-level events and substances," acquires an even more incisive meaning than previously suggested. Downward causation organizes the genetic material into those units called genes.

Nevertheless, it is vital to remember that such an adaptive difference is always fully controlled by its own genetic underpinning in the first place. Nowhere in this paper am I departing from a rigidly Darwinian framework. Yet this doesn't make the argument circular for the very reason that such a genetic underpinning is originally scattered across the whole genome (or large parts of it). It is most important for avoiding a tautology to realize that this adaptive difference is during its inception *genomic*, not *genic*. To say such an adaptive difference causes the origin of a new gene is to say that a genomic overall difference *between* individuals causes via its adaptive effect and the good offices of natural selection the emergence of a gene *inside* individuals in a process of downward causation (or genomic compartmentation, or genic individuation). This is how some specific genomic difference hardens into a single gene. The distinction between genomic and genic is vital for keeping within a rigidly Darwinian framework and yet avoiding a tautology.[3]

What is true of one gene, is also true of many genes. I assume in this scenario that at any one time there are many genes in a state of emergence. In different localities of the population different kinds of overall differences in adaptive performance acquire a selective premium, and their underlying genomic differences, in the process of spreading through the population and coming to be housed by the same individuals, harden into different genes. This is how an individual's genome becomes compartmented into many genes. It follows also that many different kinds of adaptations (as many as there are genomic differences) will tend to accumulate in every single individual. It is reasonable to assume that in the beginning each adaptive difference – though always referring to some *specific* net difference in performance among individuals – is, for instance in terms of morphological distinctness, as diffuse as is the scattering of its underlying genetic variations across the genome. But because phenotypic space is limited it is not unlikely that in a process reminiscent of character displacement (Brown and Wilson 1956) those adaptive differences also become compartmented – or harden into traits. Any one trait, in the sense I am using this term here, is then the individualized outcome of some specific difference in overall performance among individuals (and may or may not also be well-demarcated in terms of human perception). Genomic differences between individuals become compartmented into genes inside individuals, and as a corollary, differences in adaptive performance between individuals harden into traits of individuals.

Natural selection, then, builds genes and traits in populations. However, from the perspective of individual development, traits are products of genes, and this allows a recasting of the process described in the last paragraph in somewhat different terms. When natural selection turns a genomic difference into a compact gene and the diffuse adaptive difference caused by this genomic difference into a well-demarcated trait, this whole process also may be portrayed more fundamentally as one in which natural selection turns a diffuse difference in development between individuals into a minute and distinct *developmental pathway* or ontogenetic trajectory inside individuals. The intriguing point, when viewed in this perspective, is that none of the three components – the gene, its pathway, or the trait – comes first. This pathway actually *is* the gene-and-its-trait, or the streamlined outcome of a difference among individuals in adaptive performance across development. In such a perspective, the gene no longer appears solely as the cause of a minute developmental pathway (and a trait) but can be construed in the populational context with equal justification as a pathway's appendix that comes into being through natural selection building such a pathway.

In the Mendelian context, a condition in which well-demarcated genes give rise to specific traits was taken as self-evident and warranting no further explanation. This was so as long as the existence of genes was ascertained by the Mendelian traits they gave rise to. Given the present scenario, it can be inferred that cases of well-established gene-trait relations may result only from a long prior evolutionary history during which an adaptive difference in development condenses into such a minute ontogenetic trajectory. Or to borrow Gilbert's apt terminology (this volume): Adaptations have a long prior history of "arrival" before we dare apply traditional population genetic schemes in which "survival" is everything.[4] It is, then, realistic to assume that well-established gene-trait pathways form only the tip of an iceberg. Many more developmental differences never harden properly into genes and their developmental pathways, and therefore never surface as well-demarcated traits, and are continually returned to the genomic sink. At any one time there should be a continuity between genes that have hardly lifted off from their genomic background through all possible intermediate stages of distinctness to a minority of full-blown genes of the Mendelian sort.

Against this background, we can also render mutations as historically evolved products. In a state of nonexistence of genes, organisms must be quite messy things in which single nucleotide upsettings regularly pass unnoticed (and, indeed, may not even deserve to be called mutations). But to the extent that genes have formed, such genetic variations may also affect a *genic* sequence of DNA and thus a long-selected, tightly-knit minute pathway leading to a trait. This radically enhances and magnifies the effects of a mutation, and this is why in such condition a mutation very often gives rise to an alternative trait. This provides an historical explanation of the condition observed by Mendel and Johannsen. Neither genes nor mutations "just happen"; both are products of a long history of evolution of such minute ontogenetic trajectories during which mutations gradually acquire this miraculous property of nonneutrality. Or in a terminology I would prefer (but have not followed in this paper), genetic variations begin to qualify as mutations only as a result of a long period of prior evolution of gene-trait relations. A slight upsetting in nucleotide sequence begins to qualify as a mutation proper only to the degree that such a developmental pathway begins to function also as an amplification of the effects of such an upsetting. "Proper" mutations are then in a sense parasitic on gene-trait pathways.

We have already noted the bearings of the present gene concept on some of the recent findings of molecular biology, and here is the place to add a word on overlapping genes and alternative splicing. We know of *intragenomic conflict* from sociobiology. Whatever the merits of this term, in the present context it is possible to say that various kinds of differences among individuals that are present in a population may, in the process of spreading and being turned into genic differences or minute ontogenetic pathways inside individuals, compete for, partly, the same arrays of DNA variations – and finally compromise. The outcome would be overlapping genes. Different developmental stages may also do so. The outcome would be alternative splicing. Would it then be too sweeping to claim that the assumed cleverness of the genome is simply brute recapitulation of phylogeny by ontogeny? Genes are not created by the genome, but the genome is, to some extent, a condensed history of what has been going on in populations. Thus, the genome may recapitulate the partitioning of DNA strings into genes in various ways. This, then, would be an answer to the secret of how the gene was absorbed into the genome's regulatory apparatus in the first place. And this, I claim, is how the various forms of genes observed by the molecular biologist begin to fall into place.

When then may genes have arisen in the course of evolution? As just noted, present forms of noncompact genes might be frequently an ontogenetic recapitulation of past events (though, for obvious reasons I do not wish to share company with anybody searching for the "Urgen" as the beginning of life). Nevertheless, going by all we know about the flexibility and fluidity of the genome and about gene shuffling, the assumption would seem most unlikely that modern populations no longer tend to give rise to new genes in various ways.

IMPLICATIONS FOR REDUCTIONISM AND THE AUTONOMY OF BIOLOGY

In a particular sense, genes are last particles and do control organisms. From another and more comprehensive perspective, it is the other way around: Genes are the product of the evolutionary dynamics going on in populations and are brought into being by downward causation. Accordingly, downward causation and the central dogma of molecular biology may be viewed as counterparts. The central dogma that has so often reinforced reductionistic evolutionary explanations states that genetic information flows from DNA to RNA to protein and not the other way around. *Counterpart* is not quite the right word because these two processes are not only different in direction, but also in kind. The central dogma builds on molecular biological principles and was derived by Crick partly by stereochemical arguments (see Crick's 1970 retrospective). It has often been hailed as the final proof that acquired characteristics cannot be inherited, meaning that there is no downward causation in a naive Lamarckian sense. In contrast, downward causation properly understood is neither molecular biological nor physiological, but is an across-population process (and thoroughly Darwinian at that): *Only in the process of selection spreading genetic material in populations is there any justification to talk of adaptive differences materializing into genes and thus of downward causation.* Downward causation is not reducible to processes going on inside individuals taken one by one, but nevertheless takes place inside individuals-hanging-together in the evolutionary context of populations.

This element of downward causation on the population level also suggests an answer to the question asked earlier on: By which standards do a gene's regulatory and accessory sequences count as involved in gene function and thus as part of the gene? Such a sequence is part and parcel of the gene in question as long as genetic variation in this sequence affects the gene's expression in such a way that this sequence emerges on the population level as part of the same unit of selection. This is also true of barely detectable variations in regulatory sequences. Looking at single individuals, functional interdependencies fade away into the genomic horizon, and there is simply no yardstick by which to define the limits of a gene. But whenever this variation is such that it emerges in the population as part of the same unit of selection, it is, irrespective of location, also part of the gene in question. Having the same reproductive rate defines the fringes of a gene and may eventually turn it into a physically compact entity. Take the Jacob and Monod regulatory model and assume a mutation occurring, for example, in the regulator sequence with the effect of increasing the rate of transcription initiation. This may lead to the synthesis of a larger quantity of one of the proteins in question that in turn causes some difference in adaptive performance among individuals large enough for selection to detect. This reinforces a selective bias in the genetic underpinning of this adaptive difference. Because the mutation occurred in the regulator sequence, the whole regular circuit, including the operon (together with the promoter) will be affected as a single unit of selection. But maybe also some other DNA variations nearby or far away in the genome happen to contribute to the more efficient working of the mutant circuit than alternative variations at the same site and therefore also acquire the same specific reproductive rate. In any event, this whole unit as defined by a specific reproductive rate qualifies then as the gene.

Genes are (in lucky cases) the most elementary particles or ultimate downward extensions of life; they are life's atoms and come into being through a specific reproductive rate. But may not something smaller than a gene serve as such a unit? The answer is that the gene, as the smallest unit of selection, is a self-defining object. Because it includes as many bits of DNA as necessary for providing a genetic underpinning to a difference detectable by natural selection, no half or quarter gene can take its place, or when it does, it is by definition a full gene. To be sure, even mutations a level lower may radically affect a phenotype and differential reproduction, but they are not units of selection. Mutations do not compete in populations for a site in genes like genes do for a locus on a chromosome. Natural selection is then in an important sense blind for anything below the level of genes. Accepting Dobzhansky's (1973) dictum that nothing makes sense in biology except in the light of evolution, the conclusion is that mutations, which involve a shuffling among molecules too small for natural selection to detect, make sense in the evolutionary context (or in the eyes of natural selection) solely as *properties of genes.* Remember the important sense in which a mutation qualifies as proper only after a long history of evolution of the gene, that is, only when it begins to qualify as a genic property. Downward causation in the realm of biology then truly comes to an end at the level of genes and their properties, and this also ensures the autonomy of biology.

A contrasting view has been defended by Kenneth Schaffner for some decades (Schaffner 1974, 1993). After contending that a gene is satisfactorily characterized only if it is replaceable by a particular DNA (or RNA) sequence, and all the entities of the accompanying theory are completely specifiable in chemical terms, he goes on to say with respect to the Watson-Crick model of DNA replication:

The helices are held together by weak hydrogen bonds which are explicable by quantum mechanics, and the purines and pyrimidines – the nucleotide bases – are fully and completely characterizable in terms of physics and chemistry. (Schaffner 1974, 128) and more generally:

> ... since biological systems are thought by molecular biologists to be nothing but chemical systems, in the long run detailed investigations of such systems will be in full accord with the dictates suggested by the general reduction model. (Schaffner 1974, 139)

This opinion arises because Schaffner, like most molecular biologists, is looking only at processes going on inside single individuals (many such individuals, of course), but not at individuals hanging together in populations, and therefore sees only DNA and the like. My claim that genes come into being only in the context of selection-in-action-in-populations is, in turn, equal to saying that no piece of DNA displays in and of itself the capacity to form a gene. A piece of DNA, being inherently chemical, might be conferred the function of a gene by imposition from above: It begins to qualify as a gene when it becomes a unit of selection, and this is not an inborn characteristic of DNA. But when genes are seen only as pieces of DNA, a successful reduction in terms of physics and chemistry is question-begging because it is already contained in the premises.

Finally, I return to Fisher's cogency argument mentioned in the beginning that seemed the ultimate source of all reductionism in population genetics. Sewall Wright's (1931, 1932, 1982) way to overcome Fisher's cogency argument was to assume that a gene's selective value, owing to inadaptive factors in local populations like genetic drift, need not be determinate from the outset, but as it were, emerged from interactive processes in and between local demes throughout the species population. The way out suggested in this paper is more radical: Not only are the selective values of genes emergent properties, but the genes themselves are emergent units resulting from interactive processes in populations. Once it is possible to show that genes are products of populations, reductionism becomes evidently emptied of all content.

Fisher was forced by the cogency of his own argument to admit only as genes as causal factors in the evolutionary context, never as phenotypes, and so he never could have conceived of a process in which genes emerge from downward causation, even if he had wanted to. With Dawkins' (1976, 1982) reductionism the situation is more complex. To see why, it is important to consider how Dawkins defines his "egoistic gene." This definition contains, at least implicitly, an important innovation: While Fisher takes genes for granted in his population genetic foundation of the evolutionary process, Dawkins takes Fisher's foundation of evolution for granted and infers from this context what it takes to be a gene. Taking as a premise Fisher's cogency argument (that the spreading of self-contained genes is the essence of the evolutionary process), he concludes that a specific rate of differential reproduction, in fact, defines when a stretch of DNA counts as a gene. This is of course just what I have been arguing, and so it is important to point out similarities and differences. Dawkins suggests that we look for "pieces of chromosome of indeterminate length which become more or less numerous than alternatives of exactly the same length" when inquiring about genes (Dawkins 1982, 90). This is a version of the gene definition proposed by G. C. Williams: "In evolutionary theory, a gene could be defined as any hereditary information for which there is a favorable or unfavorable selection bias equal to several or many times its rate of endogenous change" (Williams 1966, 25). As I have argued, whichever piece of DNA undergoes differential reproduction is a gene by definition. This populational characteristic of the gene is for Dawkins sufficiently overriding for it to replace conventional molecular biological definitions of the gene. He contends that such a gene "is just a length of chromosome, not physically differentiated from the rest of the chromosome in any way" (Dawkins 1976, 30). It may cut across what is conventionally called a gene by the molecular biologist. For instance, it may begin in the middle of one cistron and end in the middle of another one. Any such arbitrary stretch forms a potential gene and qualifies as an actual gene as soon as it becomes a unit of differential reproduction (Dawkins 1982, 87–88). Given these insights, the question arises whether Dawkins is also aware of the role played by downward causation in the emergence of genes? Granted that a stretch of DNA qualifies as a unit of selection irrespective of internal structure, no other alternative seems to be left than by imposition from above.

Unfortunately, Dawkins' insight is immediately blurred by the selfishness that he ascribes to the gene. Dawkins, of course, never saw genes as products of adaptive effects, but solely as replicators that lever themselves into future generations by the goodness of their own effects. I take it that the very reason for him to define genes in terms of their changing frequencies was to safeguard their purported selfishness from being contradicted by empirical fact. Had he resorted to cues from a gene's internal structure, it would have always been possible to invoke empirical conditions in which such a gene's reproductive rate is causally determined by factors other than itself (by local interaction, linkage disequilibrium, group selection, etc.). However, by redefining the gene as that piece of DNA that actually does change in frequency, he turns any potential criticism of this sort into a defense of the selfish gene. Dawkins introduces the population genetic dimension only to uphold selfishness as the gene's distinctive property. Though his populational dimension is suggestive of how genes truly are brought into being, and though he makes use of this dimension as a device for defining what is a gene, he stops halfway and continues to view genes as the sole cause of their own change in frequency. This rules out the slightest possibility to conceive of downward causation. Genes are then unmoved movers; they "just are," he says (Dawkins 1976, 25), and thus he turns them into deeply ahistorical entities.

But no one else has shown that genes have a deep history of "arrival" in populations. Isn't it possible that his critics inadvertently share grounds with him in ways that outweigh their criticism? Protests emerge because Dawkins holds up a mirror more consistently than others do. Be this as it may, if it can be shown that the real course of reductionism in evolutionary theory is that it has gone unnoticed how genes are brought into being through populations, then the implementation of such a program would turn Dawkins' view of the gene upside down, or maybe the right way up.

CONCLUSIONS

What, then, is a gene? Arrays of accumulated genetic variations inside genomes that are caused by repeated shuffling of DNA strings make for some *one* difference in adaptedness during development. Various such overall differences among many individuals in turn induce selective discrimination, over the generations, within their genetic underpinnings, and such smallest units of selection we call genes. Thus the evolutionary dynamics of populations leads to the compartmentation of DNA strings into genes. Molecular biology is important not only for knowing a gene's internal structure, but also for appreciating the holistic, or across-population dimension of the gene. Matter is as much coherent and continuous as it is particulate. This is a nonpluralistic yet, hopefully, comprehensive view of the gene within a staunch Darwinian framework. We do then have last organic particles called genes; yet there is something behind them, i.e., the whole world of individuals-interacting-in-populations. Genes are not the product of human abstraction, but they become real in a process of material condensation taking place in populations of evolving organisms. This also provides an answer to the long-standing problem of reductionism in population genetics.

NOTES

1. Thanks are due to Raphael Falk and Sahotra Sarkar for many helpful comments and criticisms.
2. There may be some interesting overlap between the present gene concept and some aspects of Wimsatt's proposal to define a gene as "any element which plays a significant role in generating a large structure whose degree of adaptation determines the number of replicates or close variants of that element that are produced in the next generation" (Wimsatt 1993, 426).
3. It may be likely that the distinction made by Gifford (this volume) between "gene" and "genetic" will turn out to be a reflection of the same problem.
4. Though, I am editing slightly Gilbert's terms: In Gilbert's scheme the "arrival" metaphor underscores the specific contribution to the evolutionary process made by development while population genetics is left to handle the "survival" part. In the present scheme, "arrival" results from a process in which the developmental and population genetic dimension are all along on a par.

REFERENCES

Avery, O. T., C. M. MacLeod, and M. McCarty. 1944. Studies on the chemical transformation of pneumococcal types. *Journal of Experimental Biology and Medicine* 79: 137–158.

Barrell, B. G., G. M. Air, and C. A. Hutchinson III. 1976. Overlapping genes in bacteriophage φX174. *Nature* 264: 34–41.

Beadle, G. W., and E. L. Tatum. 1941. Genetic control of biochemical reactions in neurospora. *Proceedings of the National Academy of Sciences USA* 27: 499–506.

Benzer, S. 1955. Fine structure of a genetic region in bacteriophage. *Proceedings of the National Academy of Sciences USA* 41: 344–354.

Blake, C. C. F. 1978. Do genes-in-pieces imply proteins-in-pieces? *Nature* 273: 267.

Britten, R. J., and D. E. Kohne. 1967. Nucleotide sequence repitition in DNA. *Carnegie Institution Washington Year Book* 66: 78–106.

Brosius, J. 1991. Retroposons – seeds of evolution. *Science* 251: 753.

Brown, W. L., and E. O. Wilson. 1956. Character displacement. *Systematic Zoology* 5: 49–64.

Campbell, D. T. 1974. "Downward causation" in hierarchically organized biological systems. In *Studies in the Philosophy of Biology*, edited by F. C. Ayala and T. Dobzhansky. Berkeley: University of California Press, pp. 179–186.

Carlson, E. A. 1991. Defining the gene: An evolving concept. *American Journal of Human Genetics* 49: 475–487.

Crick, F. H. C. 1970. Central dogma of molecular biology. *Nature* 227: 561–563.

Dawkins, R. 1976. *The Selfish Gene*. Oxford: Oxford University Press.

Dawkins, R. 1982. *The Extended Phenotype*. Oxford: W. H. Freeman.

Dobzhansky, T. 1955. A review of some fundamental concepts and problems of population genetics. *Cold Spring Harbor Symposia on Quantitative Biology* 20: 1–15.

Dobzhansky, T. 1973. Nothing in biology makes sense except in the light of evolution. *American Biology Teacher* 35: 125–129.

Doolittle, R. F. 1995. The multiplicity of domains in proteins. *Annual Review of Biochemistry* 64: 287–314.

Doolittle, W. F., and C. Sapienza. 1980. Selfish genes, the phenotype paradigm and genome evolution. *Nature* 284: 601–603.

Dorit, R. L., L. Schoenbach, and W. Gilbert. 1991. How big is the universe of exons? *Science* 250: 1377–1382.

Falk, R. 1986. What is a gene? *Studies in the History and Philosophy of Science* 17: 133–173.

Fisher, R. A. 1930. *The Genetical Theory of Natural Selection*. Oxford: Clarendon Press.

Fogle, T. 1990. Are genes units of inheritance? *Biology and Philosophy* 5: 349–371.

Gale, J. S. 1990. *Theoretical Population Genetics*. London: Unwin Hyman.

Gilbert, W. 1978. Why genes in pieces? *Nature* 271: 501.

312

Gould, S. J. 1997. Darwinian fundamentalism. *New York Review of Books* 44 (10): 34–37.

Jacob, F., and J. Monod. 1961. Genetic regulatory mechanisms in the synthesis of proteins. *Journal of Molecular Biology* 3: 318–356.

Jacq, C., J. R. Miller, and G. G. Brownlee. 1977. A pseudogene structure in 5S DNA of *Xenopus laevis*. *Cell* 12: 109–120.

Johannsen, W. 1909. *Elemente der exakten Erblichkeitslehre*. Jena: Gustav Fischer.

Keller, E. F. 1987. Reproduction and the central project in evolutionary theory. *Biology and Philosophy* 2: 383–396.

Kimura, M. 1968. Evolutionary rate at the molecular level. *Nature* 217: 624–626.

Kimura, M. 1983. *The Neutral Theory of Molecular Evolution*. Cambridge: Cambridge University Press.

Lewin, B. 1980. Alternatives for splicing: Recognizing the ends of introns. *Cell* 22: 324–326.

Maynard Smith, J., and E. Szathmáry. 1995. *The Major Transitions in Evolution*. Oxford: W. H. Freeman.

Mayr, E. 1954. Change of genetic environment and evolution. In *Evolution as a Process*, edited by J. Huxley, A. C. Hardy and E. B. Ford. London: Allen and Unwin, pp. 157–180.

McClintock, B. 1951. Chromosome organization and genic expression. *Cold Spring Harbor Symposia on Quantitative Biology* 16: 13–47.

Muller, H. J. 1922. Variation due to change in the individual gene. *American Naturalist* 56: 32–50.

Muller, H. J. 1966. The genetic material as the initiator and the organizing basis of life. *American Naturalist* 100: 493–517.

Schaffner, K. F. 1974. The peripherality of reductionism in the development of molecular biology. *Journal of the History of Biology* 7: 111–139.

Schaffner, K. F. 1993. *Discovery and Explanation in Biology and Medicine*. Chicago: University of Chicago Press.

Schrödinger, E. 1944. *What Is Life?* Cambridge: Cambridge University Press.

Singer, M., and P. Berg. 1991. *Genes and Genomes: A Changing Perspective*. Mill Valley, CA: University Science Books.

Smith, C. W. J., J. G. Patton, and B. Nadal-Ginard. 1989. Alternative splicing in the control of gene expression. *Annual Review of Genetics* 23: 527–577.

Waring, M., and A. J. Britten. 1966. Nucleotide sequence repetition: a rapidly reassociating fraction of mouse DNA. *Science* 154: 791–794.

Watson, J., and F. H. C. Crick. 1953a. Molecular structure of nucleic acids. *Nature* 171: 737–738.

Watson, J. D., and F. H. C. Crick. 1953b. Genetical implications of the structure of deoxyribose nucleic acid. *Nature* 171: 964–967.

Williams, G. C. 1966. *Adaptation and Natural Selection*. Princeton: Princeton University Press.

Wimsatt, W. C. 1993. [Interview] In *Taking the Naturalistic Turn*, edited by W. Callebaut. Chicago: University of Chicago Press.

Wright, S. 1931. Evolution in Mendelian populations. *Genetics* 16: 97–159.

Wright, S. 1932. The roles of mutation, inbreeding, crossbreeding and selection in evolution. In *Proceedings of the Sixth International Congress of Genetics* 1: 356–366.

Wright, S. 1982. The shifting balance theory and macroevolution. *Annual Review of Genetics* 16: 1–19.

Final Review

The Gene – A Concept in Tension

RAPHAEL FALK[1]

> That was the gene that was.
> (rephrasing Stent, 1968)
>
> A gene is a gene is a gene.
> (Rheinberger, rephrasing Lwoff)

The gene has become a household icon of our end-of-the-century culture. There is hardly a news-story or a newspaper issue in which genes are not prominently mentioned: Scientists declare to have isolated a gene for some specific trait; a sexual practice that is claimed to deviate from the norm, or merely an unpopular literary style, is attributed by the press to be "in our genes" (see e.g., Nelkin and Lindee 1995). Geneticists, whether in the field of molecular biology or population dynamics, who should know better, often tumble and implicitly, if not explicitly, nourish and propagate this image of genes as determinants of traits. Although this sentiment is related to the dramatic developments in molecular biology of the last two decades, these notions are not new. Overwhelming as this fad may appear, the confrontation with the essence of our biological nature has always been in the center of interest. The time honored desire of humans to control their fate, and the frustration of the inadequacy of such efforts (in spite of all the progress of science and technology) have engendered the tensions that shaped much of human traditions and religions, as well as science. Francis Galton has most candidly formulated this confrontation as the conflict of nature and nurture, alluding to the words of Prospero in Shakespeare's *The Tempest* (Act 4, Scene 1):

317

A devil, a born devil, on whose nature
Nurture can never stick; on whom my pains,
Humanely taken, all, all lost, quite lost.

It is this tension of nature versus nurture that has been perpetuated in the popular concept of the gene. The gene became the ultimate entity of nature, on which "Nurture can never stick" that attained renewed prominence with the promise of genetic engineering, a promise that, unfortunately, has been often heralded by sensation-thirsty media and upheld by grant-starved scientists. As Gifford (t.v.)* puts it, it is not the relation between genes and traits, nor that of cause and effect, neither that of entities and properties that gives the gene its prominence today. Speaking of genetic traits is relating to "whether the cause is (or the causes are) genetic *as opposed to environmental.*" Genetic became the adjective for inherent, in the sense of doomed, unavoidable, and/or fatalistic. Recent achievements in genetic engineering imputed a still wider sense of determinism on genetic – that of deterministic controllability. Whether it is homosexuality, musical creativity, or criminality, the mention of genes in all these contexts indicates the assumed deterministic unavoidability of these traits. When neurobiologists wish to point out that some brain functions are, or must be, hard-wired, they assert that the only way to do this is to write the instructions for these in the genes, implying in one sense or another that flexibility of brain structure and function is *not* in the genes. Gene-transplantation became the promise of the ultimate cure.

It is, however, not this aspect of the gene concept that has been the main concern of the present volume. Science is, after all, involved in the pursuit of knowledge. Much as this is affected by and affects the social contexts in which it is carried out, a scientific tradition has emerged that claims to effectively deal in observations and concepts with its own specific tools. The notion of the gene, as it has developed recently in Western, science-oriented culture, provides an opportunity to challenge the adequacy and efficiency of some of these tools of science. It is these aspects that concern most authors of this volume. As could be expected, the advances in molecular biolog-

*References to papers included in *this volume* are mentioned with (t.v.) following the author's name.

ical research often exacerbated conflicting epistemological notions among the scientists involved. At the same time, such an epistemological scrutiny did not avoid the need to consider its historical roots, and, of course, also its social contexts.

Fifteen years ago I examined whether the *gene* may be considered to be a "concept in flux" (Falk 1986), as defined by Elkana (1970) in his analysis of the concept of *Kraft*. A concept in flux describes a situation "in which scientific concepts develop, . . . while the scientist is struggling to clarify his thoughts, that is while the discovery is made." I noticed that the situation is different for the concept of the gene: "The theoretical concepts have already been developed, so that whenever there is a conflict of experience no more than readjustments in the interior of the field are required in the Quinean model" (Falk 1986, 167). Confronted with the evidence of the papers presented in this volume, I wish to modify my previous conclusion that, whenever there is a conflict between theory and experience, no more than mere adjustments are required. I suggest that for the gene a more appropriate description would be a concept in tension. Throughout its history the gene has been defined by biologists on theoretical grounds, or on the basis of experimental and methodological considerations, as either a concrete material entity of living beings, or as an empirical instrumental one.[2] It has been conceived as an entity of developmental (physiological) or as one of evolutionary function. The gene has been instituted as the atom of life that maintains the inherited continuity of the living system versus the vagaries of environmental impacts, but also as a factor of accumulated experience that by its interaction with other environmental impacts provides ontogenetic as well as phylogenetic variability. It is these conflicting notions that make the gene concept that popular; and it is these tensions that make it so problematic.

In the first sections of this review I examine three steps in the path of reducing biology to physics and chemistry, from the acknowledgment of a particulate nature of heredity, through the attempts to characterize these particulate entities, to the structural resolution of the molecules of heredity. The conceptual impasses of the ultimate molecular resolutions could be resolved, however, only by changing directions and viewing the molecules as servants, rather than merely as components of wholes – this is discussed in the fourth section. In the last section I examine the normative meanings of the gene concept, in view of its past and with respect to the present-day social-cultural dimensions it obtained.

HEREDITY BECOMES PARTICULATE

Jean Gayon (t.v.) in his tracing of a philosophical scheme for the history of the concept of heredity notes the birth of the science of genetics as a move from heredity as a force, whose symbols *denote* magnitudes, to that of particles that are *examined* by magnitudes, that is from phenomenology to operationalism and eventually to realism. It is ironic that the person who introduced the term *gene* and opened the instrumental era of the theory of heredity was more a phenomenologist than an operationalist, according to Gayon's classification. In 1909 Wilhelm Johannsen introduced the concepts of *phenotype* and *genotype,* explicitly for "upsetting of the transmission-conception of heredity." The terms *heredity* and *inheritance* were borrowed from everyday language, and "the transmission-conception of heredity represents exactly the reverse of the real facts. . . . The *personal qualities* of any individual organism do not at all cause the qualities of its offspring; but the qualities of both ancestor and descendant are . . . determined by the nature of the 'sexual substances' – i.e., the gametes – from which they have developed. Personal qualities are then *the reactions of the gametes* . . . but the nature of the gametes is not determined by the personal qualities of the parents or the ancestors in question" (Johannsen 1911, 130). Johannsen severed the conceptual Gordian knot between traits and deterministic inputs by distinguishing between the reactions of the gametes and the personal qualities. His characterization of the genotype (and of the phenotype) was in quantitative terms, based to a large extent on Karl Pearson's (the most outspoken phenomenological philosopher) biometric analysis. The term *gene* was only introduced as a derivative of genotype, in a concession to the reductionist notion of the Mendelian theory of heredity. Superficially Johannsen's gene was just one more successor in the tradition of theoretical corpuscular entities of heredity, from Darwin's *gemmules,* through Weismann's *biophores* to de Vries's *pangenes.* However, being derived from *genotype,* it did not imply any corpuscular entity as many of the former ones did, but rather the important insight that genes are entities that are involved in the unfolding of traits but are not identical with traits. It took another ten years until Ronald A. Fisher (1918) finally toppled the remaining impact of the biometricians' phenomenological theory of heredity by providing a Mendelian foundation also to the theory of evolution by natural selection.

One immediate impact of Johannsen's genotype/phenotype disjunction was that the embryologist T. H. Morgan was able to overcome his aversion toward Mendelism as a preformist theory of heredity (Falk and Schwartz 1993). For Morgan it provided the instrumental framework to conceive genes as particles that directed the development of "unit characters" but were not identical with them (Allen 1979; Churchill 1974). The differentiation between the trait and its determinant engendered, however, a new tension, regarding the nature of the genotypic determinants and their relation to traits. For geneticists, who were primarily interested in breeding problems, like Edward M. East, the way out sounded simple: The gene provided a convenient notational concept for capturing the facts recorded in breeding and hybridization experiments. However, even for East, the instrumental conception of the genes may have been merely a means of referring to what he believed to be "real" entities, that would eventually be characterized in material terms. The dialectic between the instrumentalist and the realist treatments of the gene erupted in full in Morgan's notion, and in that of his pupils. It dominated the genetic landscape until the establishment of the "genetical implications of the structure of deoxyribose nucleic acid," as was the title of the paper by Watson and Crick in 1953.

East's instrumental conception of the gene helped to convince his friend, William E. Castle, to abandon the notion of the unit character, and to adopt, on the pragmatic ground of greater empirical import, the assumption of genetic units dissociated from the (phenotypic) unit character, even at the price that it left unresolved the problem of the interaction of those genes with environmental factors in development and differentiation. It should be kept in mind, however, that Castle was one of the most extreme operationalists in the early years of transmission genetics (comparable only to L. J. Stadler's operational concept of the gene, in the 1950s). Morgan, who was an avid empiricist (Allen 1979), was much more concerned with the wider conceptual consequences of his findings.

Transmission geneticists could infer genes only by following quantitative results of breeding experiments. From tools of statistical distribution, increasingly genes became entities in their own right, even if detectable only indirectly through their effects. However, the experimental demands of methods for individuation of specific genes on the one hand, and the elaboration of the theoretical-ontological qualities of genes on the other hand, generated tensions that have never been completely resolved. Morgan's notion that the relationship between genes and traits must be a multivariate function created a methodological difficulty for the identification of genes through their effects. The relief from this conflict, as conceived by Morgan and his coworkers, was to pick up the *difference* in many-to-many relationships between genes and traits as merely the gene's attribute (Schwartz t.v.; see also Waters 1994). As Schwartz points out, such a device for dealing with the *changes* in the attributes of genes, at a different level than that of the (ontological) nature of the entities themselves, has been a common practice in science: Experimentalists needed specific effects, markers, or Insufficient but Necessary properties, to identify the genes. Thus, instead of dealing with the Unnecessary but Sufficient ensemble for the alternative appearance of the trait, they pointed at an isolated Necessary but Insufficient component as the cause (Mackie 1965). The *proper individuation* criterion (PI), that Gifford (t.v.) provides for the explication of a genetic trait, stating that a trait is not a genetic trait unless "it is this trait *specifically* that is *caused* by the relevant gene or genes," by focusing on a single cause and screening out all the rest, is very much such a Necessary though Insufficient condition. This differential concept of the gene distracted, however, precisely from the very point that Morgan wished to stress, namely, to divert attention from the mutated factor to the remaining ones that were involved in the character. It left unresolved the nature of the gene, although it strongly suggested the gene to be a discrete material entity, rather than an abstract theoretical notion.

THE HARDENING OF THE GENE

The intensive construction of linkage maps, as well as the experimental evidence of the chromosomal correlates of genetic events (nondisjunction, deficiencies, crossing-over) by Morgan's students firmly established the chromosomal theory of heredity. But while many geneticists went a step further and imputed to the genes material characteristics, Morgan stressed that for resolving the questions of transmission genetics the material nature of these entities was irrelevant. For all practical purposes, all that the material basis of the hereditary traits implied could be ascribed to elements of the phenotype of, say, metabolic and developmental pathways, as long as *the concept of the genotype could be maintained as distinct from that of the phenotype*. Contrary to this pragmatic approach, it was primarily Herman J. Muller who conceived of genes, on a purely theoretical reductionist interpretation of the experimental results, as the material atoms of development and metabolic function.

[T]he fundamental contribution which genetics has made to cell physiology . . . consists in the demonstration that . . . there are present within the cell *thousands* of distinct substances – the "genes"; these genes exist as ultramicroscopic particles. (Muller 1922, 32)

Consequently, Muller embarked on a research program aimed at specifying the genes' physicochemical properties (Muller 1922; see also Muller 1947). For Muller, the genotype, and its entities the genes, acquired a material status of their own, distinct in principle from that of the phenotype, at whatever level of resolution the phenotypic properties are described (Falk 1986; 1997). Muller attributed to these atoms of heredity three properties (Muller 1922; 1947):

1. Autonomous self-replication;
2. Initiation of specific products that effect the development and performance of organisms' properties; and
3. The faculty to mutate (in spite of their great stability), so as to change their product/effect without losing their capacity for autonomous self-replication.

It is important to note that Muller's theoretical move had far-reaching empirical consequences. Muller realized from early on the need for a powerful experimental method for fine quantitative analysis. The *ClB* method[3] for the detection and quantitative analysis of recessive lethals in *Drosophila melanogaster* rightfully became a model of genetic analysis (even if its effects on present-day molecular techniques is less conspicuous. See, however, the selective amplification principle of molecular techniques like the PCR[4]). No wonder that the empirical success of what for most geneticists was a working hypothesis became implicit proof for the existence of genes as discrete material entities of structure, function, and mutation. Even potential evidence to the contrary, such as position effect, was interpreted by Muller (e.g., Raffel and Muller 1940) as supporting evidence at the level of functional interaction between discrete genes. It was this materialization of the genotypic level that provided the crucial lever for the study of gene function, as exemplified by the work of Beadle and Tatum in the 1940s, that culminated in the inference based on the "faith in the unproved assumption that a given gene has a single primary action" (Beadle and Tatum 1941, 115).[5] Significantly, this careful utterance was soon turned into the "one-gene – one-enzyme" doctrine. By the end of the 1930s, the concept of the discrete material gene was hardened, presumably against any opposition.

The most consistent opposition to the gene as the material particle of heredity was that of Richard Goldschmidt. If Castle's opposition to Muller's "use of the word gene in this sweeping statement" (Castle 1915) was sequestered by operationalist arguments for which he himself preached, the opposition of Goldschmidt to the gene as the atom of heredity needed other means to overcome. Goldschmidt's opposition was grounded in a complex combination of scientific and sociopolitical reasons (Harwood 1993; Sapp 1987). There is no doubt that the competition between the German and the American schools of genetics, as well as that of German unitary tendencies in the sciences, played an important role in Goldschmidt's drive "to build a unified understanding of vast arrays of biological phenomena" (see Harwood 1993). Although it is difficult to summarize Goldschmidt's views in one sentence, I would suggest that while the American school viewed genetics from the bottom up, Goldschmidt and much of the German biologists' community juxtaposed a position from the top down. While Muller accepted, on theoretical grounds, the existence of genes as the atoms of inheritance, Goldschmidt inquired whether, on theoretical grounds, genes as atoms of inheritance were required. Goldschmidt believed that by advocating a genetic hierarchy it would be possible to maintain a unified understanding of genetics and to develop a theoretical genetics in which results from cytogenetics, transmission genetics, and physiological genetics were integrated (Dietrich t.v.). Thus, he challenged, on theoretical grounds, the empirical method that inferred atomic genes from their presumed specific effects. From the fact that it was possible to localize the production of a phenotypic effect, it did not follow that there was a wild-type allele corresponding to the site of the mutation that produced the phenotypic effect. On the contrary, the accumulation of related functions in certain chromosomal segments indicated to the integrative organization of these chromosomes.

The conclusion, then, is that gene mutation and position effect are one and the same thing. This means that no genes are existing but only points, loci, in a chromosome which have to be arranged in a proper order or pattern to control normal development. Any change in this order may change some detail of development, and this is what we call a mutation. (Goldschmidt 1938, 271)

Although it is hard to avoid the conclusion that Goldschmidt's conception inevitably curbed much of his capacity as an experimentalist to take advantage of the reductionist heuristics that was so productive for the instrumentalists (Falk 1986), it was more than his experimental methodology that generated the strong criticism against him, later converting to a polite but deafening silence about him. Obviously, it was the concept of the gene as a materially discrete entity of the chromosomes that had been hardened to such an extent that any nonreductionist opposition to a particulate atom of heredity was turned down without a fair chance to examine it. Goldschmidt came into vogue again only with the introduction of the developmental notion of genes as entities in a hierarchical organization in the chromosomes (see Dietrich t.v.). However, it must be admitted that Goldschmidt's hierarchical arrangement of hereditary material seems to correspond more to a taxonomy of genetic entities than to an integrative interaction between the levels of the hierarchy (see e.g., Salthe 1985). In a sense, in his attempt to present the organismic point of view differently from the favorite particulate material *gene* conception, Goldschmidt paid the price of forsaking the advantages of reductionism, without being able to develop a proper theory of complexity.

It was, however, within the very same heuristics that served to establish the Mullerian gene concept, where doubts were eventually heaved on the gene's integrity. With the increase of the experimental resolution power in *Drosophila* itself (Lewis 1951), and even more so, with the introduction of other organisms, such as *Neurospora* and *Aspergillus* (Pontecorvo 1952), in which much higher resolution could be achieved, it became evident that genes cannot be maintained as the genetic singularities of structure, function, and mutation (see Holmes t.v.).

THE GENETICAL IMPLICATIONS OF THE STRUCTURE OF DEOXYRIBONUCLEIC ACID

According to Gayon (t.v.), while biologists were careful instrumentalists in the first half of the century, they came of age in the second half of the century: It was only with molecular biology that a realistic interpretation of hereditary material emerged. In this process of the maturation of scientific epistemology, Gayon perceives that realism became manifest by the molecular genetics approach to heredity, where the main features were "the elucidation of the physicochemical nature of the gene, and the machinery that allows it to replicate and code for polypeptides." It is, however, important to note that it was the elucidation of the physical-chemical nature of *heredity*, not of the *gene*, though many people thought so at the time, that was achieved in 1953.

In retrospect, the profound epistemological change that Watson and Crick's presentation of the model of the structural organization of DNA heralded was not the claim of the material or molecular concept of the gene – this had been taking place within the instrumental conception for years – but rather in the heuristics. It was no longer necessary to accept the existence of genes and ask about their physicochemical structure, as was Muller's research program, but rather to start with the physicochemical structure and ask what the genes were. As is well known, the significance of the structural model of DNA as that of the hereditary matter did not escape the notice of Watson and Crick even in their first publications, when they ascribed to their model the properties that Muller imputed to genes in 1922. The model allowed self-replication; it could provide enough variability needed for the coding of many and specific products; and it allowed for structural changes without losing these unique properties (Watson and Crick 1953). "Many lines of evidence indicate that [DNA] is the carrier of a part of (if not all) the genetic specificity of the chromosomes and thus of the gene itself" (Watson and Crick 1953, 964).

Thus, it is remarkable that although at least two of the genes' qualities, self-replication and mutability, were expropriated from them in favor of the DNA molecule already in the 1950s, they have been mentioned long thereafter as characterizing properties of genes. As a matter of fact, they are still frequently associated with the gene concept of the molecular biologist, as revealed even in the present volume. It is true that in the early years of molecular genetics there was an implicit hope among most geneticists that these two concepts, DNA entities and genes, would again merge. As noted by Gayon (t.v.), with the advent of the molecular phase of the science of heredity there was a feeling that "genes can be identified as fragments of chromosomal DNA, these fragments being both replicating units and functional units coding for amino-acid sequences. . . . Genes, on this account, turn out to be robust 'natural kinds'." We now know that advances in molecular biology make such a representation impossible: The qualification of some sequence of DNA as a gene depends on how the experimenter chooses to manipulate the genome. This amounts to saying that in the framework of reductionist heuristics the recent developments of genetic engineering have led to a restoration of the instrumentalist view of genetics. But, as noted by Fogle (t.v.), although the empirical details are much elaborated today, molecular genes retain the imprint of the past.

It was Seymour Benzer whose genetic analysis formalized the fact that the instrumental units of function were not necessarily identical with those of recombination or those of mutations. By calling them *cistrons*, *recons*, and *mutons* respectively, he remained noncommittal with respect to the ontological status of the gene. Although Benzer never managed to show directly the colinearity of the DNA sequence and the genetic map,[6] his research program was explicitly designed to extend transmission genetic studies to the molecular, meaning to the nucleotide level (Holmes t.v.). In this sense his ontology was molecular whereas his epistemology was that of transmission genetics.

Benzer's model soon became the pedestal to be challenged by the tensions of the empirical results and the conceptual image of the gene, or the retreat of Mendelian epistemology in confrontation with the molecular ontology. When it seemed that the units of recombination and the units of mutation converge on single nucleotides (which, with time, proved not to be the case at all), and the genetic map segment of the cistron converged on the entity of transmission genetics, Benzer's nomenclature seemed to become redundant. For a short while an effort was made to maintain the reduction of Mendelian, or transmission genetic concepts, to molecular terms. Frank Stahl (1961) suggested a model of eukaryotic chromosomes constructed of circular DNA molecules of prokaryotes. Each circular molecule was involved in one or a few related specific functions; mutations occurred within circles; recombination occurred either between circles – resembling classical crossing-over – or within circles – displaying intragenic crossing-over. Similarly, when Jacob and Monod (1961) presented their genetic regulation model they still hoped to maintain the distinction between *gene* and *cistron*. In allegiance to the one-gene – one-enzyme notion they noticed that in Yanofsky's work on the structural gene for the enzyme tryptophane-synthetase, the gene comprised two cistrons, each related to a different structural component of the enzyme. However, at the same time they also defined a new type of gene, that they called *regulator gene*. Unlike the cistron, "the regulator gene does not contribute structural information to the proteins which it controls. . . . In contrast to the classical structural gene, a regulator gene may control the synthesis of several different proteins: the one-gene – one-protein rule does not apply to it" (Jacob and Monod 1961, 334). A hierarchical notion emerged. The more it became evident that the chromosome of bacteria and viruses is one continuous molecule of DNA that replicates as a unit, with no structural indications that mark the delineation of one function from another, the cistrons as the DNA structures of function (as distinct from genes) became meaningless. Also, when all efforts to identify distinct structures along the DNA of prokaryotes failed (Cairns 1962), and even the chromosomes of

Drosophila (Kavenoff and Zimm 1973) were found to be composed of a structurally continuous DNA molecule (whose replication units were independent of cistronic or genic function), the gene as the material atom of heredity increasingly lost its heuristic power. Of the properties allocated to the genes, as the material entities of heredity, replication, recombination, and mutation became properties of the DNA per se, only the gene as a unit of function remained.

What, then, is the function that could define genes as discrete material units, if at all? More precisely, is there a physicochemical heterologous product (as distinct from the homologous product, which is the self-replicated sequence itself) that makes for a sequence of DNA to be uniquely identifiable? Thirty years ago Britten and Davidson (1969, 349–350) tried to save the new epistemology of the molecular gene that is independent of a phenotype by suggesting the minimalist definition of a gene: "A region of the genome with a narrowly definable or elementary function. It need not contain information for specifying the primary structure of a protein." But would there be at least a unique primary product of transcription that in compliance with the notion of the material gene as a unit of function/product, would define a consensus gene? That was, of course, exactly what Beadle and Tatum's slogan of "one-gene – one-enzyme" strove at in the pre-DNA era. After peeling off all the layers of secondary products and interactions responsible for the many-to-many relationships of genes and traits, the gene should be identified by its genuine unit character and that unit character is the one that defines the gene. This coveted genuine unit character was now the recognized primary product of the gene-as-a-DNA-sequence. Crick formulated the causal primacy of DNA in the Central Dogma (Crick 1958) and molecular research put increasing emphasis on the primacy of the gene. In adherence to it, geneticists devised an increasingly complex gene-regulatory scheme to tie all ends together. The *operon* of the 1960s was, in effect, a detailed model that allowed regulation to remain at the level of genes. This was manifested by Britten and Davidson's (1969) models that offered a "web of possible interactions" of genes, fitting the assumption that the eukaryotic program was gene-based, though more complex than that suggested by the operon model. I believe, however, that it was the Central Dogma of Genetics itself that, by indicating to the many factors that must be involved in the intermediate stages of transcription and translation, provided the background for the final *coup de grâce* to any one-to-one relationship of a material gene and its primary product.

Whether molecular biologists needed a comprehensive entity, or merely struggled to maintain the entity of Mendelians, it would be superfluous to repeat here the bewildering abundance of functions of not necessarily contiguous stretches of DNA that have been assigned to genes (see e.g., Fogle t.v. and Rheinberger t.v.). Mechanized gene-sequencers define genes as DNA sequences that have known transcription-starting sequences, open reading frames (ORFs) of coding nucleotide triplets (contiguous or interrupted), and transcription-stop signals. The internet *Primer on Molecular Genetics* gives the following official glossary definition of a gene:

The fundamental physical and functional unit of heredity. A *gene* is an ordered sequence of *nucleotides* located in a particular position on a particular *chromosome* that encodes a specific functional product (i.e., a *protein* or *RNA molecule*).[7]

The dire status of the molecular gene is highlighted by the three criteria that Fogle (t.v.) formulated for the gene adopted by molecular biologists:

1. Localization to a transcript-generating segment of DNA.
2. Physiological boundaries located at pretranslational compared to posttranslational activity.
3. An investigator-based assessment of functional divergence among products.

As we saw, the discriminating power of the first two criteria is very limited; they can at best suggest some wide generic reductionist definition. It is the third criterion that may lead back to an attempt to view genes as more than physicochemical entities: It was biological phenomena that defined the gene in the first place; it is biological phenomena, if at all, that should define these entities now. The two major phenomena in biology that should be considered are evolution and development.

A NEW KID IN TOWN: THE DEVELOPMENTAL GENE

The theory of evolution has been always closely linked to genetic theory. However, for many years, and especially in the heyday of the New Synthesis, it dealt with the dynamics of explicitly abstract gene frequencies in populations (Sewall Wright's interactive models notwithstanding). There was little discussion of the developmental or molecular aspects of these genes and their evolutionary dynamics. Molecular biology and evolutionary biology were "in a constant danger of diverging totally, both in the problems with which they are concerned, that is, the 'how' as against the 'why,' and as scientific communities ignorant and disdainful of each other's methods and concepts" (Lewontin, 1991, 661). The crucial change happened in 1966 when Lewontin introduced electrophoresis as a method for detecting protein polymorphisms in populations. I suggest that such studies of molecular polymorphisms eventually channeled molecular biologists to also think in terms of "why," that is to consider the holist perspective besides their traditional reductionist one of "how." It was such a conceptual change that introduced considerations of development of organisms and of the evolution of these processes, that provided new life to the gene concept. The gene's structural properties become the function of its role in development and of the evolution of that role.

Not all would agree that such a turnabout in perspective was needed. It has been proposed that the strict mechanistic talk of molecular biologists should be conceived merely as a metaphor: "No self-respecting geneticist claims that DNA is solely required for a causal explanation of cellular processes. They simply leave the rest to the protoplasm in the backdrop, a supporting character in the living drama of functionally integrated systems. Center stage is the protagonist, the DNA, manipulating and controlling cellular events" (a comment by Fogle at the second workshop on gene concepts mentioned in the introduction of this volume). However, as Keller (t.v.) notes: "Words matter." The metaphor of the gene as a determinant stifled for a long time any serious discussion of epigenetic processes among geneticists.[8] Experimental work that led to models of genetic control of embryogenesis had been going on. But because these models were formulated within the framework of the classic methodology of Mendelian genetics they proved to be largely inaccessible to molecular geneticists. It is significant that the pivotal paper of E. B. Lewis (1978), presenting an integrative model of genetic control of metamerism in *Drosophila*, is described by Morange (t.v.) as "difficult to read" and not making "any allusions whatsoever to a possible generalization of the results." Similarly, the elegant experiments of Garcia-Bellido and his associates on genes controlling compartmentalization in development that changed much of the thinking of geneticists involved in problems of development, "were forgotten" according to him. Morange seems amazed to see that these studies evoked the interest of a molecular geneticist such as F. H. C. Crick (Crick and Lawrence 1975) "in the absence of any structural [i.e., molecular] data on the selector genes" (Morange t.v.).

It was, however, within the perspective of the molecular analysis that the need for conceptual reevaluation of the gene was wanted. Jacob and Monod's (1961) model of genetic regulation at the molecular level started a course opposite to Beadle, Ephrussi, and Tatum's direction from the gene as an entity of development to an entity of biochemistry. They inaugurated a move from a molecular gene back to a gene involved in development. The model of Jacob and Monod introduced into the genetic discourse two elements needed for ontogenetic development: the notion of genes whose function was to control other genes and regulate their transcriptive activity (and eventually also other activities, like alternative splicing and translation) in time and space, and the notion of hierarchies of entities involved in structure and function of products. Within a short time McClintock (1961) and Lewis (1963) (see also Falk 1963) suggested that this molecular model of the regulation of metabolic pathways could be extended to the genetic control of development. At the time, however, the achievements of reductionist molecular genetics and the stunningly simple mechanisms of genetic transmission and translation have been perceived as the triumph of the notion of the genetic code as the source of information that instructs the functional specificity of proteins, and provides the program, according to which development obliges (see Keller t.v.). Jacob himself hoped that what he called the "paradox of development," the old "opposition between the mechanistic interpretation of the organism on the one hand, and the evident finality of . . . the development of the egg into an adult" would disappear "when heredity is described as a coded programme in a sequence of chemical radicals" (Jacob 1976, 4). Jacob, however, underwent a gradual metamorphosis already in the course of presenting his *The Logic of Life* (Jacob 1976) and even to a larger extent later (Jacob 1977). He increasingly exposed the inadequacy of the reductionist notion of the genetic program and proceeded toward the need to include an integrative view to counter the shortcomings of the particulate reductionist one. In direct correlation with this, the prominence of other cellular and extracellular components in the process of regulation increased. As the role of the genome gradually became that of provider of data, rather than being the program (Keller t.v.), the onus of genes as the determinants of development relaxed.

The shift of emphasis had to wait, however, until the introduction of a new heuristic, that of directly manipulating the DNA sequences as if they were genes that may specifically affect developmental phenotypes even far off of the primary ones – molecular methods in the service of holistic analysis. Even the paper of Nüsslein-Volhard and Wieschaus (1980) that earned its authors the cover page of *Nature*, would not have survived attention, had the tools of genetic engineering not been developed to meet the challenge of the Mendelian genes of development. The new heuristic allowed the coupling of classical Mendelian genetic methods with the methods of molecular engineering to conceive of the developmental gene. It took, however, more than a shift in heuristics to introduce the developmental gene (Gilbert t.v.; Morange t.v.).

Traditionally, development and inheritance were considered two aspects of the same problem. Heredity as an independent biological concept was established only in the eighteenth century, with the introduction of the Linnaean systematics. Embryology and the science of hybridization – that eventually became the science of heredity – diverged both in their conceptions and their heuristics throughout the nineteenth century. Formally, this differentiation reached its peak in the beginning of the twentieth century with the introduction of the new science of genetics. Although it has been attributed to Weismann's theory of the differentiation between the germ-line and the soma-line, it is more the interpretation of Weismann's notion, *Weismannism*, that provided the rationale for the alienation of embryological and genetical research during the greater part of the twentieth century (Griesemer and Wimsatt 1989; Griesemer t.v.). Weismann, as well as Morgan and Muller (Falk 1997; Falk and Schwartz 1993), Haldane and Waddington, not to mention the German school of genetics, from Correns, through Goldschmidt and Kühn, to Caspari and Stern (see Harwood 1993; see also Dietrich t.v.) always maintained the developmental aspect of the theory of genetics. They all did this, however, from a genocentric perspective: "What happened historically is that interpretation of heredity has shifted from a developmental aspect of parent-offspring relations in reproduction to genes as autonomous master molecules controlling development" (Griesemer t.v.). As Griesemer notes, a profound conceptual change was needed to revert this unyielding progression. The needed change of concepts is best illustrated by the statements of two leaders of molecular genetics: In the 1960s Gunther Stent, dazzled by the Jacob-Monod model of gene regulation, exclaimed that development was a trivial problem because all development could be reduced to that paradigm (e.g., Stent 1970). Twenty years later Sydney Brenner reflected on this tension between genetics-as-a-program and development when he stated: "The paradigm does not tell us how to make a mouse only how to make a switch" (Brenner et al. 1990). Griesemer (t.v.) suggests that for this change to be effective, not only as a philosophical heuristic but also as a prospective conceptual and methodological guideline for the biological sciences, a new paradigm should be introduced. He rejects the prevailing structural hierarchy as the starting point for interpreting the relation between heredity and development. He suggests that processes, rather than structures or functions, should be conceived as the entities of biology. Accordingly he claims that we should accept a different reductional heuristic in biology: It is reproduction that should be considered the fundamental process to which all other processes should be reduced. "Replication processes are a special case of inheritance processes, which are a special case of reproduction processes" (Griesemer t.v.).

Griesemer thus puts forward a new scientific historiography according to which "to gain heuristic purchase on a general theory of development" he views "theories of genetics as theories of development based on invariant units" (Griesemer t.v.). Once these transmission-genetics-style genes of "Developmental Synthesis" (Gilbert, t.v.) were identified as "developmental invariances" they allowed a direct experimental analysis of the genetic and epigenetic control of development by manipulation of the DNA although they were not well defined at the concrete molecular level.[9] The confinement of these genes to more-or-less specific loci has been further challenged by such elements that control their activation in time and site as the enhancers, which may reside as far away as 85 Kb upstream of the promoter (as in the case of the *Drosophila* locus of *cut*), or 69 Kb downstream of the promoter (as in the T cell receptor of a-chain gene enhancer in man). However, what gave these genes special significance was their surprising conservation both in many sites within the genome and in the genomes of other species (see Morange t.v. and Gilbert t.v.). Contrary to the genes of the Modern Synthesis that were manifest only by their difference (which is the basis for evolution by selection; see Beurton t.v.), the genes of the Developmental Synthesis are manifest by their similarities (which is an indication of conserved pathways during former macroevolution, Gilbert t.v.). The work of these genes in "context-dependent networks" (Gilbert t.v.) provides insights into the evolution by natural selection of developmental pathways.

Indeed, Beurton (t.v.) suggests that genes should be viewed as integrated entities of the process of evolution: It is natural selection that creates genes as efficient entities of determination of the proteinacious products exposed as phenotypes to natural selection, and it is the role of these products in ontogenetic development that imputes to them their selective value. Viewing natural selection as repeated "trial and error" of random mutations, he suggests that "nature in and of herself turns DNA strings into discrete and well-established entities." Thus, the gene becomes "the genetic underpinning of the smallest possible difference in adaptation that may be detected by natural selection." Genes are the ontological products of natural selection.[10] "[D]ifferential reproduction is a statistical bias among *arrays* of alternative DNA variations that, through their joint action, produce a difference among individuals large enough for selection to detect." For reasons of economy, natural selection works on dispersed sequences that happen to support some *one* minute adaptive difference among individuals, so that they might come to occupy a single location and acquire a certain degree of physical integrity. Contrary to other views of the gene (e.g., that of Dawkins 1976), Beurton asserts that a gene is not "what makes a difference" between individuals, rather, *difference in the reproduction of organisms is "what makes genes" and maintains their stability*. In a sense, Beurton revitalized the Darwinian theory of organic evolution by incorporating in its entities the dynamic dimension that views them as the consequence of evolution by natural selection, rather than merely its building blocks. Location integrity, or at least relative physical coherence, is thus not a prerequisite of a gene but simply a consequence of selection for efficient units that make for specific differentiation in reproduction. The gene "is not a unit *encountered* by natural selection. Rather, it is an emergent unit or one *generated* in the process of natural selection from a background of never-ceasing variation contained in the genome." It is natural selection that also shapes its structural properties.

The discovery of the conservation of "developmental genes" should thus be considered an important milestone in linking the concept of the gene as the product of evolution to the needs of development. The new branch of science, the evolution of development (*evo-devo*), forces upon the molecular biologists the perspective "from above," that is, from the perspective of the complex system of organisms and populations, that asks the classical biological questions of "why." The evidence-from-embryology of Darwinian evolution turns now to the evidence-from-DNA sequences. The detection of consensus DNA sequences both within the genome of the same species and between that of other, even presumably very far related species, has already led to unexpected reevaluations of the evolutionary and developmental relations at the anatomical and physiological levels. It threatens to further reshuffle many of our previous concepts of the evolution of developmental processes. Thus, anatomically and embryologically different systems like those of eyes of insects and vertebrates are, on the new molecular evidence, a rather divergent end-product of the same common structure(s) than independent structures of evolution that functionally converged (see e.g., Tomarev et al. 1997). However, this does not turn the evolving consensus sequences into genes. The very fact that DNA stretches are evolving indicates that they should be considered to be traits, or phenotypes, at a still more basic level than that of the protein or the RNA-transcript. "The 'phenotype' of an organism is the class of which it is a member based upon observable physical qualities of the organism, including its morphology, physiology, and behavior at all levels of description. The 'genotype' of an organism is the class of which it is a member based upon the postulated state of internal hereditary factors, the genes" (Lewontin 1992, 137). DNA sequences constitute phenotypes just like any other evolving trait. Here is the essence of the tension: The biological perspective identifies the gene as an entity of function that was conserved and changed in evolution, the data of which happen to be materially engraved in DNA sequences that are loosely definable from that perspective, whereas molecular biologists, especially those of the age of the mechanized Human Genome Project, are eager to define strict criteria for automatically and universally identifying DNA sequences as genes.

THE ILLUSION OF THE GENE?

If Gregor Mendel was the one who provided the acuity of the methodology that generated a theory of heredity, it was Wilhelm Johannsen who established the profoundness of the theory that guided the methodology. The gene became the central concept of the modern theory of genetics. As such, it has been torn between conflicting methodological and theoretical demands, or, put differently, between the material needs of the experimentalists and the abstract considerations of the theoreticians. The science of genetics, that struggled at the beginning of the century for its independence (Falk 1995), has since then incorporated, or was integrated in adjacent disciplines, notably evolution, embryology, cell biology, and molecular biology. Although it was only gene differences that could be followed, the consensus was that genes were concrete and discrete material entities arranged along the chromosomes. The success of Muller's working program convinced most researchers that the genes were the atoms of the material basis of heredity. Eight years before the Watson-Crick papers, Muller reiterated the foundations of his conception of the genotype and its distinct particles, the genes, in his 1945 Pilgrim Trust Lecture:

The gene has sometimes been described as a purely idealistic concept, divorced from real things, and again it has been denounced as wishful thinking on the part of those too mechanically minded. . . . However, a defensible case for the existence of separable genetic material might have been made out on very general considerations alone. . . . [I]n the organism it would be inferred that there exists a relatively stable controlling structure, to which the rest is attached, and about which it in a sense revolves. . . . [T]he material furnishing the frame of reference, whatever it is, itself underwent . . . doubling or reproduction, and . . . this too must have taken place under its own guidance. It is for this reason that it may be called the *genetic* material. (Muller 1947, 1)

Arranged in fixed linear order in each long thread is a multitude of distinctive parts, each with its characteristic chemical effect in the cell. . . . [Research] has given incontrovertible evidence of the existence of definite genetic material, of a particular nature, which certainly has, in each part, the property of self-determination in its own duplication. . . . [T]his self-duplication is not, primarily, a resultant of the action of even the genetic material as a whole, but of each part of it separately. That is, the material is potentially particulate, and each separate part, which determines the duplication of just its own material, may be called *a gene*. (Muller 1947, 3–5)

The establishment by Watson and Crick of the DNA structure as the model of the material basis of heredity and Benzer's experimental evidence of the organization of the functional genetic entities apparently upheld Muller's vision: The gene is a distinct sequence of DNA that comprises "relatively stable controlling structure, to which the rest is attached, and about which it in a sense revolves." The Central Dogma of Genetics as formulated by Crick actually canonized DNA-sequences in the role of the "unmoved movers" of metabolism and development. The special status of DNA as the genetic material to which "all other material in the organism is made subsidiary" (Muller, 1947 2), was largely maintained in the laboratories. The experimental method of reducing attention to single variables, while striving to keep all the others as constant as possible, was over-powering, and radiated on the theoretical concept, the many conceptual difficulties notwithstanding. Muller's logical conclusion for the existence of genes was maintained by imputing to what is actually a phenotype – the DNA sequence – the power of genes as the entities of the nature in contradistinction to nurture. It needed Beurton's (t.v.) assertion that a gene is not "what makes the difference" between individuals, rather, difference in the reproduction of organisms is "what makes genes," to remind us that it was actually Muller himself who had introduced, already in the 1920s, this tension between the reductionist upward conception of the gene and the biologist downward notion by seeking the answer to the "why" through the investigation of the "how."

> The subject of gene variation is an important one . . . because this same peculiar phenomenon that it involves lies at the root of organic evolution . . . . Thus it is not inheritance *and* variation which bring about evolution, but the inheritance *of* variation, and this in turn is due to the general principle of gene construction which causes the persistence of autocatalysis despite the alteration of the gene itself. Given, now, any material or collection of materials having this one unusual characteristic, and evolution would automatically follow. (Muller 1922, 35)

The triumph of the reductionist path, from the instrumental particularization of heredity, through the hardening of the particles as material genes, to the resolution of the heredity material in molecular terms, could not, in the final analysis, provide the answer to the plight of inheritance. Heredity is a property immanent to living systems and needs the perspective of the life sciences. The conflict between the perspectives is unavoidable: The notion elaborated by one perspective may consider that of the other to be an illusion. But certainly the conflict between these illusions made the gene a most fruitful concept.

I wish to contend that there are four ways genes may be discerned:

1. Genes are abstract entities. Such a notion may still be used profitably by population geneticists. The gene is an abstract variable ruled by the basic principles of segregation, recombination, and mutation, to which some additional principles of molecular genetics have been added, such as intragenic recombination, transformation, and so on. All these are of course in addition to the principles of evolution like adaptation, selection, migration, and so on. This would comprise the Modern Synthesis gene mentioned by Gilbert (t.v.), but was not further discussed in this volume.

2. Genes are material structural entities. This would comprise basically versions of Benzer's interpretation of the Mullerian particulate gene. The gene is a discrete stretch of the DNA sequence,

which has some integrity and continuity in function and/or history. The more molecular biologists tried to elaborate on such a conception the more it proved to be inefficient. Still genetic engineers cling to it in an effort to identify a universal, meaningful, structurally definable entity. Beurton's (t.v.) evolutionary approach as well as the notion of the developmental gene elaborated here by Gilbert and by Morange (t.v.) show that such a structural concept cannot be maintained.

3. Genes are functional biological entities. The concept of the gene should be reconsidered together with that of the notion of reduction in the life sciences. Living systems are singular in their functional organization; their structural organization is only a secondary consequence of the historical constraints. Accordingly, the gene is a functional entity, a derivative of reproduction as the basic process of living matter. This is the concept developed by Griesemer (t.v.).

4. Genes are generic operational entities. Living systems are essentially complex and integrative systems. It is meaningless and misleading to identify entities of such systems on an ontological basis. The gene is a generic term. This is the pragmatic approach adopted by many practicing molecular biologists (see Fogle t.v. and Rheinberger t.v.).

Modern science has been conceived and shaped along the lines of the physical sciences. The explanatory power of the reductionist methodology has been demonstrated beyond reasonable doubt. Chemistry and physics were successful in reducing all processes to ever more basic units of matter. Analysis, or methodological reduction, is part of biology as much as of any empirical science. This, however, is not what makes the study of living systems distinct. It is their specific structural integrity and complexity that makes them unique. Consequently, it is incumbent on the biologist, as distinct from the physicist and chemist, to take notice of the history, the long-range evolutionary history and the short-range developmental one. This is the reason why biologists must also provide top-down explanations, as functions of the properties of the systems in which they participate, instead of merely bottom-up explanations, as the function of the properties of the units that compose them. It is on this background that Griesemer (t.v.) suggests a wider reexamination of reduction in biological research. Instead of examining merely the customary *mapping* of elements of the reducing domain on the reduced domain, he examines the *image* of reduction. According to him, the traditional attempts to reduce development to genetics and classical genetics to molecular genetics were based on an image of reduction of the more complex and specific structures to the more basic and generalized ones. The direction of the reduction relation between genetics and development is reverted when genetic and developmental processes are analyzed in relation, not to structural levels, but to the process of reproduction and evolution. Eventually "a theory of development comparable in scope, detail and power to existing theories of genetics" should emerge. The theory should not "be constructed in isolation from genetic theories if it is to contribute to a theory of reproduction that can integrate phenomena of heredity, development and evolution" (Griesemer t.v.). Once we adopt such an image of genetic entities we may view DNA sequences in many contexts, rather than just as a program or a causative agent of inheritance.

Griesemer suggests that the continuity of living systems has been maintained by the possibility of evolving by means of reproduction composed of *progeneration* and of *development*. Progeneration is a transmission process of organisms that involves multiplication by a mechanism of material overlap. The capacity for *re*production is a prerequisite for evolution, and to count *as* reproduction, a progeneration process must result in individuation. In such a theory of functional reduction the gene as an entity of the theory of inheritance is identified and characterized as an entity that is invariant in inheritance and development. "Factors are not altered by passage through different kinds of hybrids . . . . changes in expression with change of developmental context do not constitute changes in the gene" (Griesemer t.v.).

Whether such a philosophically guided theory of inheritance and development as instantiations of reproduction will prevail must be left for the future. For a present-day experimental biologist the establishment of a material molecular baseline provides a much more pragmatic approach. Rheinberger (t.v.) goes even further and suggests that the effort to define a gene is actually self-defeating for the research effort proper. He points out the significance of the "epistemology of the imprecise" in practice of scientific research. There is a paradox in the wish to define precisely something that we do not know yet but wish to discover. "The practices in which the sciences are grounded engender epistemic objects. . . . Despite their vagueness, these entities move the world of science." And: "It is the historically changing set of epistemic practices that gives contours to these objects" (Rheinberger t.v.). Some experimentalists accept that working with an ambiguous or fuzzy term is preferable to adopting a precise definition that will need continual revision as knowledge advances (Waters 1994). Others, like François Gros or Roger Levine, did not conceal their disappointment that after all efforts of molecular genetics to identify a gene directly in the DNA the only thing that specified genes was "the products that result from its activity." Their seemingly desperate epistemological assertion that it is not "one-gene – one-enzyme" that we are left with but rather "one-enzyme – one-gene" (Rheinberger t.v., also Falk 1986) appropriately reflects the dilemma we are in. For the molecular reductionist what remains of the material gene is largely an illusion. As I pointed out, once self-replication became a property of the DNA molecule, if we wish to maintain the term at all the definition of the gene through its function/product is the only option left for us. If we cannot recognize the function from the structure, what is left is to recognize the structure (or a structure?) from the function; the problem is that it is not possible to define a function as a "kind" that refers unequivocally to specific structures (see also Griesemer t.v.). For many practical experimentalists the term gene is merely a consensus term or at most one of generic value. It may relate to a short segment of the DNA sequence coding for a polypeptide, to a chromosome segment involved in affecting a quantitative character (QTL), in the integration of a developmental process (*integron*), in replication (*replicon*) or, as a matter of fact, even to a whole chromosome or to the whole genome, depending on the function or the product intended.[11] Gilbert (t.v.) tries to delimit quite explicitly the generic gene of the Developmental Synthesis. Fuzziness, on the other hand, may pose difficulties to the experimentalist who would grope for new, epistemologically more helpful terms. Rheinberger (t.v.), after he examined the idea of the generic gene, concludes that "it has become more reasonable . . . to speak of genomes, at least 'genetic material' . . . instead of genes, in the developmental as well as the evolutionary dimension." In the footsteps of Jacob he suggested that we should speak of integrons instead of genes. Fogle (1990; t.v.) examines the proposal to focus on "domain sets for active transcription" (DSATs) in their various configurations, as more refined means to represent regions of expression, but returns to genes. "The gene has been a powerful epistemic entity in the history of heredity, in all the vagueness that is characteristic for such entities" (Rheinberger t.v.). Rheinberger too maintains, at the end, a notion of the primacy of the gene, believing that what still seems to give genes some robustness is the Central Dogma of molecular biology that puts DNA (in the context of living cells) in a unique position of unidirectionality. It leads to products – i.e., phenotypes – that are the components of the organisms, thus leading to integrated entities of long processes of constrained evolution. Waters (1994) suggests a similar pragmatic definition of the experimentalists' molecular gene concept, namely that of "a linear sequence in a product at some stage of genetic expression." Such definitions allow for structurally continuous or discontinuous sequences, and functionally for whatever form of expression of the sequences, whether transcribed or not, whether translated into a proteinaceous product, or regulating such translation. "[T]he molecular gene concept specifies what genes are for. Genes are for linear sequences in products of genetic expression" (Waters 1994, 178). The tragedy is, however, that such a pragmatic concept of the gene has been canonized by the media and the public-at-large as the biological determinants of traits and characters, and in consequence turned the genetic engineers into the sorcerers that can shape our personal and social well-being, for better or worse.

In retrospect, we see that the triumph of the concept of the gene as the material unit of inheritance was also the crucial step of its downfall. With the Watson and Crick model of the structure of DNA, the need for a distinct atom of heredity, as envisaged by Muller, became redundant. The role of the "unmoved mover" of the classical theory of inheritance was bestowed in the molecular era on DNA. Experimental results, however, only sharpened the old tension inherent in the concept of the gene as it became embodied in a sequence of DNA. Empiricists reluctantly gave in by relaxing the concept of the gene as a privileged determinant of development and inheritance, to become merely a generic term, denoting operationally defined DNA sequences having some consensus properties (ORF, initiation and termination signals, introns, etc.). DNA sequences themselves are data, rather than information, i.e., phenotypes the organization of which is amenable to evolutionary forces that interact mutually with epigenetic and environmental factors rather than being their determinants. Yet, biological considerations, especially those of the evolution of integrated development of cellular organisms, or, even more compellingly, of reproduction by overlapping individuation, suggest to define genes as meaningful universals, the phenotypic referents of which happen to be sequences of DNA.

My take-home lesson from the papers presented in this volume is that we should accept that genes are entities that refer to the hereditary input of traits (phenotypes), irrespective of whether these traits are defined in terms of behavior or of the physicochemical properties of a segment of the DNA molecule. We are back to Johannsen's original notion of the gene (or, more precisely the Morgan reinterpretation of it – see Churchill 1974):

Only the simple conception should be expressed that a trait of the developing organism is conditioned, or may be partly determined, through something in the gametes. (Johannsen 1909, 124)

NOTES

1. In the writing of this review I was helped greatly by oral and written comments from many of the authors of this volume, not all of whom I can explicitly mention. The constant advice, criticism, and, especially, support of Peter Beurton and Hans-Jörg Rheinberger were invaluable to me. I am grateful to all.
2. It should be noted that by empirical instrument I do not mean necessarily an antimaterialist approach (see a different view in Vicedo 1991, 206).
3. A sex-chromosome (X-chromosome) carrying the three genetic factors: C – cross-over inhibitor (later it turned out to be a chromosomal inversion), l – for recessive lethality, B – for dominant narrow, bar-shaped eyes allowed mass screening methods for new mutations of many genes that affect the viability of the fly (the mutations cause lethality) by simply examining the absence of males in culture vials, each derived from a single female.
4. PCR – Polymerase Chain Reaction a technique for enzymatic *in vitro*, amplification of specific DNA sequences.

5. The concept of primary reaction (*Primärreaktion*) was already introduced in Kühn, A., E. Caspari, and E. Plagge. 1935. Über hormonale Gen-wirkungen bei *Ephestia kühniella z. Nachrichten der Gesellschaft der Wissenschaften zu Göttingen, Mathematisch-Physikalische Klasse* 2: 1–29. See Hans-Jörg Rheinberger (in press), Ephestia: The Experimental Design of Alfred Kühn's Physiological Developmental Genetics.
6. F. H. C. Crick (1958, 145) wrote, however, on gene-protein colinearity: "Finally there is the recent evidence [that the nucleic acids are in some way responsible for the control of protein synthesis], mainly due to the work of Benzer on the *r*II locus of bacteriophage, that the functional gene – the 'cistron' in Benzer's terminology – consists of many sites arranged strictly *in a linear order* . . . as one might expect if a gene controls the order of amino-acids in some particular protein."
7. See *gene expression*, Human Genome Project Information 1998 http://www.ornl.gov/hgmis.
8. Jablonka and Lamb by subtitling their book *Epigenetic Inheritance and Evolution* (1995) as *The Lamarckian Dimension* used another extreme, pro-vocative metaphor to counter the determinism of the gene inherent in the metaphor of the program.
9. This approach is often referred to as *reverse genetics*, to indicate that analysis starts from the genotype to the phenotype, i.e., in a reverse logic to that of classic genetics. In reality most of these studies do start from the phenotype, but after the relevant DNA sequences are identified, it is possible to link them to phenotypes independently of the original phenotype.
10. See also Griesemer's reference (t.v.) to Van Valen's aphorism that evolu-tion is the control of development by ecology.
11. Such an interpretation dangerously approaches Richard Goldschmidt's so-called hierarchical notion of the units of heredity (see above and Dietrich t.v.). Could this development be part of Goldschmidt's resurrec-tion in recent years?

REFERENCES

Allen, G. E. 1979. Naturalists and experimentalists: The genotype and the phenotype. *Studies in the History of Biology* 3: 179–209.

Beadle, G. W., and E. L. Tatum. 1941. Genetic control of developmental reactions. *The American Naturalist* 75: 107–116.

Brenner, S., W. Dove, I. Herskowitz, and R. Thomas. 1990. Genes and Development: Molecular and logical themes. *Genetics* 126: 479–486.

Britten, R. J., and E. H. Davidson. 1969. Gene regulation for higher cells: A theory. *Science* 165: 349–357.

Cairns, J. 1962. Proof that the replication of DNA involves separation of the strands. *Nature* 194: 1274.

Castle, W. E. 1915. Mr. Muller on the constancy of Mendelian factors. *The American Naturalist* 49: 37–42.

Churchill, F. B. 1974. William Johannsen and the genotype concept. *Journal of the History of Biology* 7: 5–30.

Crick, F. H. C. 1958. On protein synthesis. *Symposia of the Society for Experimental Biology* 12: 138–163.

Crick, F. H. C., and P. A. Lawrence. 1975. Compartments and polyclones in insect development. *Science* 189: 340–347.

Dawkins, R. 1976. *The Selfish Gene*. Oxford: Oxford University Press.

Elkana, Y. 1970. Helmholtz' 'Kraft': An illustration of concepts in flux. *Historical Studies in the Physical Sciences* 2: 263–298.

Falk, R. 1963. A search for a gene control system in Drosophila. *The American Naturalist* 97: 129–132.

Falk, R. 1986. What is a gene? *Studies in the History and Philosophy of Science* 17: 133–173.

Falk, R. 1995. The struggle of genetics for independence. *Journal of the History of Biology* 28: 219–246.

Falk, R. 1997. Muller on development. *Theory in Biosciences* 116 (4): 349–366.

Falk, R. 1999. Can the norm of reaction save the gene concept? In *Thinking About Evolution: Historical, Philosophical and Political Perspectives*, edited by R. Singh, C. Krimbas, D. B. Paul and J. Beatty. New York: Cambridge University Press.

Falk, R., and S. Schwartz. 1993. Morgan's hypothesis of the genetic control of development. *Genetics* 134: 671–674.

Fisher, R. A. 1918. The correlation between relatives on the supposition of Mendelian inheritance. *Transactions of the Royal Society of Edinburgh* 52: 399–433.

Fogle, T. 1990. Are genes units of inheritance? *Biology and Philosophy* 5: 349–371.

Goldschmidt, R. 1938. The theory of the gene. *Science Monthly* 46: 268–273.

Griesemer, J. R., and W. Wimsatt. 1989. Picturing Weismannism: A case study of conceptual evolution. In *What the Philosophy of Biology Is, Essays for David Hull*, edited by M. Ruse. Dordrecht: Kluwer Academic Publishers, pp. 75–137.

Harwood, J. 1993. *Styles of Scientific Thought: The German Genetics Community. 1900–1933*. Chicago: University of Chicago Press.

Jablonka, E., and M. J. Lamb. 1995. *Epigenetic Inheritance and Evolution: The Lamarckian Dimension*. Oxford: Oxford University Press.

Jacob, F. 1976. *The Logic of Life*. New York: Vanguard.

Jacob, F. 1977. Evolution and tinkering. *Science* 196: 1161–1166.

Jacob, F., and J. Monod. 1961. Genetic regulatory mechanisms in the synthesis of proteins. *Journal of Molecular Biology* 3: 318–356.

Johannsen, W. 1909. *Elemente der exakten Erblichkeitslehre*. Jena: Gustav Fischer.

Johannsen, W. 1911. The genotype conception of heredity. *The American Naturalist* 45: 129–159.

Kavenoff, R., and B. H. Zimm. 1973. Chromosome-size DNA molecules from *Drosophila*. *Chromosoma* 41: 1–27.

Lewis, E. B. 1951. Pseudoallelism and gene evolution. *Cold Spring Harbor Symposia on Quantitative Biology* 16: 159–174.

Lewis, E. B. 1963. Genes and developmental pathways. *American Zoologist* 3: 33–56.

Lewis, E. B. 1978. A gene complex controlling segmentation in Drosophila. *Nature* 276: 565–570.

Lewontin, R. C. 1991. Twenty-five years ago in *Genetics:* Electrophoresis in the development of evolutionary genetics: Milestone or millstone? *Genetics* 128: 657–662.

Lewontin, R. C. 1992. Genotype and phenotype. In *Keywords in Evolutionary Biology,* edited by E. Fox Keller and E. Lloyd. Cambridge, Mass.: Harvard University Press, pp. 137–144.

Mackie, J. L. 1965. Causes and conditions. *American Philosophical Quarterly* 2: 245–255, 261–264.

McClintock, B. 1961. Some parallels between gene control system in maize and in bacteria. *The American Naturalist* 95: 265–277.

Muller, H. J. 1922. Variation due to change in the individual gene. *American Naturalist* 56: 32–50.

Muller, H. J. 1947. The gene. *Proceedings of the Royal Society London. Series B* 135: 1–37.

Nelkin, D., and M. S. Lindee. 1995. *The DNA Mystique: The Gene as a Cultural Icon.* New York: W. H. Freeman.

Nüsslein-Volhard, C., and E. Wieschaus. 1980. Mutations affecting segment number and polarity in Drosophila. *Nature* 287: 795–801.

Pontecorvo, G. 1952. Genetic formulation of gene structure and gene action. *Advances in Enzymology* 8: 121–149.

Raffel, D., and H. J. Muller. 1940. Position effect and gene divisibility considered in connection with three strikingly similar scute mutations. *Genetics* 25: 541–583.

Salthe, S. N. 1985. *Evolving Hierarchical Systems: Their Structure and Representation.* New York: Columbia University Press.

Sapp, J. 1987. *Beyond the Gene: Cytoplasmic Inheritance and the Struggle for Authority in Genetics.* New York: Oxford University Press.

Stahl, F. 1961. A chain model for chromosomes. *Journal de Chemie Physique* 58: 1072–1077.

Stent, G. S. 1968. That was the molecular biology that was. *Science* 160: 390–395.

Stent, G. S. 1970. DNA. *Daedalus* 99: 909–937.

Tomarev, S. I., P. Callaerts, L. Kos, R. Zinovieva, G. Halder, W. Gehring, and J.

Piatigorsky. 1977. Squid Pax-6 and eye development. *Proceedings of the National Academy of Sciences USA* 94: 2421–2426.

Vicedo, M. 1991. Realism and simplicity in the Castle-East debate on the stability of the hereditary units: Rhetorical devices versus substantive methodology. *Studies in History and Philosophy of Science* 22: 201–221.

Waters, K. 1994. Genes made molecular. *Philosophy of Science* 61: 163–185.

Watson, J. D., and F. H. C. Crick. 1953. Genetical implications of the structure of deoxyribose nucleic acid. *Nature* 171: 964–967.

Glossary

adenine	$\Rightarrow$ DNA.
allele	One of two or more alternate forms of a gene, occupying a given $\Rightarrow$ locus on a $\Rightarrow$ chromosome. Only one allele of each gene is inherited from each parent. As a rule different alleles are due to $\Rightarrow$ mutations at some sites of the respective gene, they may be referred to as "mutants" of the corresponding gene. Alleles occurring in natural populations (rather than having been found in the laboratory) are traditionally referred to as "wild-type" alleles.
alu-sequence	A family of interspersed $\Rightarrow$ repetitive sequences each of circa 300 bp ($\Rightarrow$ base pair), found with high frequency in the human genome.
amino acid	The building blocks of proteins. Any of a class of 20 naturally occurring amino acids (e.g., leucine, arginine, phenylalanine, tyrosine, proline, etc.) may be

---

A word or phrase immediately following an arrow in the glossary also appears as a main glossary entry.

combined via peptide bonds to form ⇒ proteins (⇒ polypeptides). The sequence of amino acids in a polypeptide, and ultimately the protein function, are determined by the ⇒ genetic code and according to the principle of ⇒ colinearity.

amplification

The production of an increased number of copies of a specific DNA fragment. Amplification may occur in vivo or in vitro. It is used intensively in cloning techniques and in the ⇒ polymerase chain reaction (PCR). In evolution, ⇒ repetitive sequences may be viewed as resulting from amplification processes.

autocatalysis

A process by which the production of an entity is enhanced by the presence of that same entity. This is an intrinsic characteristic of the gene as the hereditary material. The multiplication of the complementary double-stranded ⇒ DNA molecule (⇒ double helix) is catalyzed by ⇒ semiconservative replication.

autosome

A ⇒ chromosome that is not a sex-determination chromosome. A diploid cell has two sets of autosomes. The diploid human genome consists of 46 chromosomes, 22 pairs of autosomes, and one pair of ⇒ sex chromosomes (the X- and Y-chromosomes).

bacteriophage

⇒ Phage.

base pair (bp)

A pair of nitrogenous bases (adenine and thymine or guanine and cytosine) on the paired strands of the ⇒ DNA ⇒ double helix, held together by weak chemical bonds (H-bonds). Distances along DNA are measured in bp. (In double-stranded

⇒ RNA uracil instead of thymine is paired with adenine.)

cell cycle   A sequence of cytoplasmic and nuclear events that involves the periodic replication of the DNA and the multiplication of other cell components, usually leading to the production of two cells from one.

cell-line   A pedigree of cells related through asexual division.

central dogma of molecular genetics   The hypothesis formulated by Crick in 1958, according to which "once information is passed into protein it cannot get out again." Essentially, the information for the sequence of amino acids of a ⇒ polypeptide flows from the DNA to the RNA and further on to the protein.

centromere   A specialized-chromosome region that includes the site of attachment of the spindle fibers involved in the accurate segregation of chromosomes during cell division (⇒ mitosis and ⇒ meiosis).

chromatin   The material of which ⇒ chromosomes consists. It is a complex ensemble of DNA and proteins, primarily histones, that together maintain the orderly folding of the DNA strand of the chromosomes, and secure its segregation during replication and cell division. Chromatin also contains some RNA molecules and other proteins involved in the metabolic activity of the chromosomes.

chromosome   Discrete self-replicating genetic structure of cells containing the cellular DNA that bears in its nucleotide sequence the linear array of genes together with various proteins (nucleoproteins) (⇒ chromatin). Genes may be conceived as structures occupying specific chromosomal sites (⇒ locus). While in ⇒ prokaryotes the DNA molecule is usually circular, and the entire ⇒ genome is carried on one chromosome, in ⇒ eukaryotes the genome of each species is distributed into a unique pattern of chromosomes that may be distinct by their number and shape. The chromosomes usually become condensed structures only before cell division (⇒ mitosis and ⇒ meiosis), when they may be discernible under the microscope. During most of the cell cycle chromosomes are only loosely packed.

cis-trans **test (complementation test)**

Test for normal ("wild-type") function of a cell (or cellular system, organism) when two DNA segments with two mutations are present in the same cell. The mutations may be both in the same segment, while the other segment is nonmutated (*cis* configuration), or – more often – one in each segment (*trans* configuration). If the system functions normally in the *trans* configuration the mutations are in different ⇒ cistrons. If it functions abnormally (has a mutant phenotype) in the *trans* configuration the mutations reside in the same cistron.

cistron

A segment of the DNA double helix that is defined by complementation-tests (⇒ *cis-trans* test) to have a single function. A cistron is comprised of many sites at which mutations might occur (⇒ mutons), and of many segments that may cross over (⇒ recons).

clone

A group of cells or organisms derived by cell division (⇒ mitosis) from a single ancestor. Cloning is the asexual production of a group of cells, all genetically identical, from a single ancestor. In recombinant DNA technology, the use of DNA manipulation procedures to produce multiple copies of a single gene or segment of DNA is referred to as DNA cloning. Thus, a group of DNA molecules identical with a single ancestral molecule may also be termed a clone.

codon

⇒ Genetic code.

colinearity

The correspondence between the order of codons (⇒ genetic code) along the nucleotide sequence and the order of the ⇒ amino acids along the polypeptide chain.

compartmentalization

The limitation of the spread of ⇒ clones of cells by boundaries within the organism during development. When the cells of a clone are marked, the marked area does not cross the compartmental boundaries, as determined at sequential stages of development, even when the cells of the clone acquired much faster replication rates than the remaining surrounding cells. Subcellular compartmentalization refers to the organelles of a cell.

complementary base pairing

⇒ DNA.

complementary DNA (cDNA)

DNA that is synthesized from a single-stranded ⇒ RNA molecule by reverse transcriptase (⇒ reverse transcription), one strand of which is thus complementary in its sequence to that of the RNA molecule and the other identical to it, except for the Us that are replaced by Ts in DNA.

complementation test ⇒ *Cis-trans* test.

consensus sequence An idealized nucleotide sequence that represents a class of DNA sequences that are similar to each other. The consensus sequence is that sequence in which each position represents the base or base-pair most often found when many actual sequences of the given class of genetic elements are compared. The elements of the class may have a related function, or a common evolutionary origin.

conserved sequence A base sequence in a DNA molecule (or an amino acid sequence in a protein) that has remained essentially unchanged throughout evolution.

crossing over The "breaking" during ⇒ meiosis of one maternal and one paternal ⇒ chromosome at homologous sites, and the reciprocal exchange of corresponding sections of the chromosomes. This process may result in an exchange of ⇒ alleles between homologous chromosomes (⇒ recombination). The breakage and the resulting exchange is controlled by complex mechanisms involving many enzymes.

cytosine ⇒ DNA.

differentiation The process whereby cells (usually of multicellular organisms) acquire different structures and functions during embryonic development. Differentiation may be intracellular or extracellular. Often the differentiated state is transmitted from a cell to its progeny through ⇒ epigenetic inheritance. Unicellular organisms may also differentiate, for example, in the formation of spores of bacteria like those of *Bacillus subtilis*.

diploid

Possessing a double set of chromosomes (one paternal and one maternal). Most sexually reproducing organisms are diploid. ⇒ Meiosis secures that the ⇒ gametes receive only a single set of chromosomes, i.e., that they are ⇒ haploid. The diploid state is reestablished by the fusion of two gametes during fertilization. In many cases, such as those of fungi, ferns and mosses, extensive phases of the life cycles are haploid.

disjunction

The separation of members of a chromosome pair to opposite poles during cell division (⇒ cell cycle; chromosome). During ⇒ mitosis and the second meiotic division disjunction applies to daughter chromosomes (chromatids that turn out to be chromosomes); at the first meiotic division it applies to sister (homologous) chromosomes; this segregation of homologues secures the orderly transition from the ⇒ diploid state to ⇒ haploid state.

DNA (deoxyribonucleic acid)

The molecule of heredity. In the long double-stranded molecule (⇒ double helix), the strands are held together by weak bonds between the complementary bases of four nucleotides (H-bonds). The nitrogenous bases of the four nucleotides of DNA are: adenine (A), guanine (G), cytosine (C), and thymine (T). Base pairs usually form only between A and T and between G and C; thus the base sequence of each single strand can be deduced from that of its partner. The sequence of base pairs contains the genetic ⇒ information and may be transcribed into a sequence of RNA complementary to one of the strands of the double helix which is later (partly or completely, ⇒ genes-in-pieces) translated into a sequence of ⇒ amino acids (⇒ genetic code). Other sequences that do not code for ⇒ polypeptides may be involved in various regulatory activities, or even may comprise "junk DNA." During DNA replication (⇒ autocatalysis) the two strands separate and a new strand is synthesized by complementary base pairing on each of the preexisting ones as a template (⇒ semiconservative replication).

domain

A site or section of a ⇒ protein molecule that interacts coherently and specifically with its surroundings. DNA sequences whose function depends on specific interaction with protein molecules are also called domains.

dominance

The phenotypic expression of only one of the two ⇒ alleles in a heterozygote (⇒ heterozygosity). As a rule, the phenotype of a dominant allele is similar in heterozygotes and homozygotes of that allele. The allele that is not phenotypically expressed in the heterozygote (but is expressed in the homozygotes for that allele) is recessive.

double helix

The shape that Watson and Crick suggested for the two antiparallel complementary strands of ⇒ DNA when bonded together by hydrogen bonds (H-bonds) between the bases and wound up in a helical fashion.

downstream

Position of a molecular signal (or occurrence of a molecular event) usually non-contiguous to a reference point of a DNA stretch and in the direction of ⇒ transcription (⇒ upstream).

downward causation

Natural selection, when working on higher levels of organization, may determine lower-level phenomena. For instance, the efficacy of a protein may determine in the long run which DNA templates are present. Likewise, selection on the level of the organism will favor in the long run the assembly of gene combinations that through their interactions support the favored phenotype. Downward causation is an instance of a systemic process in which the whole determines the quality of the parts.

electrophoresis

A biochemical technique by which a mixture of molecules (proteins, nucleic acids) can be separated when an electric current is run through a medium in which the mixture had been positioned.

endonuclease

An ⇒ enzyme that cleaves nucleic acid bonds (of DNA or of RNA) at internal sites of the nucleotide sequence (⇒ exonuclease).

enhancer

A type of regulatory element (a DNA ⇒ domain) positioned in *cis*-position to a transcribed region that increases the rate of ⇒ transcription. The enhancer may be quite far away from the transcribed sequences.

enzyme
A protein that acts as a biocatalyst. It speeds up the rate at which a biochemical reaction proceeds without altering the nature of the reaction. Enzymes may consist of more than one peptide chain. Their catalytic properties derive from their three-dimensional conformation, i.e., from the tertiary and quaternary structure of $\Rightarrow$ proteins.

epigenetic inheritance
Inheritance of a trait for one or more cell generations that is not transmitted (or not mainly transmitted) by the genetic material proper (cell heredity). A case of epigenetic inheritance is the transmission of $\Rightarrow$ methylation patterns of the DNA: Some specific nucleotide sequences get methylated, and these methylations are maintained throughout mitotic divisions. They are "erased" upon gamete formation and meiosis.

epistasis
Originally, the expression of one gene that obscures the expression of the product or trait of another gene (at another $\Rightarrow$ locus). Today, this term is used in population genetics for all kinds of interactions between genes at different loci (nonallelic interaction).

eukaryotes
Organisms that have a nucleus bounded by a membrane, within which most of the cellular DNA is organized. These include all the multicellular organisms (plants and animals) as well as the protozoans (fungi, algae, diatoms, amoebas, paramecia, etc.), but not the bacteria, blue-

green algae and archaebacteria which are ⇒ prokaryotes.

exon ⇒ Gene-in-pieces.

exonuclease An enzyme that cleaves nucleotides specifically from one of the free ends of a linear nucleic acid substrate (⇒ endonuclease).

expression ⇒ Gene expression.

gamete Mature male or female reproductive cell (sperm or egg). Gametes are usually ⇒ haploid, and the fusion of two gametes of opposite sex gives rise to a zygote (a new ⇒ diploid cell). In diploid organisms the production of gametes is preceded by the sequence of two meiotic divisions (⇒ meiosis). In haploid organisms that reproduce sexually meiosis occurs shortly after gamete fusion ("zygote production").

gene expression The process of phenotypic manifestation of a gene. In this process, a gene's coded information is converted into the structure and/or function present and operating in the cell or in the organism as a whole (the phenotypic trait).

gene family Groups of genes closely related in their DNA sequences, apparently sharing a common ancestor. Members of a gene family usually code for similar products.

gene-in-pieces In eukaryotic cells most of the protein coding ⇒ DNA sequences are interrupted by noncoding sequences. These sequences have been termed *exons* and *introns*, respectively. The uninterrupted sequence is transcribed from the DNA into pre-mRNA ($\Rightarrow$ RNA) which in turn is spliced by special mechanisms to a continuous stretch of exons before translation. The introns may serve other functions, such as regulation of the splicing and translation processes, or even serve as exons of other genes. Moreover, for many DNA sequences alternative splicing patterns may occur in the cells of an individual: Under different circumstances (in cells of different destination) a different array of exons may be spliced (skipping in the process one or more exons) thus giving rise to different proteins from the same DNA stretch.

gene product

In molecular biology: The biochemical material, either RNA or protein, resulting from $\Rightarrow$ gene expression. Abnormal amounts or qualities of gene products can be correlated with disease-causing alleles.

genetic code

The principles by which genetic $\Rightarrow$ information is stored in and translated ($\Rightarrow$ translation) from a nucleotide sequence of $\Rightarrow$ DNA and $\Rightarrow$ RNA into a specific sequence of $\Rightarrow$ amino acids which, in turn, determine the specificity of the $\Rightarrow$ polypeptide ($\Rightarrow$ protein). The code is read on the $\Rightarrow$ ribosomes sequentially in groups of three nucleotides. Each of these triplets or codons contains the information necessary for specifying one amino acid. Since any one of the four kinds of nucleotides of which a DNA strand is built can occupy each of the three positions of a codon, there is a totality of $4^3 =$ 64 different codons as compared to only

20 available amino acids. Therefore most amino acids are encoded by more than one codon (the genetic code is degenerate or redundant). Three codons function as ⇒ termination codons.

genetic map   ⇒ Linkage map.

genome   All the genetic material in the chromosome set of a particular organism. In ⇒ diploid organisms there are two genomic complements.

genotype   The genetic constitution of an organism, or the genetic constitution (the ⇒ allele composition) of a specific gene locus, as opposed to the organism's ⇒ phenotype which is partly the product of the interaction of the genotype with environmental conditions during development.

germ line   The lineage of cells, continuous across the generations, that gives rise to gametes in every generation.

germ plasm   A term in use at the end of the nineteenth century and the first half of the twentieth century, designating the cellular components that carry the hereditary material.

guanine   ⇒ DNA.

haploid   A single set of chromosomes, in contradistinction to the double (⇒ diploid) set. The egg and sperm cells (⇒ gametes) of diploid organisms are haploid. Bacteria and many plants like ferns and mosses, as well as some animals (e.g., male bees) are haploid also in their soma cells.

heritability   The proportion of the phenotypic variability in a population that may be attributed to (additive) genetic variability.

heterocatalysis

The process in which an entity catalyzes the construction of a product other than itself. This has been denoted as one of the necessary characteristics of the gene as the basis of heredity, besides that of ⇒ autocatalysis. The mRNA (⇒ RNA), and eventually the ⇒ polypeptides and ⇒ proteins synthesized according to the DNA sequence (⇒ genetic code) stand for such heterocatalytic products of DNA.

heterozygosity

The presence of different versions of a gene (⇒ allele) at one or more loci of the same organism.

homeobox

A short stretch of some 180 nucleotides whose base sequence is nearly identical in all the genes that contain it (⇒ consensus sequence). It has been found in many organisms from fruit flies to mice and human beings and is consequently highly conserved. It is usually located near the ⇒ downstream end of the genes, and codes for a protein ⇒ domain of about sixty amino acids that is engaged in regulating other genes' expressions by binding specifically to their ⇒ upstream regulatory DNA sequences. Homeoboxes appear to determine when and where particular groups of genes are expressed along the embryonic axis during development.

homeotic gene

Genes with the ability, when mutated, to transform one organ into the likeness of another, usually homologous, organ such as an antenna of Drosophila into an extra leg, or a thoracic segment into an abdominal segment (⇒ master gene).

homologous chromosomes   Chromosomes containing the same array of genes, though not necessarily the same ⇒ alleles. Sometimes a homologous chromosome may carry the same array of genes rearranged in their sequential order (e.g., through an inversion). In ⇒ diploids homologous chromosomes pair during ⇒ meiosis and exchange segments by ⇒ crossing over.

information   The potential of a ⇒ DNA strand, owing to the particular sequence of its nucleotides and the principles of the ⇒ genetic code, to give rise to a well-specified ⇒ gene product.

initiator codon   A codon (⇒ genetic code) in the mRNA that directs initiation of polypeptide synthesis (⇒ translation) in the presence of a complex set of initiation factors (initiator complex). A common initiator codon is AUG, which codes for the amino acid methionine.

interphase   The stage between one cell division and the following one (⇒ cell cycle).

intron   ⇒ Gene-in-pieces.

karyotype   A presentation of an individual's (or of a related group of individuals') entire complement of chromosomes arranged in a standard format showing the number, size, and shape of each chromosome type. Usually a photomicrograph used in cytological mapping. The karyotype is often helpful in correlating gross chromosomal abnormalities with the characteristics of specific diseases.

kilobase (kb)   Unit of length for DNA fragments equal to 1,000 nucleotides (1,000 base pairs).

leader sequence
A sequence of nucleotides on the transcribed mRNA that precedes the ⇒ initiator codon, involved in regulating ⇒ translation.

linkage
The association in inheritance of certain genes as a consequence of their location on the same chromosome. Hence, genes come in linkage groups each of which corresponds to a specific chromosome. Linkage is broken by ⇒ crossing over (⇒ recombination).

linkage disequilibrium
Association of alleles at different loci at frequencies that deviate from the random combination expected at equilibrium of these alleles in a large population. This may be a result of natural selection favoring particular combinations of alleles at different loci (⇒ epistasis) or due to stochastic processes in a small population that has not yet reached equilibrium.

linkage map
A linear plot of the relative positions of genes on a chromosome, determined on the basis of how often the loci are inherited together. The closer together two genes are on a chromosome, the lower the probability that they will be separated through ⇒ recombination.

locus (pl. loci)
The position on a chromosome of a gene or a distinct ⇒ marker. In molecular biology the use of locus is sometimes restricted to mean regions of DNA that are expressed.

macromutation
A heritable change in the genetic material that produces a large phenotypic effect. It has often been hypothesized in the history of biology that macromutations may

lead to the saltational origin of new species or larger taxa.

marker
A genetically controlled phenotypic difference used to keep track as reliably as possible of the genotypic characteristics of an individual, a tissue, or a cell. A marker may be a physical location on a chromosome (e.g., restriction enzyme cutting site) whose inheritance can be monitored, indentifiable repeats of several nucleotides of the DNA, or a visible trait like eye color.

master gene
A DNA sequence that controls the activity of a whole battery or cascade of genes of a developmental pathway.

meiosis
Occurs in all sexually reproducing ⇒ eukaryotes: Two successive cell divisions following only one chromosome duplication. Meiosis of a ⇒ diploid cell ends in four ⇒ haploid daughter cells (⇒ gametes). During meiosis ⇒ homologous chromosomes pair and recombine (⇒ crossing over) before segregating.

Mendelism
The theory of particulate inheritance based on the interpretation of Mendel's theory. It is studied mainly through hybridization experiments and is concerned primarily with the transmission of traits by independent factors obtained from both parents in the hybrids (⇒ transmission genetics).

methylation
The attachment of methyl groups to organic molecules. In DNA, methylation occurs mainly at the cytosine (C) of C-G sequences. Patterns of methylation are involved in ⇒ epigenetic inheritance.

mitosis   The process of nuclear division (usually followed by cell division) giving rise to two daughter nuclei that are identical in chromosome numbers to one another and to the parent nucleus.

mRNA   ⇒ RNA.

mutation   Heritable change in the genetic material. Depending on the units involved, there are gene mutations (any change in the nucleotide sequence of a gene), ⇒ chromosome mutations (involving the gain, loss, or relocation of chromosomal segments), and ⇒ genome mutations (changes in the number of one or more chromosomes or of the complete chromosome set contained in the cell).

muton   ⇒ Cistron.

negative interference   ⇒ Crossing over at one site of a chromosome usually interferes with another crossing-over event in its vicinity. This phenomenon has been known as chromosomal interference. However, the fine analysis of crossing over reveals that there is an increased probability for more exchanges to occur extremely close to an exchange event. This coincidence of exchange events is called negative interference.

norm of reaction   The range of potential phenotypes that a single genotype may develop if exposed to a variety of environmental conditions.

nucleic acid   ⇒ DNA, ⇒ RNA.

nucleotide   Building block of the polynucleotide chains of ⇒ DNA and ⇒ RNA. Each nucleotide is composed of a purine or pyrimidine base bound to a sugar residue (deoxyribose in DNA and ribose in RNA) and a phosphoric acid residue.

operator

A short specific sequence of DNA to which proteins bind that either activate or repress the transcription of the associated coding sequences.

operon

A group of contiguous, often functionally related genes organized into a single transcription unit, the expression of which is controlled by a single regulatory region ($\Rightarrow$ operator). Operons are common in prokaryotes.

ORF (open reading frame)

DNA sequence presumed to carry the information for an as yet unidentified gene product because the sequence is "open," i.e., not interrupted by a termination codon.

overlapping genes

Functionally defined genes that share some portion of their DNA sequence.

phage

A $\Rightarrow$ virus for which the natural host is a bacterial cell. It is also called bacteriophage. In the laboratory phages are usually detected by the changes that they induce in bacterial cultures. Thus, on an agar medium on which a "lawn" of bacteria were grown, the form of the "holes" in the lawn, the plaques, are indicators of the phages involved. Phage mutants are named according to the host cells they attack and the form of the plaques single phages produce. Thus, T4 and T2 are phages attacking different host strains of *Escherichia coli*, and *r*II mutants of T4 phages represent the form of the plaques produced by mutant phages grown on specific *E. coli* strains.

phenotype
The appearance, or characteristic of an organism, resulting from the interaction of its ⇒ genotype and the conditions (environmental and others).

point mutation
Mutation that does not affect the linkage relationship of the genes. In molecular genetics, a mutation that changes a single base pair, or deletes or inserts one.

polyadenylation
A posttranscriptional addition of a sequence of up to 200 adenine nucleotides to the trailer (⇒ downstream) end of a mRNA transcript.

polymerase
An enzyme that catalyzes the synthesis of nucleic acids (DNA or RNA) on preexisting nucleic acid templates, assembling DNA or RNA from the corresponding ⇒ nucleotides.

polymorphism
Existence of multiple alleles of given genes (differences in DNA sequences) among individuals of a population. Genetic variation occurring in more than 1% of a population would be considered polymorphism maintained by forces other than random mutation (e.g., by selection or migration).

polypeptide
A sequence of ⇒ amino acids that folds into a three-dimensional ⇒ protein structure that may interact with similar or other polypeptides to form quatenary complexes. Polypeptides can act as structural and as functional entities (predominantly enzymes in the latter case).

position-effect
A change in the phenotypic effect of one or more genes due to a change in their

location on the chromosome in relation to other genes.

posttranslation processing

Biochemical changes to a polypeptide after the completion of translation. This may include enzymatic cutting of the polypeptide or the formation of special bonds (like disulfide bridges) as a prerequisite for the folding of the ⇒ protein into its active tertiary structure.

prokaryote

Organism lacking a membrane-bound, structurally discrete nucleus and other subcellular compartments, e.g., bacteria and blue-green algae (⇒ eukaryote).

promoter

A site on DNA to which RNA ⇒ polymerase will bind and initiate ⇒ transcription. It usually includes some ⇒ consensus sequences, for example, the ⇒ TATA box.

protein

A major macromolecular constituent of cells, composed of one or more ⇒ polypeptides. Proteins are required for the structure, function, and regulation of the body cells, tissues, and organs, and each protein has unique functions. Examples are hormones, enzymes, antibodies, and fibers. A protein's primary structure is given by the sequence of ⇒ amino acids (⇒ genetic code) forming the polypeptide. The characteristic three-dimensional folding of the polypeptide is referred to as the protein's tertiary structure. The association of two or more folded polypeptides into functional units leads to the quarternary structure of proteins.

pseudoallele

Any of two (or more) mutants that do not complement in *trans*-position, i.e., are ⇒ alleles, but do complement each other in the *cis*-position (⇒ *cis-trans* test).

pseudogene

DNA sequence similar to that of well-characterized genes that is unable to be transcribed or is transcribed only at very low rates and is presumed to have no effect on cellular function.

recessiveness

⇒ Dominance.

recombination

The process by which progeny derive a combination of alleles different from that of either parent. In diploid organisms recombination occurs through independent assortment between chromosome pairs and ⇒ crossing over within chromosome pairs during ⇒ meiosis.

recon

⇒ Cistron.

repetitive DNA sequence

A set of homologous DNA sequences that are found in the genomes of multicellular eukaryotes. The repetitive sequence may be short (a couple of base pairs each) or long (e.g., the rRNA genes), dispersed or clustered in specific chromosome regions. Some clusters comprise up to a million repeats, and may be discernible during DNA preparations as satellite DNA.

restriction enzyme

A protein (⇒ endonuclease) that recognizes specific, short nucleotide sequences and cuts DNA at those sites. Different bacteria contain altogether over 400 such enzymes that recognize and cut over 100 different specific DNA sequences. Restriction enzymes are major tools in recombinant DNA technology. They can also be used to uncover DNA ⇒ polymorphism in the form of Restriction Fragment Length Polymorphism (RFLP), i.e., variation between individuals in DNA fragment sizes cut by specific restriction enzymes. Polymorphic sequences that result in RFLPs are used as markers on both physical maps and genetic linkage maps.

reverse transcription Synthesis of DNA on a template of RNA by the enzyme reverse-transcriptase, contrary to the "normal" path from DNA to RNA to protein (⇒ central dogma of molecular genetics).

ribosome A small cellular organelle composed of two subunits, one small and one large, each containing different ribosomal RNAs (rRNA) and proteins; sites of protein synthesis. In eukaryotic cells ribosomes are usually concentrated along the endoplasmic reticulum of the cell, giving it a "rough" pattern (rough endoplasmic reticulum). The transcribed (⇒ transcription), or posttranscriptionally spliced (⇒ gene-in-pieces) mRNA is "threaded" through a groove between the two ribosomal subunits and tRNAs with anticodon triplets complementary to the codon sequence of the mRNA and the corresponding amino acids attached to the tRNAs to the nascent polypeptide chain, one at a time.

RNA (ribonucleic acid) A nucleic acid found in the nucleus and cytoplasm of cells. The structure of RNA is similar to that of ⇒ DNA, however the sugar residue is ribose instead of deoxyribose, and of the four bases three (adenine, guanine, cytosine) are identical to those of DNA, while uracil replaces the thymine of DNA. There are three major classes of RNA: messenger RNA (mRNA), transfer RNA (tRNA), and ribosomal RNA (rRNA; ⇒ ribosome). The genetic information contained in DNA, after having been transcribed (and spliced ⇒ gene-in-pieces) into mRNA (⇒ genetic code), is transferred to the ribosomes for ⇒ translation into proteins. Translation is mediated by tRNAs that serve as adaptor molecules to position each amino acid in proper alignment on the mRNA running through the ribosome for polymerization into a ⇒ polypeptide. There are other small RNAs, each serving a different purpose. Some viruses have RNA instead of DNA as their genetic material.

RNA polymerase ⇒ Polymerase.

RNA splicing ⇒ Genes-in-pieces.

rRNA ⇒ RNA, ribosomes.

selector gene ⇒ Master gene at a "developmental crossroads" with the ability to select between alternate developmental pathways by turning on or off the activities of many genes of the developmental pathways.

semiconservative replication Mode of replication of the double helix of the ⇒ DNA molecule. The two strands separate and two new double helices are produced, each containing one original strand and one newly syn-

thesized, complementary to the "conserved" one.

sequence
The linear order of nucleotide bases in a
DNA or RNA molecule (base sequence).
The linear order of amino acids in a $\Rightarrow$
polypeptide molecule. Sequencing is the
procedure of determining the order of
nucleotides in a DNA or RNA molecule
or the order of amino acids in a
polypeptide.

sex chromosome
The chromosome that differs in males
and females of bisexual organisms (contrary to the $\Rightarrow$ autosomes that are the
same in both sexes). Usually females
have two similar chromosomes designated X-chromosomes, and males have
one X- and a Y-chromosome.

soma, somatic cells
All the cells of a multicellular organism,
except for those belonging to the $\Rightarrow$ germ
line.

splicing
$\Rightarrow$ Gene-in-pieces.

start codon
$\Rightarrow$ Initiator codon.

stop codon
$\Rightarrow$ Termination codon.

suppressor
A mutation that suppresses the effect of
another mutation. A suppressor mutation may be intragenic, compensating for
the first mutation, e.g., by restoring the
reading frame of the sequence ($\Rightarrow$ ORF),
or occur in other genes.

targeted gene mutation
Mutations induced in a predetermined
base or sequence: a technique of genetic
engineering.

TATA box
A $\Rightarrow$ consensus sequence common for the
$\Rightarrow$ upstream regulative sequences ($\Rightarrow$

promoters) of eukaryotic cells. It is essential for the initiation of ⇒ transcription and may be involved in the binding of the RNA polymerase (⇒ polymerase) to the DNA sequence.

termination codon   Three codons (out of the 64 codons; ⇒ genetic code) that signify the end of the translation of the RNA transcript of the ⇒ ribosomes.

thymine   ⇒ DNA.

trailer sequence   A sequence following the termination signal at the end of messenger ⇒ RNA.

transcription   The synthesis of an ⇒ RNA copy from a sequence of ⇒ DNA.

transformation   Transfer and incorporation of foreign DNA into a cell and the subsequent recombination of at least part of it into the cell's genome.

translation   The synthesis of a polypeptide chain on the ⇒ ribosomes according to the sequence of triplets of a mRNA molecule (⇒ RNA).

transmission genetics   The study of the mechanisms involved in the passage of a gene from one generation to the next (⇒ Mendelism).

triplet   A set of three nucleotides that form a codon (⇒ genetic code) that codes for an ⇒ amino acid.

tRNA   ⇒ RNA ⇒ ribosome.

untranslated sequence   A sequence of the transcribed mRNA that is not translated to a polypeptide sequence. Usually these sequences are involved in some regulatory activity.

upstream	Position of a molecular signal (or occurrence of a molecular event) usually noncontiguous to a reference point of a DNA stretch and in the direction opposite of $\Rightarrow$ transcription ($\Rightarrow$ downstream).

uracil	$\Rightarrow$ RNA.

virus	An infectious noncellular biological entity that can reproduce only within a host cell. Viruses lack protein-synthesizing machinery and energy-conversion systems, and need the host's cellular system to grow and multiply. They consist of a core of nucleic acid (DNA or RNA) and some proteins, covered by protein shell, that includes cell-penetration mechanisms. Some animal viruses are also surrounded by a lipid membrane. Viruses that infect bacterial cells are known as bacteriophages, or, for short, $\Rightarrow$ phages.

X-ray spectrography	A physical experimental procedure that allows one to determine the three-dimensional structure of molecules or molecular $\Rightarrow$ domains.

Index

Index

Index